THE CAUSES OF TROPICAL DEFORESTATION

The Causes of Tropical Deforestation

The economic and statistical analysis of factors giving rise to the loss of the tropical forests

Edited by

Katrina Brown
David W. Pearce

CSERGE
University of East Anglia
University College London

UBC PRESS / VANCOUVER

First published in 1994 by UCL Press Limited
The name of University College London (UCL) is a registered trade mark
used by UCL Press with the consent of the owner.

Published simultaneously in Canada by
UBC Press, University of British Columbia,
6344 Memorial Road, Vancouver, British Columbia V6T 1Z2

Canadian Cataloguing-in-Publication Data

Main entry under title:

The Causes of tropical deforestation

Includes bibliographical references and index.
ISBN 0-7748-0511-0

1. Deforestation—Tropics. 2. Forest policy—Tropics. I.
Brown, Katrina, 1960–   II. Pearce, David W.
SD418.3.T76C38 1994   333.75'137'0913   C94-910463-9

Typeset in Bembo.
Printed and bound by
Biddles Ltd, Guildford and King's Lynn, England.

Contents

CONTENTS

ix

Preface and acknowledgments

The idea for this volume came from teaching the master's course in Environmental and Resource Economics at University College London and as a result of our work on the Centre for Social and Economic Research on the Global Environment (CSERGE)'s biodiversity and global warming research programmes. Part of the master's course is devoted to environment and development problems in the developing world, and the issue of tropical deforestation was, and is, a focal point for that discussion. Perhaps the most interesting feature of the debate has been the extent to which various factors and institutions have been blamed for deforestation. It is regularly alleged that, like other environmental problems, deforestation is due to poverty, massive external debt, greedy multinational logging companies, inept governments, the World Bank, the International Monetary Fund, population growth, the non-sustainable practices of agricultural colonists, and so on. The debate seemed more a matter of emotional heat and hobby horses than scientific investigation. We therefore began to collect together the literature that sought to investigate the problem in a more rigorous fashion. The result is the current book.

A few of the essays in the book are reprinted from other sources. Most are printed here for the first time and emanate either as new essays written especially for the volume, or as updates of previous "grey" papers circulating in the world of pre-prints and discussion papers. They generally share the same concern: the application of statistical technique to the analysis of the problem of deforestation, a form of "environmetrics". The result, we think, is illuminating even if some readers will have difficulty identifying with the idea of applying statistical technique in this area.

However, seeking statistical explanations for deforestation is only part of the story. As we discuss in our introductory chapter, deforestation, like most environmental problems, arises from the absence of a "level playing field" for conservation and economic development. Simply put, conservation loses out because (a) so many of the economic services it provides for the human species do not have markets, and (b) the resource and land-uses that it competes with secure unfair economic advantage through being subsidized. The absence of conservation markets means that the economic value of conservation tends to zero. The subsidies to land conversion, and hence deforestation, mean that there is financial advantage to deforestation even though, properly accounted for, a lot of that deforestation yields no profit for society. In economic value terms, this "zero plays plus" game therefore inevitably means that deforestation wins over conservation. In a nutshell, this is the main underlying cause of deforestation. It is mixed up with the factors that give urgent stimulus to non-level playing fields: population and poverty in particular. Those interactions are explored in the various studies in the book, but the phenomenon of "missing markets" is not explored in most of the essays since it is not amenable to statistical testing.

We are very much indebted to our colleagues as CSERGE – the Centre for Social and Economic Research on the Global Environment. CSERGE is core-funded by the UK Economic and Social Research Council and we are indebted to it, and especially to its Chairman – Howard Newby – for having the vision to see the need for CSERGE and for having confidence in us to develop it on two sites – University College London and the University of East Anglia. Though separated by a hundred miles or so, we have developed a close, collaborative style of working, and this book is one of the products of that collaboration. Special thanks go to our colleagues, Neil Adger, Camille Bann, and Leon Clarke for help in its preparation, and to Kerry Turner and Tim O'Riordan – co-directors of CSERGE – for their continuous support and encouragement. Also special thanks to our contributors who responded with enthusiasm to the idea and who have survived fairly constant pestering from us as editors.

Individual contributors wish to acknowledge help as follows:

Kahn and McDonald (Chapter 4): financial support through a Cooperative Agreement between SUNY-Binghampton and the Climate Change Division, Office of Policy Analysis, US Environmental Protection Agency, who are not responsible for the position taken in the paper; D. W. Pearce, S. Ferry, W. Branson, W. O'Neill, M. J. Kealy and J. Andrasko for comments and suggestions, and Tae H. Ha for research assistance.

Nemat Shafik (Chapter 6) wishes to thank Will Cavendish, Sweder van Wijnbergen, David Wheeler and participants in seminars at Columbia and Princeton Universities, and at Swathmore College for their helpful comments. Special thanks to Sushenjit Bandyopadhyay for outstanding research assistance.

Tom Rudel (Chapter 7) acknowledges valuable comments from Helmut Anheier, Mike Gildesgane, Tim Jessup, George Morren and Andrew P. Vayda.

Southgate (Chapter 9): The research reported here was undertaken while Doug Southgate worked at the Instituto de Estrategias Agropecuarias (IDEA), which was founded with support from the U. S. Agency for International Development (AID) to analyze policies affecting rural development in Ecuador. He benefited greatly from interchange with current and former staff members of IDEA and of AID's Quito mission. Particular thanks go to Dr Morris Whitaker, of Utah State University.

David Kummer and Chi Ho Sham (Chapter 10) would like to thank Dr B. L. Turner II, Dr William Meyer and David Edmunds of the Graduate School of Geography at Clark University, and Dr Tom Stone of the Woods Hole Research Centre for comments on an earlier draft of their chapter.

Dennis Mahar and Robert Schneider (Chapter 11) wish to thank Maryla Webb for her assistance in the preparation of their chapter.

Diane Osgood (Chapter 15) would like to thank Dr John Harriss and Dr Ian Rowlands at the Development Studies Institute at London School of Economics and Political Science, and David Pearce.

Jeffrey Vincent (Chapter 19) acknowledges financial assistance from the Biodiversity Support Program (USAID, World Wildlife Fund, and Conservation Foundation) and the Harvard Institute for International Development, and C. Binkley, M. Pullenbeck, T. Panayotou and R. Sedjo for comments on earlier drafts of the chapter.

We would like to thank *Science* for their permission to reprint Jeffrey Vincent's paper (Chapter 19).

The book is a modest contribution to an improved understanding of the forces at work in global environmental degradation. If we are to save something resembling the beauty of the world we have, and pass it on to the next generations, then we have to understand why environmental degradation occurs. Our belief is that much, though not all, of the explanations lie in the workings of national and international economies. This book is testimony to that belief.

K. B.
D. W. P.

Contributors

Bruce Aylward is Research Associate at the London Environmental Economics Centre (LEEC) at the International Institute for Environment and Development (IIED), 3 Endsleigh St, London, WC1H ODD, UK.

Edward Barbier is Senior Lecturer in the Department of Environmental Economics and Environmental Management at the University of York, Heslington, York, YO1 5DD, UK. This work was underatken whilst Director of the London Environmental Economics Centre (LEEC).

Richard Bilsborrow is Research Professor in the Departments of Biostatistics, City and Regional Planning, and Ecology at the University of North Carolina at Chapel Hill, University Square, Chapel Hill, NC 27516–3997, USA.

Josh Bishop is Research Associate at the London Environmental Economics Centre (LEEC) at the International Institute for Environment and Development (IIED), 3 Endsleigh St, London, WC1H ODD, UK.

Nancy Bokstael is Professor in the Department of Agricultural and Natural Resource Economics, University of Maryland, College Park, MD 20740, USA.

Katrina Brown is Senior Research Associate at the Centre for Social and Economic Research on the Global Environment (CSERGE), University of East Anglia, Norwich, Norfolk, NR4 7TJ, UK.

Joanne Burgess is Lecturer in the Department of Environmental Economics and Environmental Management at the University of York, Heslington, York, YO1 5DD, England. This work was undertaken whilst a researcher at the London Environmental Economics Centre (LEEC).

Ana Doris Capistrano is Representative at The Ford Foundation, House No. 30, Road No. 15, Dhanmondi, Dhaka 1209, Bangladesh.

Manab Chakraborty is Environmental Advisor to the India–Canada Cooperation Office, D1/56 Vasant Vihar, New Delhi 110 057, India.

Martha Geores is completing her PhD Thesis in the Department of Geography, University of North Carolina, Chapel Hill, University Square, Chapel Hill, NC 27516–3997, USA.

Rolando Guzmán is Research Assistant in the Instituto de Planejamento Econômico e Social (IPEA), Av. Pres. Antonio Carlos, 51/17° andar, 20020 Rio de Janeiro, Brazil.

James Kahn has a joint appointment as Professor of Economics, Department of Economics, University of Tennessee, Knoxville, TN 37996–0550 and at Oak Ridge National Laboratory, PO Box 2008, Bldg 4500N, MS 6205, Oak Ridge, TN 37831–6205, USA.

David Kummer is Visiting Assistant Professor and Research Associate at George Perkins Marsh Institute, Graduate School of Geography, Clark University, 950

Main Street, Worcester, MA 01610, USA.

Chiara Lombardini is a doctoral student with the Centre for Social and Economic Research on the Global Environment (CSERGE), University College London, Gower St, London, WC1E 6BT, UK.

Dennis Mahar is Chief, Environment Division, Latin America and Caribbean Division, World Bank, 1818 H Street NW, Washington DC 20433, USA.

Judith McDonald is at the Department of Economics, Rauch Business Center, 621 Taylor St, Lehigh University, Bethlehem, PA 18015, USA.

Norman Myers is an independent scientist and a Visiting Fellow of Green College, Oxford. His address is Upper Meadow, Old Road, Headington, Oxford OX3 8SZ, UK.

Diane Osgood is doctoral candidate, Development Studies Institute, London School of Economics and Political Science, Houghton Street, London WC2A 2AE, UK.

Matti Palo is Director of the Department of Forest Resources at the Finnish Forest Research Institute, PO Box 37, SF–00381, Helsinki, Finland.

Theodore Panayotou is Fellow of the Harvard Institute for International Development, One Eliot Street, Cambridge, MA 02138, USA.

David Pearce is Professor of Environmental Economics at University College London, and Director of the Centre for Social and Economic Research on the Global Environment (CSERGE), University College London, Gower St, London, WC1E 6BT, UK.

Eustáquio Reis is Senior Economist in the Instituto de Planejamento Econômico e Social (IPEA), Av. Pres. Antonio Carlos, 51/17° andar, 20020 Rio de Janeiro, Brazil.

Thomas Rudel is Professor of Human Ecology and Sociology at Rutgers University, PO Box231, New Brunswick, New Jersey 08903, USA.

Robert Schneider is Principal Economist, Environment Division for Latin America and the Caribbean at the World Bank, 1818 H Street NW, Washington DC 20433, USA.

Nemat Shafik is at the World Bank, 1818 H Street NW, Washington DC 20433, USA.

Chi Ho Sham is Senior Associate and Director of GIS for the Cadmus Group Inc., 135 Beaver Street, Waltham, MA 02154, USA.

Douglas Southgate is Associate Professor in Ohio State University's Department of Agricultural Economics. During 1990–93 he held an assignment with the U. S. Agency for International Development and the Instituto de Estrategias Agropecuarias in Quito, Ecuador in August of 1990. He is also an Associate Fellow of the London Environmental Economics Centre.

Ivar Strand is in the Department of Agricultural and Natural Resource Economics, University of Maryland, College Park, MD 20740, USA.

Somthawin Sungsuwan is a doctoral candidate in the Department of Public Policy Analysis, University of Pennsylvania, Philadelphia PA 19104, USA.

Jeffrey Vincent is an Institute Associate of the Harvard Institute for International Development, One Eliot Street, Cambridge, MA 02138, USA and Associate at the Institute of Strategic and International Studies, 1 Pesiaran Sultan Salahuddin, Peti Surat 12424, 50778 Kuala Lumpur, Malaysia.

I
The issues

1

Saving the world's tropical forests

David Pearce and Katrina Brown

1.1 Introduction

The destruction of tropical forests has received worldwide attention because of the unique rôle which they play in evolutionary and ecological terms; the diversity of functions they serve; and the accelerating threat to their existence. The purpose of this book is to explain why these forests are being destroyed by compiling recent studies which test various hypotheses as to the causes of deforestation in the tropics. Some of these studies examine deforestation as a global phenomenon, whereas others look at the issue at a national or regional level. This first chapter presents the latest data on current rates of tropical deforestation; discusses the benefits of conservation and the likely impacts of deforestation; summarizes the underlying causes of tropical deforestation and the findings of the collected studies; and makes suggestions for potential solutions to excessive deforestation. We concentrate specifically on the *global* benefits of forests, and the solutions suggested propose how these might be realized so that the tropical countries containing forests are compensated by the international community. Such mechanisms therefore provide incentives for the conservation and sustainable management of tropical forests.

1.2 Benefits of forest conservation and the impacts of deforestation

An underlying assumption of this volume, and the main reason why the causes of tropical deforestation is researched, is that deforestation has negative environmental impacts and hence leaves those humans who benefit from the use or existence of forests worse off. It is usually assumed that deforestation degrades ecosystems and that other species are put under stress. Many of the benefits of forest conservation as well as the impacts of deforestation are globally observed and well-documented, but as these benefits are not universally accepted (see Wood 1993, and riposte by Fearnside 1993), it is worth reiterating them here.

Tropical forests are home to many millions of people including numerous indigenous groups, and comprise the richest diversity of flora and fauna – biological diversity or biodiversity. Tropical forests cover some seven per cent of the Earth's surface, and contain 50 per cent of all species. This biodiversity is valuable for a number of different reasons: direct use, education, and scientific research. The tropical forests provide timber and a whole range of other products, known as non-timber forests products, such as fruit, nuts, oils, latex and other exudates, medicines, building materials, meat and many other products which meet the subsistence needs of forest inhabitants, and are traded both locally and in international markets. In addition forests provide ecological and environmental functions which include watershed and soil protection, and climate regulation. Forests are also perceived to have intrinsic value and in many cases are vested with profound spiritual and cultural, as well as economic and scientific, importance.

These benefits accrue through the existence and continuing use of forests, and so deforestation has a number of deleterious impacts on livelihoods of forest inhabitants and on the local and global environment. In addition, deforestation causes the local and global extinction of species, some of which remain unknown and unrecorded to science. This chapter concentrates on the global aspects of deforestation: not because these are any more important than either local or national impacts, but because these are well represented in the literature (see for example, Anderson 1990, Redford & Padoch 1992, Grainger 1993, Panayotou & Ashton 1993), and because the global aspects are more amenable to action at the international level. Some authors argue very strongly for focus on *local* solutions to deforestation (see Banuri & Marglin 1993, The Ecologist 1993), and we would contend that such strategies do not conflict with the global mechanisms we present here, and that both are necessary and should be complementary.

Deforestation and greenhouse gas emissions

The atmospheric concentrations of greenhouse gases, including carbon dioxide (CO_2) and methane (CH_4) have been increasing steadily in the last century or more with the likely consequences of changes in the global climate through the so-called greenhouse effect. The increases are a direct result of the burning of fossil fuels, of other industrial processes, and of other interventions in the global carbon cycle such as deforestation and other land-use changes. The Intergovernmental Panel on Climate Change's central estimate indicates that in the 1980s deforestation accounted for 1.6 billion tonnes of carbon emissions (compared to 5.4 billion tonnes from fossil fuel use) (IPCC 1990: 13). Table 1.1 shows that tropical deforestation may have contributed approximately one fifth to one quarter of the greenhouse effect in the 1980s. Deforestation is therefore a major cause of future climate change, the impacts of which will include sea-level changes; changes in the frequency of extreme events such as hurricanes and droughts; and changes in the productivity of plants, and hence of agricultural and natural ecosystems throughout the world. Forests not only play a rôle in the global carbon cycle, but are also likely to experience the major impacts of changes in the Earth's atmosphere and climate (Schneider 1992a).

Carbon is released as a result of deforestation at different rates according to the method of clearance and subsequent land-use. With burning there will be an immediate release of CO_2 into the atmosphere, and some of the remaining carbon will be locked in ash and charcoal which is resistant to decay. If forests are simply cut and allowed to decay, as they may be after clear-felling, then most of the carbon is released to the atmosphere within 10–20 years. Studies of tropical forests indicate

Table 1.1 Tropical deforestation and biotic sources of greenhouse gases.

	Annual emissions	% total emissions	% contribution to greenhouse effect in 1980s Total[a]	From deforestation
CO_2			50	
Industrial	5.6 Pg C			
Biotic[b]	2.0–2.8 Pg C			
Tropical deforestation[c]	2.0–2.8 Pg C	26–33		13–16
CH_4			20	
Industrial	50–100 Tg C			
Biotic[b]	320–785 Tg C			
Tropical deforestation	155–340 Tg C	38–42		8
N_2O[d]			5	
Industrial	<1[d] Tg N			
Biotic[b]	3–9[d] Tg N			
Tropical deforestation	1–3[d] Tg N	25–30[d]		1–2
CFCs			20	
Industrial	700 Gg			
Biotic[b]	0	0		
TOTAL			95	22–26

Notes: a) The greenhouse gases considered in this table are only those released as a direct result of human activities. Tropospheric ozone, formed as a result of other emissions, contributes about five per cent to the total. The major greenhouse gas, water vapour, is not directly under human control but will increase in response to a global warming (positive feedback); b) Biotic emissions include emissions from tropical deforestation as well as natural emissions; c) Relatively little of this CH_4 is emitted from deforestation; most of the emissions result from rice cultivation or cattle ranching, land-uses that replace forest. Additional releases occur with repeated burning of pastures and grasslands; d) Estimates. *Source: Houghton 1993.*

that significant amounts of cleared vegetation become lumber, slash, charcoal and ash. The proportion of these differs for closed and open forests as the smaller stature and drier climate of open forests result in the combustion of higher proportions of the vegetation.

If tropical forested land is converted to pasture or permanent agriculture, then the amount of carbon stored in secondary vegetation is equivalent to the carbon content of the biomass of crops planted, or the grass grown on the pasture. If a secondary forest is allowed to grow, then carbon will accumulate, and maximum biomass density is attained after a relatively short time.

Table 1.2 illustrates the net carbon storage effects of land-use conversion from tropical forests: closed primary, closed secondary, or open forests, to shifting cultivation, permanent agriculture, or pasture. The negative figures represent emissions of carbon; for example, conversion from closed primary forest to shifting agriculture results in a net loss of 204 tC/ha. The greatest loss of carbon involves change of land-use from primary closed forest to permanent agriculture. These figures represent the once and for all change that will occur in carbon storage as a result of the various land-use conversions. The data suggest that, allowing for the carbon fixed by subsequent land-uses, carbon released from deforestation of secondary and primary tropical forest is of the order of 100–200 tonnes of carbon per hectare.

Table 1.2 Changes in carbon with land-use conversion (tC/ha).

		Shifting agriculture	Permanent agriculture	Pasture
	Original C	79	63	63
Closed primary	283	−204	−220	−220
Closed secondary	194	−106	−152	−122
Open forest	115	−36	−52	−52

Note: shifting agriculture represents carbon in biomass and soils in the second year of the shifting cultivation cycle.   *Source: Brown and Pearce 1994.*

Biodiversity loss

Species loss is the other major impact of tropical deforestation, and is often estimated using the species–area model of MacArthur and Wilson (1967). This postulates a mathematical relationship between the number of species present, the area, the biogeographical region, taxon group and population density. Table 1.3 presents some estimates of global species extinctions which illustrate a range of 1–11 per cent of species lost globally per decade. Most of these studies use species–area curves to estimate numbers of species likely to become extinct. A recent study by Lugo et al. (1993) however, suggests that the use of these models overestimates

Table 1.3 Estimated extinctions as a result of global deforestation.

Estimate	% Global loss per decade[a]	Method of estimation	Reference
15–20% of species between 1980 and 2000	8–11	Estimated species–area curve	Lovejoy 1980
2000 plant species per year in tropics and subtropics	8	Loss of half the species in area likely to be deforested by 2015	Raven 1987
25% of species between 1985 and 2015	9	As above	Raven 1988a,b
At least 7% of plant species	7[bc]	Half of species lost over next decade in 10 "hot spots" covering 3.5% of forest area	Myers 1988
0.2–0.3% per year	2–3[b]	Half of rain forest species assumed lost in tropical rain forests to be local endemics and becoming extinct with forest loss	Wilson 1988, 1989
2–13% loss between 1990 and 2015	1–5[b]	Species–area curve range includes current rate of forest loss and 50% increase	Reid 1992

Notes:a) Based on total species number of ten million. Estimates in **bold** indicate the actual loss of species over that period (or shortly thereafter). Estimates in standard type refer to the number of species that will be committed to extinction during that time period as a new equilibrium is attained; b) Estimate refers to number of species committed to eventual extinction when species numbers reach equilibrium following forest loss; c) This estimate applies only to hot spot regions, thus the global extrapolation is conservative.   *Source: Reid 1992.*

likely extinctions as it fails to take account of species gains by immigration or succession. Even if this is the case, then it remains clear that tropical deforestation is likely to cause a considerable number of extinctions, and perhaps the most worrying aspect is that many of these will be of species as yet unknown to science.

Although the loss of biodiversity may be viewed as a *global* cost of deforestation, in many respects it is the costs to local populations of the loss of diversity of species and of other functions of forests which is of primary concern. The use of forests by those people dwelling in or adjacent to them for productive, non-productive and symbolic and cultural purposes is a principal factor in determining a strategy for sustainable utilization (Redford & Padoch 1992). The range of mechanisms available include protecting intellectual property rights on products developed from genetic material originating in forests (Reid et al. 1993), to enhancing the status of traditional herbal-based medicine (Brown 1994) and maximizing the value of non-timber forest products (Plotkin & Famolare 1992). All these mechanisms essentially enhance the value of conserving the forest relative to deforestation and alternative land-uses (Panyotou & Ashton 1993).

1.3 Rate of loss of tropical forests

Uncertainties in the global assessment of impacts of deforestation, for example for CO_2 emissions and biodiversity loss, are partially a function of uncertainty in the rate of deforestation itself. Although forests are a stock of wealth, they are not directly measured over time as part of agricultural or land-use censuses (see, for example, Bilsborrow & Geores, Chapter 8 in this volume). Many deforestation estimates are therefore based on survey, which in recent years utilize remote sensing techniques. Even with these the paucity of accurate data is a perpetual theme of the chapters of this book. Table 1.4 shows the current state of forest inventories in tropical countries: only two out of 40 countries in Africa, and six of 33 in Latin America have conducted more than one inventory, and no assessment was available for three countries in Africa when the survey was carried out as part of the FAO 1990 Forest Assessment.

Historical data are clearly lacking, and even when available, their accuracy may be highly suspect. Methodological issues and the lack of accurate data are highlighted by many of the authors in this volume (see particularly Bilsborrow & Geores, Chapter 8, and Shafik, Chapter 6). Grainger (1993) presents a comparison of various estimates made from 1970–90 which highlights serious discrepancies; Table 1.5 illustrates the range in magnitude of these different estimates. First, it is not clear whether the estimates refer to the same thing, as they use different definitions of deforestation – be it total clearance, selective logging, or degradation – and also different types and categories of forest. Second, methodologies and techniques of measurement differ, and remote sensing has done little to provide more accurate data. Remote sensing techniques offer conflicting estimates,

Table 1.4 State of forest inventory in the tropics.

		Africa	Asia/ Pacific	Latin America/ Caribbean	Total
Number of countries surveyed		40	17	33	**90**
Number of countries with forest resources data at national level					
	No assessment	3	0	0	**3**
Forest area information	One assessment ⎰ Before 1980	17	1	11	**29**
	⎱ 1980–90	18	5	16	**39**
	More than one assessment	2	11	6	**19**
Forest conservation and management		7	9	12	**28**
Forest plantations		5	8	4	**17**
Volume and biomass		2	8	4	**14**
Forest harvesting and utilization		4	7	4	**15**

Source: Singh 1993

Table 1.5 Estimates of rates of deforestation in the humid tropics (million ha per annum).

Source	Date	Period	Total
Sommer[a]	1976	1970s	11–15
Myers[b]	1980	1970s	7.5–20
Grainger[c]	1983	1976–1980	6.1
Myers	1989	late 1980s	14.2
FAO[d]	1990	1981–1990	16.8
FAO[e]	1992	1981–1990	12.2
FAO[f]	1993	1981–1990	15.4

Notes: a) 15 commonly quoted; b) 7.5 a later revision; c) cf. 7.3 for all tropics; d) interim estimate; e) revised interim estimate presented to UNCED; f) 1990 Tropical Forest Assessment as reported by Singh (1993): figure for all tropics.   *Source: Adapted from Grainger 1993.*

depending on resolution and sampling coverage. For example, two estimates of deforestation in Brazilian Amazonia using remote sensing provide very different figures. One estimate by a group in the Brazilian National Space Agency (INPE) estimates a rate of deforestation at 1.7 million ha/yr during 1978–88, whereas another group at the same agency estimated 8.1 million ha/yr for 1987. The different figures were produced as a result of using different resolutions, and the second measured smoke from fires as an indicator of deforestation. Accurate estimates wholly based on remote sensing measurement are not currently available, so on-the-ground monitoring of forest areas or even of stand growth would presently be the only option.

Many of the studies in this volume utilize data from the 1980 FAO/UNEP tropical forest assessment. Completed in 1982, this aimed to provide the first comprehensive and statistically consistent assessment of forest resources, deforestation and afforestation, in tropical countries. Since the completion of the 1980 project, the need for a continuous process of global forest inventory has been made apparent, and a new assessment of tropical forests was initiated in 1990 and intended to be part of an ongoing monitoring and data gathering process. The 1990 assessment consists of two phases: the first based on collation and analysis of existing information, and the second using remote sensing (multidate high-resolution satellite) data (Singh 1993). It also attempts to overcome many of the problems of comparability by issuing guidelines (published by FAO in 1990), and has developed a software programme known as Forest Resources Information System (FORIS) for use on personal computers designed for easy entry, retrieval and storage of data.

The definitions used by the 1990 assessment are as follows: *forests* are defined as ecosystems with minimum of 10 per cent crown cover of trees and/or bamboos, generally associated with wild flora, fauna and natural soil conditions and not subject to agricultural practices. *Deforestation* refers to a change of land-use with the depletion of tree crown cover to less than 10 per cent. Changes within the forest class (from closed to open forest), which negatively affect the site or stand, and in particular, lower the production capacity are termed *degradation*. Degradation is not reflected in the estimates.

Table 1.6 presents data from the 1990 assessment. This shows that total forest area in the tropics has decreased from 1910.4 million hectares in 1980 to 1756.3 million hectares in 1990; this represents a decline of 154 million hectares, or a little over 8 er cent. Annual rates of decrease between 1981 and 1990 are calculated to be 0.7 per cent in Africa, 1.1 per cent in Asia, and 0.7 per cent in Latin America, equivalent to a decline of 0.8 per cent for all tropical countries. Although comparisons are difficult because of changing definitions, deforestation rates may have been lower, probably running at annual rates of about 0.6 per cent, in the late 1970s. Norman Myers, in Chapter 2, discusses the rates and patterns of tropical deforestation in greater detail and sets the scene for the later analyses of the causes of deforestation in tropical countries.

Analysis is not yet complete, but the 1990 assessment aims to provide detailed regional breakdown of land-use change. So far, the African data has been analyzed,

Table 1.6 Estimates of forest cover area and deforestation by geographical subregions.

Geographic regions and subregions	Number of countries surveyed	Total land area (10^6 ha)	Forest area (10^6 ha) 1980	1990	Annually deforested area (10^6 ha)	Rate of change 1981–90 (% per annum)
Africa	**40**	**2236.1**	**568.6**	**527.6**	**4.1**	**−0.7**
West Sahelian Africa	9	528.0	43.7	40.8	0.3	−0.7
East Sahelian Africa	6	489.7	71.4	65.3	0.6	−0.8
West Africa	8	203.8	61.5	55.6	0.6	−0.8
Central Africa	6	398.3	215.5	204.1	1.1	−0.5
Tropical Southern Africa	10	558.1	159.3	145.9	1.3	−0.8
Insular Africa	1	58.2	17.1	15.8	0.1	−0.8
Asia	**17**	**892.1**	**349.6**	**310.6**	**3.9**	**−1.1**
South Asia	6	412.2	69.4	63.9	0.6	−0.8
Continental Southeast Asia	5	190.2	88.4	75.2	1.3	−1.5
Insular Southeast Asia	5	244.4	154.7	135.4	1.9	−1.2
Pacific Islands	1	45.3	37.1	36.0	0.1	−0.3
Latin America	**33**	**1650.1**	**992.2**	**918.1**	**7.4**	**−0.7**
Central America/Mexico	7	239.6	79.2	68.1	1.1	−1.4
Caribbean	19	69.0	48.3	47.1	0.1	−0.3
Tropical South America	7	1341.6	864.6	802.9	6.2	−0.7
TOTAL TROPICS	**90**	**4778.3**	**1910.4**	**1756.3**	**15.4**	**−0.8**

Source: Singh 1993.

and Table 1.7 presents a forest cover change matrix for Africa, showing the changes in land-use between 1980 and 1990. For example, the area of closed forest has declined from 24 per cent in 1980 to 22.6 per cent in 1990, and the area of short fallow land has increased from three per cent in 1980 to four per cent in 1990. Similar data for Latin America and Asia should be available by the end of 1993 (Singh 1993).

As for likely future rates, it is inappropriate to extrapolate rates of loss. As deforestation proceeds, the remaining forest is increasingly characterized by steep slopes or permanent and seasonal flooding which makes it unsuitable for conversion to agriculture and other uses. Sheer unsuitability for conversion may be the ultimate protector of a minimum stock of tropical forest, although even that cannot be safe from regional air pollution or global warming impacts (see Myers, Chapter 2, and Reis & Guzmán, Chapter 12, in this volume). Nonetheless, a rate of approximately one per cent per annum is an alarming loss rate and appears to be almost a doubling of earlier loss rates, although those are very much "guesstimates".

1.4 The fundamental causes of tropical deforestation

Widespread deforestation has significant economic and ecological impacts, but accurate data on precise rates, and in particular on historic rates, are difficult to obtain. The proximate causes of deforestation are the clearing of land for other uses such as agriculture, and unsustainable logging. But the underlying reasons for uneconomic and environmentally destructive decisions about forest and land-use management also need to be explored. For example, Panyotou & Sungsuwan, in Chapter 13, make the distinction between activities such as logging, fuelwood collection, and land clearing for agriculture as *sources* of, as distinct from the *causes* of deforestation. To abstract from specific causes such as population pressure, indebtedness, or the structure of world timber trade, all of which are discussed in later chapters, it is useful to define the *fundamental* forces giving rise to deforestation. These arise from two factors:

- Competition between humans and non-humans for the remaining ecological niches on land and in coastal regions. In turn, this competition reflects the rapidly expanded population growth of developing countries.
- "Failures" in the workings of the international and national economic systems. "Failure" in this sense means the failure of these economic systems to reflect the true value of environmental systems in the working of the economy. Essentially, many of the functions of tropical forests are not marketed and, as such, are ignored in decision-making. Additionally, decisions to convert tropical forest are themselves encouraged by fiscal and other incentives for various reasons.

Table 1.7 Forest cover change matrix for the tropical African region.

Classes at year 1980	Area of classes at year 1990									Total at year 1980 ('000 ha)	(%)
	Closed forest	Open forest	Forest + shifting cultivation	Fragmented forest	Shrub	Short fallow	Other land cover	Water	Planted woody vegetation		
Closed forest	16781	382.1	82.6	291.8	9.5	524.3	247.5	–	–	18319.3	24
Open forest	23.6	10049	48.3	371.2	12.7	117.8	397.3	0.1	1.4	11021.8	14
Forest + shifting cultivation	7.7	14.6	556.8	1.6	4.4	51.7	28.5	–	–	665.40	0
Fragmented forest	24.1	40.0	1.0	8088.8	7.7	5.8	293.5	0.1	–	8461.1	11
Shrub	0.8	10.8	–	1.1	3877.9	–	164.3	0.1	–	4055.1	5
Short fallow	7.6	10.9	9.6	2.1	–	2254.8	53.3	0.4	–	2338.6	3
Other land cover	16.9	38.2	11.0	63.1	86.6	34.3	26452	51.2	–	26752.8	35
Water	0.5	–	–	0.5	0.1	3.2	81.5	2960.1	–	3045.9	4
Plantations	–	–	–	–	–	0.4	0.4	–	4.6	5.3	0
Total 1990	16863	10546	709.2	8820.2	3999.0	2992.4	27718	3011.9	6.0	74665.2	
Percentage of total land area	22.6	14.1	0.9	11.8	5.4	4.0	37.1	4.0	0.0		100

Source: Singh 1993.

Other factors are at work as well. In a comprehensive review we would need to add misdirected past policies by bilateral and multilateral aid agencies, corruption, the indifference of much big business to environmental concerns, the results of international indebtedness and poverty itself. But, contrary perhaps to widespread opinion, these factors are not well understood and a popular mythology has emerged about them, usually more in keeping with preconceived political agendas than with any respect for proper research and judgement. The remaining essays in this volume directly address these issues in an attempt to divorce myth from analysis.

Competition for space

Most of the competition for space between man and other species is demonstrated by the conversion of land to agriculture, aquaculture, infrastructure, urban development, industry and unsustainable forestry. Table 1.8 shows land-use conversions by world region between 1977 and 1987.

The loss of the world's forests – rich sources of biodiversity – is apparent, especially in Asia and South America. Unless the reasons for these conversions are understood, the outlook for the conservation of biodiversity is bleak.

Population pressure is clearly a force of some considerable importance. Whereas

Table 1.8 Land conversions 1977/9–1987/9 (million hectares).

	Cropland	Pasture	Forest	Other
Africa	+8	−4	−25	+22
N&C America	+3	+11	+7	−20
S America	+14	+20	−41	+11
Asia	+4	−2	−29	+29
Europe	−2	−3	+2	+4

Note: other land includes roads, uncultivated land, wetlands, built-on land.
Source: estimated from World Resources Institute 1992, Table 17.1.

Table 1.9 World population projections.

	1990	2100	2150
World Population (billion)	5.4	12.0	12.2
Per cent in:			
Asia/Oceania	59.4	57.0	56.8
N & S America	13.7	11.0	10.8
Africa	11.9	23.9	24.5
Europe	15.0	8.1	7.9

Source: World Bank 1992a.

humans compete only marginally for niche space in the world's oceans, they compete directly with other species for land and coastal waters space. The exponential nature of historic world population change is well known. Table 1.9 records World Bank projections for the next 160 years. World population is expected to stabilize at around 12 billion people towards the end of the next century, but this is more than twice the number of people on Earth today. The fastest growth rate is in Africa, currently growing at 2.9 per cent p. a. and heading for a population of three billion people towards the end of the next century, around five times the population of today.

These figures suggest that sheer pressure of human beings for space will displace other living species. Some authors have argued that the benefits of this process probably outweigh the costs (for example, Simon 1981, 1986), but one suspects that such views arise from a lack of appreciation of the economic values embodied in biodiversity (Pearce et al. 1992). The displacement hypothesis should, however, be tested empirically – an issue we return to shortly.

The economic theory of species extinction was developed mainly in the water context and very largely in terms of the fisheries (see Clark 1990). There it is comparatively straightforward to see that a combination of *open access property rights* (no one owning the sea) and profitability (the difference between revenues and the cost of fishing effort) does much to explain overfishing and the loss of mammalian species. Once the theory is moved to land, the additional factor is the sheer competition for niche occupancy, and the theory needs to change, as Swanson (1993) has demonstrated. Rather than saying open access explains excess harvesting effort, the question is why nation states allow open access conditions to prevail. Put another way, why don't governments invest more in conservation land-uses? There are three immediate reasons and they comprise the second major strand in the explanation for tropical forest loss: an economic theory of deforestation.

Economic failure

Local market failure

First, under-investment arises for the classic economic reason of "market failure". What this means is that the interplay of market forces will not secure the economically correct balance of land conversion and land conservation. This is because those who convert the land do not have to compensate those who suffer the local consequences of that conversion – extra pollution and sedimentation of waters from deforestation, for example. The corrective solutions to this problem are well known – a tax on land conversion, zoning to restrict detrimental land-uses, environmental standards, and so on. From this it follows that we cannot rely on market forces to save the world's environmental assets as long as those assets are dispensable because of market forces. Notice, however, that the measures needed to correct this market failure do not result in zero deforestation. In the economist's language there is an "optimal" rate of loss. It is less than what happens now, but it is not zero.

Intervention failure

Secondly, and a more recent explanation, is "intervention failure" or "government failure" – the deliberate intervention by governments in the working of market forces. As we shall see, this can co-exist with market failure: both forces are at work at the same time. The examples are, by now, well known (Pearce & Warford 1993) and include the subsidies to forest conversion for livestock in Brazil up to the end of the 1980s; the failure to tax logging companies sufficiently, giving them an incentive to expand their activities even further; the encouragement of inefficient domestic wood processing industries, effectively raising the ratio of logs, and hence deforestation, to wood product; and so on. What intervention does is to distort the competitive playing field. Governments effectively subsidize the rate of return to land conversion, tilting the economic balance against conservation.

Table 1.10 assembles some information on the scale of the distortions that gov-

Table 1.10 Economic distortions to land conversion.

Agricultural price subsidies[a]		
Mexico	mid-1980s	+54%
Brazil	mid-1980s	+10%
South Korea	mid-1980s	+55%
Sub-Saharan Africa	mid-1980s	+9%
OECD	1992	+44%
Timber stumpage fees as percent replacement costs[b]		
Ethiopia	late 1908s	+23%
Kenya	late 1980s	+14%
Côte d'Ivoire	late 1980s	+13%
Sudan	late 1980s	+4%
Senegal	late 1980s	+2%
Niger	late 1980s	+1%
Timber charges as per cent of total rents		
Indonesia	early 1980s	+33%
Philippines	early 1980s	+11%

Notes: a) Producer subsidies are measured by the "Producer Subsidy Equivalent" (PSE) which is defined as the value of all transfers to the agricultural sector in the form of price support to farmers, any direct payments to farmers and any reductions in agricultural input costs through subsidies. These payments are shown here as a percentage of the total value of agricultural production valued at domestic prices; b) A stumpage fee is the rate charged to logging companies for standing timber. It is expressed here as a percentage of the cost of reforesting and as a percentage of total rents.
Source: agricultural PSEs from Moreddu et al. (1990), OECD (1993); stumpage fees from World Bank (1992a), and Repetto & Gillis (1988).

ernments introduce. Such distortions are widespread. The general rule in developing countries is for agriculture to be *taxed*, not subsidized, but significant subsidies exist in several major tropical forest countries such as Brazil and Mexico. By comparison, OECD countries are actually worse at subsidizing agriculture. In 1992 OECD subsidies exceeded $180 billion (OECD 1993). These subsidies work in two ways. Subsidies in developing countries will tend to encourage extensification of agriculture into forested area. Subsidies in the developed world make it impossible for the developing world to compete properly on international markets, locking them into primitive agricultural practices. While the removal of OECD country subsidies would appear to be a recipe for expanding land conversion in the developing world to capture the larger market, the demands of a rich overseas market are more likely to result in agricultural intensification and hence reduced pressure on forested land. Table 1.10 also shows that many developing countries fail to tax logging companies adequately, thus generating larger "rents" for loggers. The larger rents have two effects: they attract more loggers and they encourage existing loggers to expand their concessions and, indeed, to do both by persuading the host countries to give them concessions. Persuasion involves the whole menu of usual mechanisms, including corruption.

Global appropriation failure

The rate of return to forest conservation is distorted by what economists call "missing markets". What this means in the tropical forest context is that systems of habitat and species are serving valuable functions which are not marketed. Effectively, then, no one values these functions because there is no obvious mechanism for capturing the values. Local market failure describes this phenomenon within the context of the country or local area. But there are missing global markets as well. We can illustrate this by the example of the value of carbon storage by tropical forests, which is discussed below. Other global benefits of tropical forests, such as non-use and existence values are also not reflected in global markets. Such values can be estimated using a number of different techniques including contingent valuation method (CVM), travel cost method, hedonic property price and production function approaches (Georgiou 1993). Global appropriation failure (or GAF) arises because these values are not easily captured or appropriated by the countries in possession of tropical forests.

All forests store carbon so that, if cleared for agriculture there will be a release of carbon dioxide which will contribute to the accelerated greenhouse effect and hence global warming. In order to derive a value for the "carbon credit" that should be ascribed to a tropical forest, we need to know (a) the net carbon released when forests are converted to other uses, and (b) the economic value of one tonne of carbon released to the atmosphere.

The carbon released from burning tropical forests contributes to global warming, and we now have several estimates of the minimum economic damage done by global warming, leaving aside catastrophic events. Recent work by Fankhauser (1993) suggests a "central" value of $20 of damage for every tonne of carbon released. Applying this figure to the data in Table 1.2, we can conclude that converting an open forest to agriculture or pasture would result in global warming damage of, say, $600–1000 per hectare; conversion of closed secondary forest would cause damage of $2000–3000 per hectare; and conversion of primary forest to agriculture would give rise to damage of about $4000–4400 per hectare. Note that these estimates allow for carbon fixation in the subsequent land-use.

How do these estimates relate to the development benefits of land-use conversion? We can illustrate this with respect to the Amazon region of Brazil. Schneider (1992b) reports upper bound values of $300 per hectare for land in Rondônia. The figures suggest carbon credit values 2–15 times the price of land in Rondônia. These "carbon credits" also compare favourably with the value of forest land for timber in, say, Indonesia, where estimates are of the order of $2000–2500 per hectare. All this suggests the scope for a global bargain. The land is worth $300 per hectare to the forest colonist but several times this to the world at large. If the North can transfer a sum of money greater than $300 but less than the damage cost from global warming, there are mutual gains to be obtained.

Note that if the transfers did take place at, say, $500 per hectare, then the cost per tonne carbon reduced is of the order of $5 tC ($500/100 tC/ha). These unit costs compare favourably with those to be achieved by carbon emission reduction policies through fossil fuel conversion. Avoiding deforestation becomes a legitimate and potentially important means of reducing global warming rates. The kinds of policies and mechanisms which may attempt to capture GAFs are discussed in later sections.

1.5 Testing causal explanations of deforestation

The previous sections have been concerned with explanations of deforestation which contain only *some* testable hypotheses. The link between deforestation and population change for example is inherently testable by statistical technique. That between "missing markets" and deforestation is not directly testable. The rest of this volume is concerned with the testable hypotheses linking deforestation to poverty, income growth, external indebtedness, structural adjustment, population growth and densities, and other factors. Table 1.11 summarizes the findings of many of these studies, and also presents those of other studies not included, though frequently cited in the text. (See also Chapter 10 by Kummer & Sham which reviews earlier quantitative work on deforestation). Not all the studies in this volume are directly amenable to the matrix form of the table, for example some test multiple factors affecting deforestation, while others produce inconclusive results. It is of course, essential to consult the original sources for the detail, caveats and general context.

The studies test a number of hypotheses using different econometric and statistical techniques. It is therefore difficult to generalize across a range of studies which employ different techniques and test different variables at different levels of analysis. Many of the studies test causal factors of deforestation globally, using data obtained from a number of tropical countries. These studies comprise Part II of the book. A number of the chapters examine the impact of population on deforestation. These include Chapters 3, 4, 5, 7, and 8. The studies generally indicate a positive correlation between population growth and deforestation. However, they also highlight the complex causal relationships (see Bilsborrow & Geores in Chapter 8, and Palo, Chapter 3) and the problems associated with lack of historic data, as pointed out by Chakraborty in Chapter 16. The area of forest present in a country today is the result of deforestation in the past, and if accurate data are not available on historic forest cover, then it is difficult to test the relationship between forest cover today and population increases. Cross-country studies also reveal the importance of particular "outlier" countries in influencing results (see Bilsborrow & Geores, Chapter 8), and so again, caution should be exercised in interpreting the results presented in Table 1.11. Income, particularly as an indicator of poverty, is examined by Capistrano in Chapter 5, Rudel in Chapter 7, and Shafik in Chapter 6. Macroeconomic factors, including indebtedness and foreign exchange earnings are tested by Kahn & McDonald in Chapter 4, Capistrano in Chapter 5, and Shafik, Chapter 6.

In the individual country studies and regional analyses presented in Part III, similar factors are tested; for example, Panyotou & Sungsuwan, Chapter 13, find that population, income level and price of wood were the most important causes of deforestation in northeast Thailand. The effects of infrastructural developments such as road density and kilometres of road constructed are the subject of the study by Kummer & Sham (Chapter 10) which examines deforestation in post-war Philippines. Other chapters look at agricultural land-use, productivity and prices to test their effects on deforestation. Is deforestation encouraged by higher productivity and profitability of agriculture, or by the need to expand farming on the agricultural frontier? In other words are there pull or push factors operating to encourage farmers to expand? Chapters 9, 11, and 8 examine these issues. The Latin American case contrasts with the Indian, as presented in Chakraborty's study (Chapter 16), where frontier agriculture is not a major cause of forest destruction; this is presumed to be because of strict enforcement of forest boundaries. These effects may be encouraged by government policies which fund road building, and

Table 1.11 Selected econometric studies on deforestation.

Study	Type of analysis	Dependent variable	Independent variables					Other significant variables
			Population	Population density	Income	Agricultural productivity	External indebtedness	
Shafik, this volume Ch. 6	Panel regression of the causes of deforestation, 3 models	annual rate of deforestation 1962–85, 66 countries			per capita GDP not significant		debt per capita not significant	investment rate – positive, electricity tariff – negative, trade shares in GDP – negative, political rights – positive, civil rights – positive
		Total deforestation 1961–86, 77 countries			not significant	not significant		
Burgess, 1992	Cross-sectional analysis of deforestation in 53 countries	change in closed forest area 1980–85		negative 0.01	real GNP per capita in 1980 positive 0.05			roundwood production per capita 1980 – negative 0.05, log of closed forest area as a percentage of total forest area in 1980 – positive 0.1
Burgess, 1991	Cross-sectional analysis of deforestation in 44 countries	model 1 level of deforestation	population growth negative 0.05		GDP per capita positive 0.05		debt-service ratio as a % of exports positive 0.05	total roundwood production – positive 0.05
		model 2 level of deforestation	negative 0.01				positive 0.10	food production per capita – positive 0.10, total roundwood production – positive 0.05
Southgate, this volume Ch. 9	OLS regression analysis of causes of agricultural colonization in 23 Latin American countries	growth in the area used to produce crops and livestock	population growth positive 0.01			negative 0.01		agricultural export growth – positive 0.05

Study	Model	Dependent variable	Population growth	Income per capita	Debt-service	Other variables
Kahn & McDonald, this volume Ch. 4	2-stage-least-squares model to show economic mechanisms by which debt may lead to deforestation	deforested area (1000 ha)	negative 0.017 (significant alpha level)	positive 0.01		forested land area – positive 0.01; annual change in public external debt – positive 0.05
Capistrano, this volume Ch. 5	OLS analysis of macroeconomic factors of deforestation in 45 countries 1967–85, 2 linear models	depletion of broadleaved forests during four periods, 1969–71, 72–75, 76–80, 81–85	positive 0.01 P2	GNP per capita positive 0.01 P2, P3	debt-service ratio negative	log export value – positive 0.01 P1; real devaluation rate – positive 0.01 P3; cereal self sufficiency ratio – positive 0.01 P2; arable land per agricultural capita – positive 0.01 P4
Rudel, this volume Ch. 7	decline in closed tropical forests for 36 countries across Africa, Asia and Latin America	average annual decline in hectares of a country's tropical forests during the period 1977–80	population growth positive 0.001; rural population growth positive 0.01	GDP per capita positive 0.01		forest land area – positive 0.001
Paulo, Mery & Salmi, 1987	cross-sectional test of factors influencing deforestation in 72 countries	absolute forest cover in 1980	negative 0.01			food production per capita – negative 0.01; share of forest fallow – positive 0.01; agricultural area coverage – negative 0.1

Study	Type of analysis	Dependent variable	Independent variables					Other significant variables
			Population	Population density	Income	Agricultural productivity	External indebtedness	
Allen & Barnes, 1985	deforestation during 1968–78 in 39 countries in Africa, Asia and Latin America	model 1: annual change in forest areas model 2: the decade change in forest area, 1968–78	population growth negative 0.10					logarithm of % forest cover 1986 – positive 0.10 % area under plantation crops in 1968 – negative 0.05 per capita wood fuels consumption and wood exports in 1968 – negative 0.05
Lugo, Schmidt & Brown, 1981	deforestation in all greater Caribbean countries	% forest cover		negative 0.001				energy use per unit area – positive 0.001
Reis & Guzman, this volume Ch. 12	Brazilian Amazon deforestation and its contribution to CO_2 emissions	deforestation density				positive 0.001		cattle herd – positive 0.05 logging – negative 0.01
Katila, 1992a	deforestation in Thailand	relative forest cover		negative 0.01		negative 0.05		wholesale price of construction timber – negative 0.01
Constantino & Ingram, 1990	deforestation rates in Indonesia	relative forest cover		negative 0.01	GDP per capita positive 0.05	positive 0.01 (rice production used as proxy)		time – negative 0.01

Study	Type	Dependent variable			Other variables
Kummer & Sham, this volume Ch. 10	deforestation in post-war Philippines	cross-sectional analysis for the years 1957, 1970 and 1980	negative 0.05 1970 and 1980		road density – negative 0.05 1957, 1970 & 1980
		absolute amount of forest cover per province (ha)			kilometres of road – positive 0.05, 1980
		panel analysis of absolute loss of forest cover, 1970–80 per province (ha)			forest area – positive 0.05; distance from Manila – positive 0.05; logging in 1970 – positive 0.05
Panayotou & Sungsuwan, this volume Ch. 13	deforestation in northeast Thailand	forest cover	negative 0.01	positive 0.01 (provincial income)	wood prices – negative 0.01; distance from Bangkok – positive 0.01; rural roads – negative 0.10; rice yields – positive 0.10; price of kerosesn – negative 0.01
Southgate, Sierra & Brown, 1989	deforestation in 20 cantons in Eastern Ecuador in early 1980s	deforestation	agricultural population 0.05		tenure security – negative 0.06

support colonization of border regions, and give distorting subsidies to alternative land-uses such as cattle ranching and provide incentives to deforest. Such policies are discussed by Mahar & Schneider in Amazonia (Chapter 11), Osgood in Indonesia (Chapter 15), and Southgate in Latin America (Chapter 9). Ecological impacts are also touched upon, and Reis and Guzmán model carbon releases due to deforestation in Amazonia in Chapter 12.

Aspects of the international timber trade are explored in Part IV. In Chapter 17 Barbier et al. present a partial equilibrium timber trade model to explain the effects of various policy interventions on the trade and tropical deforestation in Indonesia, a country in the process of forest-based industrialization (see also Chapter 15 by Osgood). The study also looks at the likely effects of the introduction of sustainable timber management strategy and asks how the government of Indonesia could provide sufficient incentives to make sure such a strategy is implemented. Vincent's chapter (19) demonstrates how the "boom and bust" nature of tropical countries' participation in the international timber trade is probably due to artificially low prices of tropical timbers on the international market, although unsustainable exploitation of forests has also been exacerbated by domestic policies related to timber concessions and wood processing industries (also illustrated by Osgood in Indonesia). In general the findings of these studies indicate that trade interventions, such as import bans, taxes and quantitative restrictions, as advocated by many developed country environmentalists, will actually reduce incentives for sustainable forest management and may even increase overall tropical deforestation.

1.6 Controlling tropical deforestation: creating global environmental markets

Many of the studies in this volume suggest ways in which the factors causing excessive deforestation, and the unsustainable management of forests could be combated. These include policies which attempt to remedy the economic failures which we have identified earlier in this chapter. These include recommendations on property rights and land tenure (for example, in India, Chapter 16, and Latin America, Chapter 11) and on reforms in the international timber trade (Chapters 18 and 19). Many of these policies involve enhancing the appropriation of benefits of forests which are not normally captured by either local resource managers, or nation states. In this section we will discuss one aspect of this appropriation, and continue the theme introduced earlier, and examine mechanisms for capturing some of the global benefits of tropical forests, and how these may be transferred back to the countries owning such resources.

There are several ways in which global appropriation failure can be corrected

Table 1.12 A schema for global environmental markets.

	Regulation-induced	Spontaneous market
Public/official ventures	Examples: government-to-government measures under joint implementation provisions of the Rio treaties:Norway, Mexico, Poland, GEF	Examples: government involvement in market ventures: Swiss green export aid; debt-for-nature swaps
Private sector ventures	Examples: carbon offsets against carbon taxes and externality adders	Examples: purchase of exotic capital – Merck and Costa Rica

through creating global environmental markets (GEMs). We distinguish between private and public ("official") ventures, and between those that are regulation-induced and those that are "spontaneous market" initiatives. Public regulation-induced activity arises because of international agreements, such as the Biodiversity and Climate Change Conventions. Table 1.12 sets out the resulting schema. Some of the examples are now discussed.

The first way in which markets are emerging is via the existence of regulations or anticipated regulations. In turn, these regulations are international and national but since implementation is always at the national level we can treat them together.

In terms of public or official ventures, government–government trades may take place under international regulations, such as the Climate Change Convention when it comes into force. Under the Convention each ratifying country will have an obligation to cut back on CO_2 emissions, but the Convention quite explicitly recognizes that it is often cheaper for one country to cut back on emissions in another country, besides making its own domestic effort to cut back. Similarly, it may be cheaper to create "sinks" for CO_2 in another country compared to cutting back domestically (Brown & Adger 1993). This scope for "carbon offsets" or "joint implementation" is potentially large, and the first joint implementation agreement has already been agreed between Norway, Poland and Mexico, through the medium of the Global Environment Facility (GEF). Norway agrees to create additional financing (through the revenues from its own carbon tax) for GEF carbon-reducing projects in Mexico (energy efficient lighting) and Poland (converting from coal burning to natural gas) (GEF 1992). The US Environmental Defence Fund is understood to be in the process (1993) of developing a reforestation project in Russia. The US government announced the Forest for the Future Initiative (FFI) in January 1993 under which carbon offset agreements will be negotiated between the USA and several countries, including Mexico, Russia, Guatemala, Indonesia and Papua New Guinea. The aim is for the US Environmental Protection Agency to broker deals involving the private sector.

As yet, the procedures for joint implementation under the Convention are not agreed and it is likely that more deals will develop once the ground rules have been established.

Private sector trades may also be facilitated by regulation. For example, the European Community Draft Directive on a carbon tax and other European legislation also provides an incentive to trade in this way, as does state regulation on pollution by electric utilities in the USA. While not strictly a private enterprise trade, in the Netherlands, the state electricity generating board (SEP) has established a non-profit making enterprise (FACE – Forests Absorbing Carbon Dioxide Emissions) and is planning to invest in forest rehabilitation to absorb CO_2 in Indonesia, Ecuador, Costa Rica and the Netherlands itself. The FACE Foundation already has a contract with Innoprise in Sabah, Malaysia for the regeneration of degraded forest lands.

In the US case the offset deals are currently not directly linked to legislation, but several have occurred which are clearly a mix of anticipation of regulation and "global good citizenship" (Newcombe & de Lucia 1993). These include the New England Electric System's investment in carbon sequestration in Sabah, Malaysia through the reduction of carbon waste from inefficient logging activities. New England Electric estimate that some 300,000 to 600,000 tonnes of carbon (C) will be offset at a cost of below \$2/tC. Many recent offset deals are reviewed in Dixon et al. (1993).

While these investments are aimed at CO_2 reduction, sequestration clearly has the potential for generating joint benefits, i.e. for conserving biodiversity as well through the recreation of habitats. Much depends here on the nature of the offset. If the aim is CO_2 fixation alone, there will be a temptation to invest in fast growing species which could be to the detriment of biodiversity. It is important therefore to extend the offset concept so that larger credits are given for investments which produce joint biodiversity – CO_2 reduction benefits.

Several offset deals appear to have been undertaken quite independently of legislation or anticipation of regulation. Applied Energy Services (AES) of Virginia has also undertaken sequestration investments in Guatemala (agroforestry) and Paraguay and is in the process of setting up another project in the Amazon basin. The Guatemala project is designed to offset emissions from an 1800 MW coal fired power plant being built in Uncasville, Connecticut. The intermediary for the project is the World Resources Institute and in Guatemala the implementing agency is CARE. The project involves tree planting by some 40,000 farm families. Carbon sequestration is estimated to be 15.5 million tonnes of carbon. The \$14 million cost includes \$2 million contribution from AES; \$1.2 million from the government of Guatemala; \$1.8 million from CARE, with the balance coming in-kind from USAID and the Peace Corps. Table 1.13 shows some examples of carbon offset deals agreed or negotiated to date.

Financial transfers may take place without any regulatory "push". The consumer

Table 1.13 Private sector carbon offset deals.

Company	Project	Other participants	Million tC sequestered	Total cost $ million	$ tC sequestered
AES	Agro-forestry Guatemala	US CARE Govt of Guatemala	15–58 over 40 years	14	a) 0.5–2 b) 1–4 c) 9
AES	Agro-forestry Paraguay	US Nature Conservancy FMB	13 over 30 years	2	a) 0.2 b) 0.4 c) <1.5
NEES	Forestry, Malaysia	Rain Forest Alliance, COPEC	0.3–0.6, period not stated	0.45	a) n.a. b) n.a. c) <2
SEP	Reforestation, Malaysia	Innoprise	? over 25 years	1.3	a) n.a. b) n.a. c) n.a.
Tenaska	Reforestation, Russia	Trexler, Ministry of Ecology, Russian Forest Service etc.	0.5 over 25 years	0.5?	a) n.a. b) n.a. c) 1–2
PacifiCorp	Forestry, Oregon	Trexler	0.06 p.a.	0.1 p.a.	a) n.a. b) n.a. c) 5
PacifiCorp	Urban trees, Utah	Trexler, TreeUtah	?	0.1 p.a.	a) n.a. b) n.a. c) 15–30

Notes: a) assumes 10% discount rate applied to total cost to obtain an annuity which is then applied to carbon fixed per annum, assuming equal distribution of carbon sequestered over the time horizon indicated; b) assumes 4% discount rate applied to costs; c) cost per tC as reported in Dixon et al. (1993). *Sources: adapted from Dixon et al. 1993, and Bann 1993.*

demand for green products has already resulted in companies deciding to invest in conservation either for direct profit or because of a mix of profit and conservation motives. The Body Shop is an illustration of the mixed motive, as is Merck's royalty agreement with Costa Rica for pharmaceutical plants and Pro-Natura's expanding venture in marketing indigenous tropical forest products.

The deal between Merck & Co, the world's largest pharmaceutical company, and INBio (the National Biodiversity Institute of Costa Rica) is already well documented and studied (Gámez et al. 1993, Sittenfield & Gámez 1993, Blum 1993). Under the agreement, INBio collects and processes plant, insect and soil samples in Costa Rica and supplies them to Merck for assessment. In return, Merck pays Costa Rica $1 million plus a share of any royalties should any successful drug be developed from the supplied material. The royalty agreement is reputed to be of the order of one to three per cent, and to be shared between INBio and the Costa Rican government. Patent rights to any successful drug would remain with Merck. Biodiversity is protected in two ways – by conferring commercial value of the biodiversity, and through the earmarking of some of the payments for the Ministry of Natural Resources.

How far is the Merck–INBio deal likely to be repeated? Several caveats are in order to offset some of the enthusiasm over this single deal. First, Costa Rica is in the vanguard of biodiversity conservation, as its strong record in debt-for-nature swaps shows. Secondly, Costa Rica has a strong scientific base and a considerable degree of political stability. Both of these characteristics need to be present and their combination is not typical of many developing countries. Thirdly, the economic value of such deals is minimal unless the royalties are actually paid and that will mean success in developing drugs from the relevant genetic material. The chances of such developments are small – perhaps one in one to ten thousand of plant species screened (Principe 1991).

1.7 Conclusion

These international aspects of appropriating the benefits of tropical forests represent just one mechanism which can be utilized to create greater incentives for the conservation and sustainable use of tropical forests. Many other policies, at local, national and international level need to be employed (see for example, Panyotou & Ashton 1993 for a discussion of managing tropical forests for multiple use). There are many suggestions contained in the studies presented within this volume. However, it is clear that only when we understand more fully the complex causes of unsustainable exploitation and degradation of tropical forests can we formulate coherent policy to prevent destruction of these resources, the value of which we are just beginning to appreciate.

2

Tropical deforestation: rates and patterns

Norman Myers

2.1 Introduction

Tropical forests, covering around 7.5 million km^2 of the humid tropics, or roughly half of their original expanse, are being destroyed at a progressively more rapid rate. In 1979 an estimated 75,000 km^2 was lost (FAO 1981a, Myers 1980; see also Houghton et al. 1985, Lanly 1982, Melillo et al. 1985, Molofsky et al. 1986). This paper reviews the situation as of 1991, when the annual amount had risen to at least 132,000 km^2 according to a recent survey by this writer (Myers 1989, 1992a, b); or, according to an estimate of the Food and Agriculture Organization (1992a), 126,000 km^2. It would be difficult to make a country-by-country comparison of the two sets of figures because of different definitions and criteria which become unduly complex at country level.

Plainly we need a clear grasp of what forms of forest destruction are at work, how fast they are proceeding, and what we can realistically expect for the future. In this chapter I examine the present situation, consider some recent trends and patterns, and offer a prognosis for the next few decades. There is also an analysis of the main sources of deforestation, with particular reference to the slash-and-burn cultivator.

First, a few definitions. The chapter concentrates on *tropical moist forests*, these being by far the richest forests in biological senses, the most complex in ecological senses, and the most valuable in commercial-timber senses. They are defined as "evergreen or partly evergreen forests, in areas receiving not less than 100 mm of precipitation in any month for two out of three years, with mean annual temperature of 24-plus degrees C and essentially frost-free" (Myers 1980). Moreover, two-thirds of the forests in question are considered to be little if at all disturbed, meaning that they still contain their complement of biodiversity, and they offer a full range of environmental services such as watershed functions, regulation of rainfall regimes, albedo absorption and carbon storage.

Deforestation refers to the complete destruction of forest cover through clearing for agriculture of whatever sort (cattle ranching, smallholder agriculture whether planned or spontaneous, and large scale commodity crop production through e.g. rubber and oil palm plantations). It means that not a tree remains, and the land is given over to permanent non-forest purposes. There are certain instances too where the forest biomass is so severely depleted, notably through the very heavy and unduly negligent logging of *dipterocarp* forests in Southeast Asia, resulting in the removal of, or unsurvivable injury to, the great majority of trees, that the remnant ecosystem is a travesty of natural forest as properly understood. Decline of biomass and depletion of ecosystem services are so severe that the residual forest can no longer qualify as forest is any practical sense of the word. So this particular kind of over-logging is included under the term "deforestation".

As for *agents of deforestation*, it is not necessary to define the commercial logger or the cattle rancher. But it is appropriate to distinguish between various forms of slash-and-burn cultivator. Until the early 1960s the small scale farmer was usually a shifting cultivator of traditional type, one who practised a migratory mode of agri-culture. As long as there were not many of these cultivators and there was plenty of forest for them to shift around in, they made sustainable use of forest ecosystems. But during the past quarter century there has been an influx into forests of landless or otherwise displaced peasants, small scale farmers who are sometimes known as "shifted" cultivators (Myers 1984, Westoby 1989). They have formerly made a liv-ing in long-established farmlands of the countries concerned, often in territories far distant from the forests. For various reasons – population growth, maldistribu-tion of established farmlands, lack of agrotechnologies for intensive cultivation, and inadequate rural development generally, they have increasingly been squeezed out of their erstwhile homelands. Perceiving no alternative way to sustain themselves, they head for the only unoccupied lands available to them: tropical forests.

These shifted cultivators have grown exceptionally numerous in recent decades (Cruz 1988, Palo & Salmi 1988, Peters & Neuenschwander 1988, Schuman & Partridge 1989, Thiesenhusen 1989, Uhlig 1984). Advancing upon the forest fringe in large numbers, they penetrate deeper into the forest season by season.

Behind them come still more displaced and impoverished peasants, leaving the forest no chance to regenerate itself. With scant understanding of how to make sustainable agricultural use of forest ecosystems, these shifted cultivators now cause more forest destruction than all other agents of deforestation combined (Myers 1988a, 1988b, 1992c, World Resources Institute 1985).

On top of deforestation, there is a good deal of forest degradation through less-than-destructive logging, forestland farming and fuelwood gathering. While not documented in detail, this is estimated to amount to rather more than the area deforested (Myers 1989). Since this chapter is confined to outright deforestation, forest degradation is not considered.

2.2 Tropical deforestation: a summary

As noted, the deforestation total for 1991 has been estimated to be at least 132,000 km^2, or 1.8 per cent of remaining forests covering roughly 7.5 million km^2. This 1991 figure contrasts with the amount in 1979, some 75,000 km^2. In 1989 the total had risen to 126,000 km^2, meaning the annual amount had increased during the 1980s by 68 per cent. Table 2.1 shows the situation in 1989, listing the 34 countries that make up 97.3 per cent of tropical forests. The 1991 figure for biome-wide deforestation represents an extrapolation of the 1989 calculation, with some variations for certain individual countries.

A deforestation rate of 1.8 per cent in 1991 (it will have grown somewhat higher by 1994) does not mean that all remaining forests will therefore disappear in another half century at most (actually less due to the compounding effect). Patterns and trends of deforestation are far from even throughout the biome. In Southeast Asia it is likely, supposing recent land-use trends and patterns persist unvaried, that virtually all forest will be eliminated by the end of the century in Thailand and Vietnam, and virtually all primary forest in Philippines and Myanmar. Little forest of whatever sort is likely to remain in another 20 years' time in most of Malaysia, and in Indonesia outside of Kalimantan and Irian Jaya. But in Papua New Guinea with its low population pressures (fewer than four million people in an area of 461,700 km^2, almost the size of Spain or California), there could well be sizeable tracts of forest remaining for several decades into the next century.

A similarly differentiated picture emerges in Africa. If recent land-use patterns and trends persist unchanged, one can realistically anticipate that hardly any forest will remain in Madagascar, East Africa and West Africa beyond the end of the century, due to the combined pressures of population growth and peasant poverty. But in the Zaire basin, comprising Gabon, Congo and Zaire, there are only 43 million people occupying an area of three million square kilometres. Moreover, these countries are so well endowed with mineral resources that their governments sense little urgent need to exploit their forest stocks in order to fund development. However, the situation may change rapidly with the opening of the Trans-Gabon Railway, a large African Development Bank investment in saw milling in Congo, and increased foreign investment for commercial logging in Zaire. Moreover, the population total is projected to reach 105 million, for a 145 per cent increase, as early as the year 2025.

In Latin America, it is difficult to see that much forest can persist long into the next century in Mexico, Central America, the Atlantic-coast sector of Brazil and Amazonian Ecuador. The Colombian *Choco* may survive a while longer. Amazonia presents a mixed picture. The sectors in Peru and Bolivia may join that in Ecuador within another few decades, by being largely eliminated. By contrast, the Venezuelan sector is hardly affected thus far, and much of it may well remain intact for a

Table 2.1 Tropical moist forests: status in selected countries, 1989.

Country	Area (sq. km)	Original extent of forest cover (sq. km)	Extent of forest cover (sq. km)	Amount of deforestation in 1989 (sq. km)	(%)
Bolivia	1,098,581	90,000	70,000	1,700	2.4
Brazil	8,511,960	2,860,000	2,200,000	30,000	1.4
Cameroon	475,442	220,000	164,000	2,300	1.4
Central America	522,915	500,000	90,000	3,300	3.7
Colombia	1,138,891	700,000	278,500	6,500	2.3
Congo	342,000	100,000	90,000	800	0.9
Ecuador	270,670	132,000	76,000	3,000	4.0
Gabon	267,670	240,000	200,000	800	0.4
Guyanas (French Guiana, Guyana and Suriname)	469,790	500,000	410,000	500	0.1
India	3,287,000	1,600,000	165,000	4,300	2.6
Indonesia	1,919,300	1,220,000	860,000	12,000	1.4
Côte d'Ivoire	322,463	160,000	16,000	2,500	15.6
Cambodia	181,035	120,000	67,000	500	0.7
Laos	236,800	110,000	68,000	1,100	1.6
Madagascar	590,992	62,000	24,000	2,000	8.3
Malaysia	329,079	305,000	157,000	4,800	3.1
Mexico	1,967,180	400,000	166,000	7,000	4.2
Myanmar (Burma)	696,500	500,000	245,000	8,500	3.5
Nigeria	924,000	72,000	28,000	4,000	14.3
Papua New Guinea	461,700	425,000	360,000	3,500	1.0
Peru	1,285,220	700,000	515,000	3,700	0.7
Philippines	299,400	250,000	50,000	2,700	5.4
Thailand	513,517	435,000	74,000	6,000	8.1
Venezuela	912,050	420,000	350,000	1,600	0.5
Vietnam	334,331	260,000	60,000	3,500	5.8
Zaire	2,344,886	1,245,000	1,000,000	4,200	0.4
		13,626,000[a]	7,783,500[b]	120,800	1.6

Notes: a) 97% of estimated total original extent of tropical forests, or around 14 million sq. kms; b) 97.3% of present total extent of tropical forests, or 8 million sq. km.

Sources: Myers 1989, 1992a and b.

good while to come; while the Colombian government has recently assigned a large proportion of its Amazonian forest to the care of its tribal peoples, who do not generally engage in destructive forms of forest exploitation.

The Latin American situation is dominated by Brazil, with well over half of the region's forests (and well over one quarter of all tropical forests). There has been creeping attrition of Brazil's Amazonian forest throughout the past two decades, with a sharp acceleration in the deforestation rate during 1987 and 1988, albeit followed by a marked decline in the rate during 1989–92. The 1987 and 1988 rate appeared to average around 50,000 km^2 of forest burned, but by 1991 it had declined to 11,130 km^2 (Instituto Nacional de Pesquisas Espaciais 1992; Lisansky 1990). The recent decline in the deforestation rate appears, however, to have been due more to Brazil's economic recession than to enhanced legislation and better law enforcement; many observers anticipate that as Brazil's economy recovers, there will be a resumption of progressively increasing deforestation. But so vast is Brazil's expanse of forest that even the high rate of burning during the late 1980s leaves the proportionate amount of deforestation behind a good number of other countries with higher percentage rates. Fortunately we can hope that while the peripheral states along the southern, eastern and even northern borders of Brazilian Amazonia may well continue to experience extensive deforestation, the west-central bloc could conceivably survive with scant depletion for several more decades.

There is even better prospect for the Guyanas, with more than 400,000 km^2 of forest and only two million people. Until the road linking Guyana south to Brazil is built, little deforestation appears likely.

To consider the overall analysis from a different standpoint, we can note that ten countries are losing forest at a rate of 4000 km^2 or more each year. These are: Brazil, Colombia, India, Indonesia, Malaysia, Mexico, Myanmar, Nigeria, Thailand and Zaire. Their collective total in 1989 was 87,300 km^2, or 69 per cent of all deforestation.

Note also that ten countries – or, in the case of the Guyanas, a group of countries – each possess 200,000 km^2 or more of remaining forest. These are: Brazil, Colombia, Gabon, the Guyanas, Indonesia, Myanmar, Papua New Guinea, Peru, Venezuela and Zaire. Their collective total in 1989 was 6,418,500 km^2, or 80 per cent of remaining forests. Just three countries, Brazil, Indonesia and Zaire, still possessed rather more than four million km^2 altogether, or over half of the entire biome. As for the rate of annual deforestation, ten countries, or in the case of Central America, a group of countries, feature rates that are more than twice the average rate for the biome, which is 1.6 per cent. These are: Central America, Ecuador, Côte d'Ivoire, Madagascar, Mexico, Myanmar, Nigeria, Philippines, Thailand and Vietnam. (Of these, all except Central America, Ecuador, Mexico and Myanmar are three times or more above the biome-wide average; four countries are 4.5 times or more above the average: Thailand (8.1 per cent), Madagascar (8.3 per cent), Nigeria (14.3 per cent) and Côte d'Ivoire (15.6 per cent).) Six countries feature a rate that is less than half the average biome-wide rate: Gabon, the Guyanas, Cambodia, Peru, Venezuela and Zaire.

Recall, moreover, that these figures reflect the situation in 1989, the final year with detailed documentation from which the data are drawn (Myers 1989; see also 1992a and b). In some countries, deforestation turns out to have been greater than was supposed in 1989 (though in Brazil it has declined sharply); and in many of these countries, the deforestation rate since 1989 appears to have accelerated sharply. These countries include Bolivia (Government of Bolivia 1990, Southgate & Runge 1990), Congo (Hecketsweiler 1990), Gabon (Barnes 1991), Côte d'Ivoire (Jonkers & Glastra 1989), Laos (Salter 1990, Salter & Phanthavong 1989), the Sarawak state of Malaysia (Aiken & Leigh 1992, de Milde 1991), Papua New Guinea (Srivastava & Butzler 1989), Peru (de Soto 1989, Dourojeanni 1991, Serra-Vega 1990), Philippines (Basa 1991, Kummer 1992, World Bank 1990a), Thailand (Klankamsorn & Charuppat 1991), and Vietnam (Kemf 1990). So far as can be roughly determined, this additional deforestation may well have amounted to an aggregate total of as much as 25,000 km^2 per year.

To reiterate, deforestation presents a decidedly mixed picture, not only now but as concerns the likely future. This makes it all the more unrealistic to generalize from the biome-wide deforestation rate of 1.8 per cent in 1991, and postulate that all forests will therefore be eliminated in another four decades or so. Many if not most countries will surely have little forest left in half that time. A few could well expect, if we make a simple extrapolation of the recent past, to retain sizeable expanses for at least several decades more. But this raises a key question: will the future be a simple extrapolation of the recent past? Will deforestation rates continue in linear manner, or will there be a geometric increase in certain sectors of the biome – and might some areas even experience a decline in deforestation rates?

2.3 Main agents of deforestation

In 1989 commercial logging was affecting some 45,000 km^2 of new forest each year, much the same as ten years ago. Of this, roughly two-thirds, or 30,000 km^2, was in Southeast Asia, where it was so heavy and negligent that it amounted to forest destruction. Cattle ranching, almost entirely confined to Central America and Amazonia, caused 15,000 km^2 of forest to be cleared in 1989. Forest conversion to cash-crop plantations (oil palm, rubber, etc.), plus forest destruction for roads, mining and other activities of similar relatively small scale sort, amounted to perhaps 10,000 km^2. So in 1989, these three categories totalled 55,000 km^2. The rest, a little over 87,000 km^2, was ostensibly due to slash-and-burn farmers, mainly shifted cultivators, though this was, and remains, a very rough-and-ready estimate, advanced with the sole aim of gaining an insight into the proportionate share of forest destruction attributable to this agent of destruction. Shifted cultivators thus accounted for 61 per cent of all forest destruction, a proportion that since 1989 appears to have been increasing steadily.

2.4 Regrowth and secondary forests

It has been supposed by some observers (for example, Lugo 1988, Lugo & Brown 1982) that most deforestation is quickly compensated for by regrowth, and that areas cleared and then abandoned are soon covered with secondary forest. This conclusion is questionable since deforested areas are not usually abandoned nowadays. As we have seen, the bulk of deforestation is due to the activities of slash-and-burn cultivators, who, operating in large numbers (as many as 300 million by the late 1970s, according to Denevan 1980, citing a range of authorities), rarely practise a rotatory form of agriculture. Rather, vast numbers of "shifted" cultivators impose huge pressures on forest fringes, clearing extensive areas before driving deeper into the remaining forest year after year. Behind them come still more waves of shifted cultivators, allowing the forest no chance to re-establish itself (see Battjees 1988, Cruz & Zosa-Feranil 1987, Myers 1988b, Westoby 1989).

This is now the pattern in much of West Africa, East Africa, and Southern and Southeast Asia, where population pressures are greatest, although in parts of Latin America and especially in Amazonia, it is often the case that abandoned croplands and pastures become available for forest regrowth. For example, in Brazilian Amazonia in 1988, between 20 and 40 per cent of deforested lands were starting to feature secondary-forest recovery (Browder 1989; see also Uhl 1987). But this is far from the invariable outcome. Indeed there are many instances in Amazonia where regrowth does not occur at all. It may be more difficult than is sometimes supposed for successional processes to get underway in a manner that leads to secondary forest. In the Bragantina Zone in eastern Amazonia, an area of some $35,000 \, \text{km}^2$ was abandoned after a short-lived effort at agricultural settlement early this century. Fifty years later there was still little vegetation beyond scrub and brush growth (Egler 1961). There is scant forest regrowth in many parts of the Amazonian sectors of Colombia (Battjees 1988), Ecuador (Gentry 1989) and Peru (Dourojeanni 1988).

This outcome, of no secondary-forest regrowth, is even more pervasive in Asia. Indonesia features at least $160,000 \, \text{km}^2$ of *alangalang* (*Imperata*) grasslands that have colonized deforested areas; the grass proves so competitive that forest growth simply cannot establish itself. There are similar tracts of fire-climax grasslands in the Philippines ($70,000 \, \text{km}^2$) (Revilla et al. 1987), also in Thailand (Hirsch 1987), Malaysia (Chin 1989) and Papua New Guinea (Saulei 1989), making up $450,000 \, \text{km}^2$ in all. In Africa too a parallel phenomenon is widespread, due in the main to

33

sheer pressure of numbers of small scale farmers, for instance in Cameroon and Ghana (Hepper 1989), in Côte d'Ivoire (Bertrand 1983), and in Madagascar (Jenkins 1987).

Moreover, even in cases where secondary forest manages to establish itself, the forest's stature and biomass are often less than in primary forest, and this has implications for their carbon-storing capacity and hence their linkage to the carbon budget of the global atmosphere. In Malaysia, secondary forests usually achieve only half to three-quarters the height of primary forest (Ng 1983). Near Altamira in Para, Brazil, they have been found to possess less than half the original biomass (Fearnside 1986a). Secondary forest on abandoned pasture lands in Brazilian Amazonia proves much slower to take hold than would occur on cropland sites abandoned as part of shifting-cultivator cycles (Uhl 1987). Furthermore the success of secondary forest varies according to the degree of cattle use before abandonment, with only half as much biomass on heavily-used sites (the predominant type) than on more lightly used sites (Fearnside 1985; see also Lugo & Brown 1986, Fearnside 1986a, and Uhl et al. 1988).

2.5 Future outlook

Should we anticipate a simple extension of recent trends and patterns of deforestation? Or are there likely to be discontinuities such as a still greater acceleration of the deforestation rate in certain countries, or perhaps a decline in the rate in certain other countries? Clearly the 1989 rates cited for Côte d'Ivoire, 15.6 per cent, and for Nigeria, 14.3 per cent, cannot be maintained for many more years since there would simply not be any forest left. Equally clearly the rates are likely to decline before then as remaining forests are reduced to fragments in inaccessible localities such as ultra-steep hillsides or ravines. In Côte d'Ivoire the rate has declined a good deal during the 1980s; so too in Central America and the Philippines. How far, then, can we engage in a realistic prognosis of the future situation, given the many variables at work? Consider a number of salient factors.

Planned deforestation

Certain governments have formulated plans to exploit their remaining forests on a broad scale. The most notable example cited is Brazil (see Repetto & Gillis 1988). Here we shall briefly consider two other countries, Peru and Indonesia, by way of illustration of specific intent to deforest.

In Peru there is likely to be a steady stream of settler farmers into Amazonia, for reasons akin to those driving the migrants' surge in Brazil: to apply a safety valve for land hunger in established farming areas, to avoid issues of agrarian reform, and to expand agricultural production (the government still perceives Amazonia as a breadbasket for the rest of the country, despite much experience to the contrary) (Dourojeanni 1988). As in most other Amazonian countries, there could also be a motive to assert "demographic sovereignty" over remote territories bordering other countries, especially territories believed to be rich in minerals. Furthermore there could eventually (or soon?) be an extension of the Brazilian highway from Acre into Peru in order to supply a conduit to the Pacific coast for raw materials from Brazilian Amazonia. The major beneficiary would be the likely financier of the road, Japan. At the same time, the road would open up the southern sector of Peruvian Amazonia, being the sector least affected by migrant settlers to date.

Thus it is not unrealistic to suppose that an additional expanse of Peruvian Amazonia, somewhere between 110,000 km^2 (Dourojeanni 1988) and 200,000 km^2 (Salati et al. 1990), will be deforested during the period 1985–99, or an average rate of between 7330 km^2 and 13,300 km^2 of forest per year. This is more than the 1989 rate, estimated at 3500 km^2.

In Indonesia, given the current rates of logging, all timber concessions, which account for the great bulk of remaining lowland forests, are scheduled for exploitation within 25 years at most (Abell 1988, Sutter 1989). In addition, the government plans that during the period 1988–2000 a forest expanse of at least 60,000 km^2, conceivably three times as much, will be converted to planned agriculture, primarily through the Transmigration Programme (which, however, has recently slowed down, though due to lack of funds rather than doubts about its effectiveness) (Booth 1989, Sutter 1989). In addition there will be a sizeable amount of forest taken for unplanned agriculture by "spontaneous" settlers and other types of slash-and-burn cultivators. These three deforestation factors could well eliminate a total of at least 200,000 km^2 of forest during the 1990s, possibly much more, resulting in an average of 20,000 km^2 per year, by contrast with the 1989 deforestation estimate of 12,000 km^2.

Population pressures

Population pressures are often singled out as the prime factor behind deforestation. Table 2.2 shows 1992 population and projected growth. But these effects are far from simple (Myers 1990, Rudel 1989). For example, while tropical-forest countries' populations expanded by amounts ranging from 15 to 36 per cent during the 1980s, deforestation expanded by almost 70 per cent. The single largest agent of tropical deforestation is the "shifted" cultivator (Jones & Richter 1981, Schuman & Partridge 1989, Uhlig 1984). With fast-growing numbers, these people find themselves squeezed out of traditional farmlands in countries concerned, whereupon they head for the last unoccupied lands they are aware of, the forests. Driven

Table 2.2 Population growth in selected tropical forest countries.

Country	Population in 1950 (millions)	Population in 1992 (millions)	Growth of population 1992 (%)	Per cent of population in rural areas	Population projected in 2000 (millions)	Population projected in 2050 (millions)	Projected size of stationary population (millions)	Per capita GNP 1990 (US$)
LATIN AMERICA								
Bolivia	3	8	2.7	49	9	14	21	620
Brazil	53	151	1.9	26	178	237	305	2680
Central America	9	30	2.5	52	38	63	88	1350
Colombia	12	34	2.0	32	38	54	63	1240
Ecuador	3	10	2.4	45	13	18	24	960
The Guyanas	n/a	2	2.1	52	2.3	3	n/a	1050
Mexico	27	88	2.3	29	103	143	184	2490
Peru	8	23	2.2	30	27	37	50	1160
Venezuela	5	19	2.5	16	25	35	45	2560
ASIA								
India	362	883	2.0	74	1006	1383	1862	350
Indonesia	77	185	1.7	69	209	278	360	560
Cambodia	4	9	2.2	87	10	13	n/a	n/a
Laos	n/a	4	2.9	84	6	10	21	200
Malaysia	6	19	2.5	65	22	35	44	2340
Myanmar (Burma)	18	43	1.9	76	51	70	96	n/a
Papua New Guinea	2	4	2.3	87	5	7	11	860
Philippines	20	64	2.4	57	74	101	137	730
Thailand	20	56	1.4	82	64	76	105	1420
Vietnam	24	69	2.2	80	82	108	159	n/a
AFRICA								
Cameroon	5	13	3.2	58	16	36	53	940
Congo	1	2	2.9	59	3	5.5	14	1010
Gabon	n/a	1	2.5	57	1.6	1.8	6	3220
Côte d'Ivoire	3	13	3.6	57	17	39	64	730
Madagascar	5	12	3.2	77	15	32	46	230
Nigeria	41	90	3.0	84	153	216	453	370
Zaire	14	38	3.1	60	50	98	172	230

Sources: Population Reference Bureau 1992; World Bank 1992a.

significantly by population growth and sheer pressure on existing farmlands (albeit often cultivated with only low or medium levels of agrotechnology), slash-and-burn farming is the principal factor in deforestation in Colombia, Ecuador, Peru, Bolivia, Nigeria, Madagascar, India, Thailand, Indonesia and the Philippines, probably also in Mexico, Brazil, Myanmar and Vietnam.

Populations of these cultivators are often increasing at annual rates far above the rates of nationwide increase. In Rondônia in Brazil the numbers of small scale settlers grew during the period 1975–86 at a rate that surpassed 15 per cent per year for much of the period since 1975, whereas the population growth rate for Brazil as a whole averaged only 2.1 per cent (Malingreau & Tucker 1988). There are similar migrations into tropical forests, albeit at lower rates, in Colombia, Ecuador, Peru, Bolivia, Côte d'Ivoire, Nigeria, India, Thailand, Vietnam, Indonesia and Philippines. In all these instances, population growth is a significant if not the predominant factor in deforestation.

A host of related factors frequently operate in addition to population growth. They include pervasive poverty among peasant communities concerned, maldistribution of existing farmlands, inequitable land-tenure systems, lack of property rights, inefficient agrotechnologies, insufficient attention to the subsistence farming sector, lack of rural infrastructure, and faulty development policies overall. In Brazil, for example, five per cent of farmland owners possess 70 per cent of all farmlands, while 70 per cent cultivate only five per cent, a skewed situation that is growing more acute; another 1.7 million Brazilians enter the job market each year, over half of them failing to find enough employment to support themselves.

Moreover there is vast scope in population growth of the future for still larger numbers of shifted cultivators to accelerate deforestation in many sectors of the biome. Of the one billion people projected to be added to the global population during the 1990s, a full 60 per cent are expected to be in tropical forest countries. By the year 2030 or thereabouts, 80 per cent of the world's projected population of eight billion people are expected to be in tropical forest countries; this translates into 6.4 billion people, or almost one billion people more than now alive. Given the "demographic momentum" built into population growth processes in countries concerned, and even allowing for expanded family planning programmes, population projections (see Table 2.2) suggest that in those countries where economies appear likely to remain primarily agrarian, there will be progressive pressures on remaining forests, extending for decades into the future. For instance, Ecuador's population is projected to increase from 10.2 million today to 24 million (135 per cent greater) before it attains zero growth in about a century's time; Cameroon's from 13 million to 53 million (308 per cent); Côte d'Ivoire's from 13.5 million to 64 million (374 per cent); Madagascar's from 12.3 million to 46 million (274 per cent); Nigeria's from 90 million to 453 million (403 per cent); Myanmar's from

37

43.3 million to 96 million (122 per cent); India's from 883 million to 1862 million (111 per cent); Indonesia's from 185 million to 360 million (95 per cent); and Vietnam's from 69 million to 159 million (130 per cent). Unless there is a reduction of population growth together with a resolution of the landless-peasant phenomenon (a prospect that appears less than promising (Sinha 1984)), it is difficult to see that much forest will remain in just a few decades' time in most of the countries cited.

Of course we must be careful not to overstate the case on a biome-wide basis. In much of the island of New Guinea, for instance, population pressures are slight to date, and appear unlikely to become significant within the foreseeable future. Much the same applies in the countries of the Zaire Basin in central Africa; in the countries of the Guyanas (Guyana, Suriname, and French Guiana); and in the western sector of Brazilian Amazonia.

In certain countries there are exogenous factors at work. In Central America deforestation has stemmed much more from the spread of cattle ranching on the part of a relatively few large scale cattle ranchers than from population pressures in a region that is conventionally perceived to experience pronounced population problems. In turn, the spread of cattle ranching has reflected demand for "cheap" beef from North America—the so-called "hamburger connection" (Myers 1981). There is a parallel phenomenon of developed-world consumerism driving deforestation in Thailand via the cassava connection with the European Community (Myers 1986a). Similar linkages operate with respect to the developed world's rôle through demand for tropical timber and through international debt (Myers 1986b).

The rôle of the shifted cultivator

Slash-and-burn cultivation, practised much more by the shifted cultivator of recent form than the shifting cultivator of traditional type, accounts for well over half of all deforestation, a proportion that is likely to increase rapidly. However, whereas it would be fairly straightforward to relieve deforestation pressures from the commercial logger (by growing timber in plantations on deforested lands) and from the cattle rancher (by engaging in sustained-yield production of beef on established pasture lands), no such "easy fix" is available to tackle the problem of the slash-and-burn cultivator. A broad based approach is needed to overcome the economic, social, political and institutional marginalization of the shifted cultivator which would involve the redistribution of existing farmlands, reform of land-tenure systems, build-up of agricultural extension services, improvement of credit facilities, and provision of agrotechnologies. The slash-and-burn cultivator should not be viewed as some sort of "culprit", and is no more to be held responsible for felling the forest than a soldier is to be blamed for fighting a war.

Regrettably there seems limited prospect of the shifted cultivator problem being resolved within a time horizon that will assist tropical forests, unless there is much more attention directed by governments concerned, and by international development agencies to the particular development challenges posed by the phenomenon. In the first place we simply do not know how many forestland farmers there are, beyond estimates that range from 300 million to 500 million. If the latter estimate is correct (it may even be an under-estimate), then this accounts for almost one in ten of humankind. Most measures to address deforestation seek to relieve problems within the forests, but these respond to less than half the problem. Many of them (more protected areas and the like) reflect efforts to tackle symptoms rather than sources of problems, given that the ultimate source of the biggest problem lies way outside the forest, a dimension beyond the purview of the great majority of tropical forest analysts.

Additional sources of future deforestation

There are further possible sources of deforestation, these being atmospheric and climatic in nature. True, they are not likely to make their impact felt until early next century, whereas the factors described above are already powerful causes of deforestation. But these additional factors could eventually prove to be some of the most harmful of all. They include region-scale climatic feedbacks, global-warming changes, and acid rain, all of which could induce pronounced depletion of tropical forests (Galloway 1989, Rodhe & Herrera 1988).

2.6 Likely acceleration in the deforestation rate

On the basis of the problems presented by the shifted cultivator alone, there is solid reason to suppose there will be a continued acceleration in deforestation rates in much if not most of the biome. That is to say, deforestation will not increase in linear manner, rather the rate will accelerate exponentially. But by how much? Could we reasonably anticipate that whereas there has been an almost 70 per cent increase in the annual deforestation rate during the 1980s, the 1990s could well see the rate soar as high as 150 per cent, meaning the amount of forest then being lost per year would expand to almost 190,000 km^2? Meantime the total of forest remaining would have declined from today's 7.5 million km^2 to only a little over six million km^2. So a loss in the year 2000 of 190,000 km^2 of forest would work out at three per cent of remaining forests. In other words, it takes far less than a doubling in the annual rate of loss to produce almost a doubling of the expanse lost.

This finding applies of course at the level of the individual country (though with a qualifier – see below). Even were the depletive pressure exerted on a forest stock to remain unchanged, the proportion of remaining forest being eliminated each year would increase due to a "tightening" effect. But the depletive pressure is likely to increase if only through growing human numbers and growing human demands (in tropical forest and non-tropical forest countries alike). So the tightening effect becomes compounded. This all means that the last tracts of tropical forest could be eliminated far faster than one might expect.

At the same time, it is possible that deforestation will slow down under certain circumstances. As in the case of Madagascar, remnant forests can ultimately be confined to steeply sloping hillsides, to ravines and other inaccessible places. In this situation, relict fragments of forest can persist for lengthy periods. Note too, however, that desperation born of poverty can eventually eliminate these last hold-outs of forest, as witnessed in El Salvador where even in broken terrain the last vestiges of forest are about to disappear.

2.7 Conclusion

This chapter presents two main findings. First is that the deforestation rate in the humid tropics has expanded by 68 per cent during the 1980s; and this rapid increase in the deforestation rate looks likely to expand still further during the foreseeable future if only because of the phenomenon of the shifted cultivator.

Secondly, most efforts to reduce deforestation thus far have concentrated on the commercial logger and the cattle rancher, who actually account for under 40 per cent of all current deforestation. Were we to redirect our attention to the primary source of deforestation, and undertake the broad range of policy and programmatic measures needed to address the particular though diverse problems presented by the shifted cultivator, we could still do much to counter the threat of accelerating deforestation before most remaining forests are eliminated within just a few more decades. But this would require a recognition that the source problem is an amalgam of non-forestry problems, including population growth, maldistribution of farmlands, inadequate rural infrastructure, and lack of government attention to subsistence agriculture; hence the overall problem must be tackled largely through non-forestry measures. We shall not achieve success in safeguarding remaining forests until we direct much more emphasis at the root causes of deforestation, which generally originate in lands far removed from the forests. There is still time, though only just, to confront the challenge, and thus to convert a profound problem into a splendid opportunity.

II
Explaining global deforestation

3

Population and deforestation

Matti Palo

3.1 Introduction

Different schools of thought have existed concerning the rôle of human population pressure on economic development, environment and tropical deforestation. Two centuries ago Malthus introduced his ideas on population growing exponentially and food production only linearly. His scenario of consequent human misery has been overcome by the progress of technology, sectoral transformation and rising real incomes. However, the neo-Malthusian tradition has remained alive, particularly among biologists and ecologists.

By contrast, the neoclassical school of economics has considered population pressure either as neutral or beneficial to development. More people has meant more demand, more labour force, more skills and better technology, economies of scale and lower production and distribution costs. Deforestation and other environmental deterioration were assumed to be taken care of by competitive markets: if natural forests, for instance, become economically scarce, their increasing real prices would induce more investments in forestry. Hence their sustainability is maintained automatically (Woods 1987, Birdsall 1988, Boserup 1990, Keyfitz 1991, Hyde & Seve 1991). The evidence for this view is limited however.

Some econometric evidence has been found in Malawi to the effect that smallholders with secure tenure have responded to higher fuelwood prices by increasing afforestation (Hyde & Seve 1991). The authors point out one remaining market failure: watershed management and hence downstream siltation. This study suggests that market forces will contain the deforestation rate, but this finding can be criticized. First, in most other parts of Africa the substitution of plantations for fuelwood from natural forests does not necessarily work, due to insecure tenure conditions and existing restrictive forest legislation. The plantations can never fully substitute for the mix of hardwood logs or for full erosion protection *in situ*, or biodiversity, carbon sequestration, medicinal plants and other non-timber goods. The missing *in situ* markets inside the natural forests – for instance, competitive stumpage markets – is the main factor that has severely handicapped successful functioning of this economic substitution process in the tropical forests. Even in the Malawi case there was no effective control to prevent the full deforestation of all accessible natural forests.

3.2 Modelling deforestation

Whereas the economic substitution process may work to some extent with fuelwood and timber it does not with threatened species, biodiversity, carbon sequestration and other non-priced commodities. Accordingly, economic substitution theory needs to be complemented by a system causality model (Figure 3.1). The two frameworks jointly explain the terms of transition from deforestation into sustainable forestry.

A fundamental feature of excess deforestation is that the causal factors of deforestation are linked together as various chains or mechanisms into a causal system (Palo 1987 and 1990). The mechanisms comprise mostly positive feedback loops which tend to accelerate the process of deforestation. The only effective negative feedback loop is caused by the inaccessibility of forests and by the successive reduction of remaining forest areas. Sustainable forest management has been so far successful on only a small scale in the tropics (ITTO 1990). Accordingly, excess deforestation has prevailed. More advances have been gained by creating forest plantations, but they only rarely decelerate deforestation in natural forests or substitute for them. Inaccessibility has been gradually overcome by technological and medical progress and appears to be a braking force only in the later stages of the deforestation process.

Such reasoning leads to a question of system causality connected with complex, intersectoral transformation processes with positive feedbacks prevailing over the negative ones (Forrester 1969, Ozbekhan 1969, Toffler 1980, Capra 1982). Complex systems are counter-intuitive because cause and effect are remotely located in time and space. They also tend to be insensitive to changes, even to purposeful policy changes. Those few points which are more sensitive to changes are not evident; they have to be discovered through careful examination. Symptoms abound in complex systems, but they are hard to change if underlying causes remain. This kind of holistic system causality applies most appropriately to the intersectoral, accelerating, socio-economic deforestation process in the tropics. The structure of system causality has been viewed as hierarchic in terms of international, national and local factors. In this way system causality can be interpreted as having a consistent global structure and functioning, although a high geographical variation may prevail

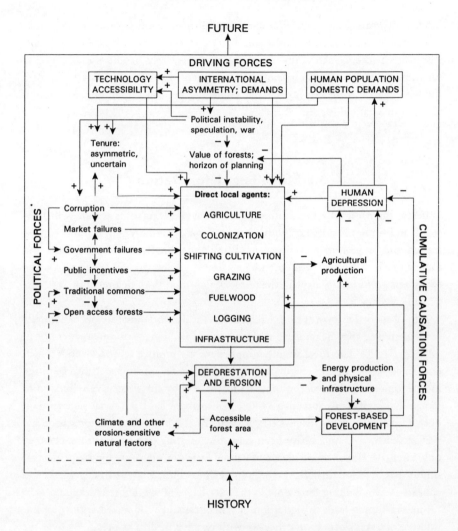

Figure 3.1 The system causality model of deforestation.

regarding the intensity of individual causal factors. Many authors tend to consider the causal factors of deforestation only nationally and locally from an atomistic view of science. In most cases, as long as the remaining accessible forest area is large enough, the deforestation process is continuously stimulated by the driving forces. No equilibrium is reached until all accessible forests have been cleared due to the fact that both control by the market ("the invisible hand") and control by the government fail to decelerate or halt deforestation in tropical countries.

In a quantitative multiple variable statistical modelling experiment (Palo et al. 1987) population density was considered as a proxy for domestic demands of forest and food products. Growing human populations, *ceteris paribus*, create increasing demands for food, clothing, energy, shelter, and services of health and culture which again add pressures on the local agents of deforestation. Population density appeared to be a key independent variable in "explaining" forest cover as a dependent variable, and as a negative proxy for deforestation. The human population impact on deforestation is indirect (see Figure 3.1). Therefore, the ultimate effect of population density on deforestation is strongly linked with the other variables of the system causality model and with the level of technology in particular.

Increasing agricultural production can take place via either extensification or intensification. Intensification has often been hindered in the tropics by the marginal farmers who are risk averters rather than profit maximizers, by an insufficient credit supply, and by traditional rural cultures which are not sensitive to novel innovations. Also, an unequal income distribution has frequently prevented the diffusion of new technologies (Palo 1990). The remaining response is the extensification process which has mostly meant deforestation of new marginal land badly suited to agriculture. The migrating people have also usually brought along tilling methods not appropriate to the new environment. On the marginal soils more land is needed to feed a family than on the previously worked, more fertile soils, and at the margin the soils are mined in a shorter time frame.

Based on the system causality model of deforestation a theoretical inference about the functional form between natural forest cover and population density was deduced as a decreasing logistic function. The parallel function of non-forest area has been defined as analogous to the Chapman–Richards function for biological growth (Scotti 1990, FAO 1992b). Deforestation is assumed to start slowly, then gradually accelerate, and finally slow down in those cases where inaccessibility and/ or effective policies prevent further deforestation. When the latter does not occur, the collapse function is relevant. Here only natural forest cover is considered. If forest plantations are also included, the functional form has a tendency to turn upwards towards the end.

3.3 Empirical findings

Using multiple regression analysis of 60 tropical countries, population pressure was related to the extent of forest cover. The total population density correlated somewhat more strongly than rural population density, although their mutual correlation was high ($R = 0.97$). Population growth was found to have zero correlation with forest cover. The original forest cover variable correlated more strongly with the logarithm of the population density than with population density. The logarithms of both variables gave weaker correlations than the first combination. Total

	1 Forest cover	2 Total population density	3 Rural population density	4 Share of forest fallow area	5 GNP per capita	6 GDP per land area	7 Relative industrial roundwood production	8 Forest product exports per forest area	9 Agricultural area coverage	10 Food production per capita	11 Livestock production index	12 Population growth	13 Relative fuelwood production
1	1.00												
2	-0.78	1.00											
3	-0.77	0.97	1.00										
4	-0.38	0.32	0.28	1.00									
5	-0.14	-0.08	-0.16	-0.06	1.00								
6	-0.55	0.64	0.51	0.32	0.36	1.00							
7	-0.48	0.55	0.58	0.19	-0.01	0.26	1.00						
8	-0.00	0.06	0.03	0.25	0.22	0.25	0.24	1.00					
9	-0.71	0.44	0.46	0.01	0.07	0.28	0.24	-0.17	1.00				
10	-0.48	0.32	0.30	0.07	0.48	0.36	0.18	0.21	0.27	1.00			
11	-0.20	0.12	0.12	0.02	0.34	0.23	-0.02	0.13	0.26	0.58	1.00		
12	0.00	-0.20	-0.17	0.10	0.07	-0.13	-0.01	-0.02	0.00	0.08	-0.07	1.00	
13	-0.51	0.48	0.54	0.14	-0.12	0.12	0.77	-0.06	0.39	0.11	-0.04	-0.01	1.00

Figure 3.2 Correlation matrix of deforestation in 60 tropical countries in 1980. *Source: Palo et al. 1987.*

population density correlated strongly negatively with forest cover ($R = -0.78$), but while it had multicollinearity with most other of the 13 variables (Figure 3.2), it was concluded that population density was a key variable, which also reflected the effects of some other variables in the causality system. Population density had close to zero correlation with GNP per capita, the forest products exports per forest area, and the livestock production index (Palo et al. 1987).

The system causality model of deforestation has so far been partly supported by empirical research (Palo 1984, 1987, 1988, Palo & Mery 1986, Palo et al. 1987, Repetto 1989), although the integrated empirical testing remains problematic. However, the most essential inference of the model, the accelerating nature of deforestation, is more liable to empirical testing. The non-linear relationships between these variables appeared stronger than the linear ones. In general, the correlations have been comparatively strong, particularly within the more homogeneous country groups (Figures 3.3a and 3.3b), but also in the set of districts at the subnational level in Latin America (170 observations in 1966–91) and Asia (143 observations in 1956–89) (Figures 3.3c and 3.3d).

Repetto & Holmes (1983) argued that population growth jointly with open access, asymmetric tenure, and commercialization with increasing international demands led to considerably faster, even accelerating, deforestation than with population growth alone. Barbier (1989a: 14–15) concluded as follows: "The socio-economic factors that induce rational, individual household decisions to expand populations can easily lead to cumulative and unsustainable demographic pressures on a fragile resource base. Once the threshold level of carrying capacity is breached, the 'cumulative causation' of poverty and environment degradation is hard to reverse."

The final results of phase 1 of the FAO Tropical Forest Resources 1990 Assessment Project indicate that the pace of global tropical deforestation has been about 15.4 million hectares per annum in the 1980s, which is 36 per cent faster than that of the late 1970s (FAO 1993). This result indicates support for the projection of the non-linear, accelerating nature of tropical deforestation. The same assessment tested the population pressure hypothesis with a fresh and large data set, particularly at the subnational level of provinces and states. Strong correlations varying according to the ecological zones were discovered. In fact, the Project decided to apply the regression coefficient linking of deforestation to population in updating the national forest areas from the random years of the past national assessments into the baseline years of 1980 and 1990 (FAO 1992b and 1993, Scotti 1990).

Latin America had a total tropical forest area of 918 million hectares in 1990. During 1980–90 this area was annually deforested by 7.4 million hectares, a relative rate of -0.8 per cent from the initial forest area. Latin America plays a key rôle concerning the future of the tropical forests, having nearly half of both the remaining total tropical forest area and the recent deforested area (FAO 1993.)

a)

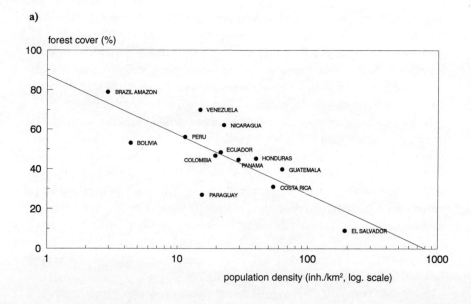

b)

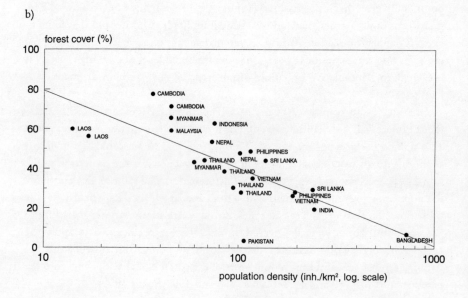

Figure 3.3 a) Correlation (R = −0.77) between forest cover and population density in 13 tropical Latin American countries during 1875–1989; b) correlation (R = −0.76) between forest cover and population density: 23 observations in 13 tropical Asian countries during 1956–89.

c)

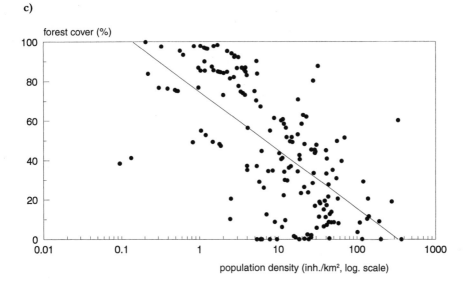

d)

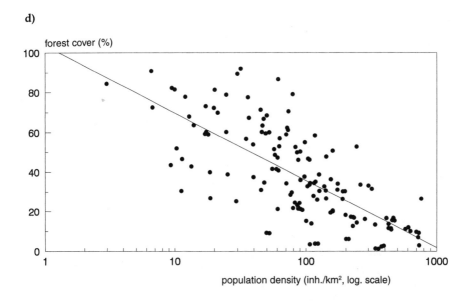

Figure 3.3 c) correlation (R = −0.67) between forest cover and population density: 170 observation during 1966–91 of 104 subnational districts in 13 Latin American countries; d) correlation (R = −0.73) between forest cover and populaation density: 143 observation during 1956–89 of 86 subnational districts in 13 tropical Asian countries.

Tropical Asia had a total forest area of 311 million hectares in 1990 and the fastest relative rate of deforestation (1.2 per cent) among the three tropical continents, while its absolute rate was lowest, owing to the smallest initial forest area. Continental Southeast Asia had the fastest relative rate of annual deforestation (1.6 per cent) and insular Southeast Asia the largest absolute rate (1.9 million ha) within tropical Asia. (FAO 1993.)

In 1990 Africa had a total tropical forest area of 528 million hectares. Africa has had the second largest annual deforestation (4.1 million hectares) among the continents. The most rapid rate of deforestation (1.0 per cent annually) has been occurring in West Africa and the highest absolute deforestation (1.3 million hectares) has occurred in southern Africa (FAO 1993). Only FAO/UNEP 1980 Assessment data were available for national observations (FAO 1988a). The 38 African countries were grouped into the categories of the moist, semi-moist and dry countries (Figures 3.4a–c). The correlations and regressions between forest cover and population density were declining from the moist to the dry countries. FAO (1992b: 5) provides similar results: the deforestation rate was found to vary with the ecological zone for the same population density.

It can be observed that two countries may have the same population densities but different rates of deforestation, or vice versa. Such cases, the argument goes, falsify the rôle of the population pressure as a causal factor of deforestation. Those situations, however, may logically appear as a consequence of the variations of the ecological zones or in the intensities of the socio–economic variables of the deforestation process (see Figure 3.1), or because of variations in the historical progress of the deforestation process (Palo 1990).

The system causality model of deforestation resembles what John Stuart Mill one and half centuries ago called "chemical causation". Comparative sociologists refer to it as conjunctural or combinatorial causation, which refers to causal complexity (Ragin 1987). A causal argument cites a combination of conditions and is concerned with their interaction. "It is the intersection of a set of conditions in time and space that produces many of the large-scale qualitative changes, as well as many of the small-scale events, that interests social scientists, not separate or independent effects of these conditions." (Ragin 1987: 25). The conclusion is that the association of the population pressure with deforestation is bound up in context. It cannot be classified either as necessary or sufficient, or both necessary and sufficient.

3.4 Comparative country cases

The hypothesis of a negative relationship between forest cover and population density was confirmed in the above analyses of tropical countries. The same hypothesis was tested in a parallel way for a group of 35 industrialized countries. Medium strong, negative and non-linear correlations were found for most of the countries, provided a few clear outliers were excluded (see Figure 3.5 and Palo & Mery 1986). The industrialized OECD-member countries have experienced since 1965 a declining trend in the average annual growth of population (from 0.8 per cent to 0.5 per cent) as well as high incomes and standard of living (life expectancy at birth is 77 years) (World Bank 1992a).

More than half (56 per cent) of the total land area in Latin America is covered by forest (with a crown density of more than 10 per cent; FAO 1993). Guyana, Surinam and French Guyana have by far the highest forest covers (85–90 per cent), but also the lowest population densities (1–4 inhabitants per square kilometre). In Puerto Rico and Trinidad and Tobago relatively high forest covers (30–40 per cent) co-exist with relatively high population densities (220–370 inh./sq. km). In the former case, the intimate linkage with the United States has provided stable politics, a capable administration, high level of living (life expectancy at birth is 76 years) and a release from additional population pressure via open access for emigration to continental USA. In the latter case, industrialization based on oil and alternative income sources as well as the relatively low rate of the average annual growth of population (a decrease from 1.2 per cent to 1.0 per cent since 1965) may provide some explanation for decelerated deforestation.

In tropical Asia only the little oil-rich state of Brunei provides a case of a relatively high forest cover of 61 per cent and of medium population density of 37 inhabitants per square kilometre in 1980 (FAO 1988a). The alternative incomes created by oil and the consequent high standard of living is one explanation for the prevailing situation. Tropical Africa, on the other hand, does not provide any single case of sustaining forest cover under population pressure, reflecting the absence of alternative income prospects and limited migration potential.

About half of the world's total forest area is located in the industrialized countries. Denmark and the United Kingdom are examples of countries that were once totally forest covered, but by the 19th century were practically totally deforested. Most other present-day industrialized countries were heavily deforested under socio-economic conditions which resembled those in the developing tropical countries today. Large-scale erosion and other problems were encountered mostly only in the Mediterranean region, the USA, Australia, South Africa and southern Russia. The more northerly temperate countries have been less erosion-sensitive after deforestation (Albion 1926, Pinchot 1947, Thirgood 1981). Also, in the cases of Denmark and the United Kingdom the transformation of the forests was allocated primarily to the intensification of agriculture and the industrialization, whereas in the tropics the respective allocation concerns mostly capital outflows or luxury consumption.

Japan appeared to be the most remarkable case among all the countries studied:

a)

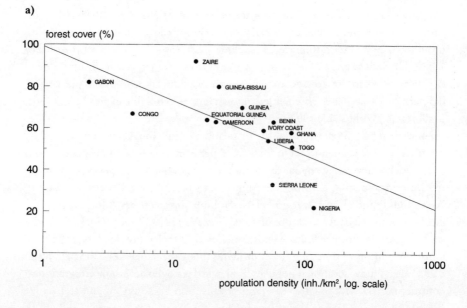

b)

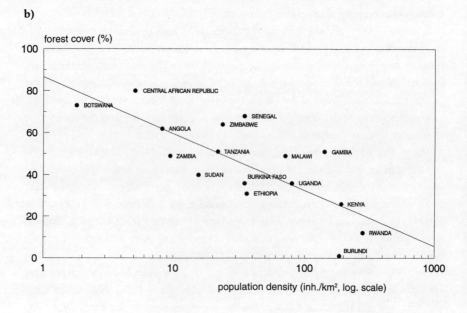

Figure 3.4 a) Correlation (R = −0.69) between forest cover and population density in 14 moist African countries in 1980; b) correlation (R = −0.77) between forest cover and population density in 16 semi-moist African countries in 1980.

c)

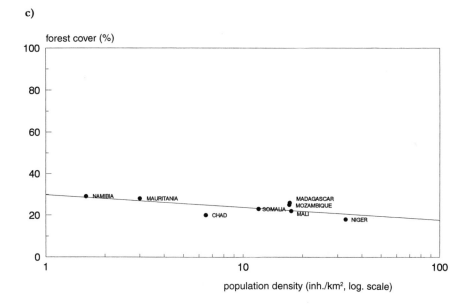

Figure 3.4 c) correlation (R = −0.70) between forest cover and population density in 8 dry African countries in 1980.

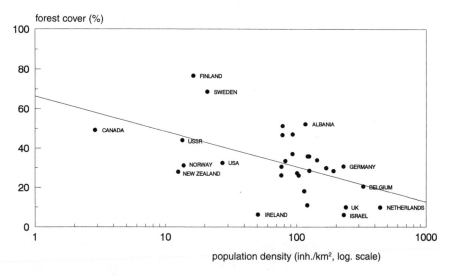

Figure 3.5 Correlation (−0.52) between forest cover and population density in 31 industrialized countries in 1990.

despite a high population density of 329 inhabitants per square kilometre Japan has been able to maintain one of the largest forest covers (68 per cent) in the whole world. Japan's achievement is accentuated by its highly erosion-sensitive natural environments, where mountains and heavy precipitation prevail. Several factors may have contributed to this, but rather stable political conditions, the prevailing clear and secure private tenure of the forests (and hence comparatively efficient roundwood markets), effective public forest policies (Shioya 1967), easy accessibility for roundwood imports and a high status of trees in religious life could be the most relevant factors.

The Republic of Korea is among the few Asian countries which have recently been able to go from a deforestation stage into a sustainable forestry stage. The forest cover of the country was 66 per cent in 1988. The amount of growing stock is projected to increase strongly during the next decades. Korea is a country where industrialization, substitution of fuelwood and rapid sectoral transformation have taken place, releasing population pressure on the forests, although the country is even more densely populated than Japan (440 inh./km^2). Both Japan and Korea have been able to decrease population growth. The respective average annual rates of population growth since 1965 have been from 1.2 per cent to 0.3 per cent, and from 2.0 per cent to 0.9 per cent (World Bank 1992a).

The industrialized countries have 90 per cent of the total global coniferous forest area, but only 26 per cent of the total human population. Although population density seems to correlate moderately with forest cover within this region also, it no longer plays such a rôle in deforestation as in the tropics. The correlation reflects a situation already reached some time ago. The different socio-economic context and the level of technology have affected the industrialized countries in their transition from deforestation to sustainable forestry. In fact, the European (excluding the former USSR) forest area was reported as having increased by 28 million hectares during 1950–90. However, the great majority of this increase is in the form of forest plantations. The deforestation of the majority of the natural forests, particularly in Europe, has been irreversible.

3.5 Discussion

The empirical findings of this chapter can be criticized for their elementary nature. Two-variable correlations are notorious in the analysis of causal connections. The first question is the direction of the causal effect. Closed tropical natural forests certainly hinder human settlements and accordingly a high forest cover may both affect and be a cause of low population densities. However, there are strong theoretical arguments for supposing that the causality of population pressure on deforestation remains as the stronger effect. One further aspect is whether a third factor exists and is causal to both of these factors. Certainly both forest cover and population density are strongly correlated with time which has often been interpreted as the progress of technology and know-how in econometric studies. This possibility is eliminated at the present phase of the economic development in the contemporary tropics by the Boserup (1990) hypothesis which suggests that denser populations induce technological progress to overcome the adverse effects of population pressure on development.

Another question is whether the two-variable correlation remains significant in multiple variable analysis and with the same sign. There are some results where the sign changed – the higher the population pressure the less deforestation (for example, Burgess 1991). Burgess used the absolute deforested area, whereas we have used forest cover (forest area as percentage of the total land area) as a negative proxy for deforestation. The latter is more accurate since sampling theory shows that changes are always more difficult to estimate than the stocks at one point of time (Palo et al. 1987). This aspect is most relevant because tropical deforestation statistics are in general notoriously inaccurate (FAO 1992b and 1993). One more critical aspect concerning Burgess's (1991) findings is that the deforestation data used were based on the FAO/UNEP 1980 assessment, the accuracy of which has later been shown to be rather poor (FAO 1993). Forest area data in this chapter are based on the 1990 assessment (FAO 1991a, 1992b and 1993), which are more accurate than the 1980 data.

There are other multiple variable analyses where the population variable has received theoretically correct signs. In our previous 11-variable regression model both population density and growth had correct signs but only the former was statistically significant (Palo, et al. 1987: 103). Reis & Guzman (1992) discovered that population pressure was, among other variables, statistically significant in explaining deforestation in the Brazilian Amazon. There exist a few other supporting findings (Repetto & Holmes 1983, Panayotou & Sungsuwan 1989, Grepperud 1992, Katila 1992b). During 1993 the new final data of both phases I and II of the 1990 assessment (FAO 1993) have become available. Accordingly, more theoretically and methodologically sound multivariate analyses will then be possible.

Westoby (1978a and 1989) has been among the strongest critics against population pressure being regarded as a real cause of tropical deforestation. His criticisms are valid, if a population variable is used independently and as a direct factor. The system causality model of deforestation of this paper includes Westoby's focal factors, such as unequal tenure and poverty. Population pressure as a causal factor of deforestation has to be considered only in the context of the other relevant factors.

In general, the stages of transition from deforestation to sustainable forestry, and the specific rôle of population, can be analyzed by applying economic substitution theory complemented with the system causality model of deforestation in order to identify the significance of the various assumed causal factors. Natural forests have primarily been deforested both in the tropics and in Europe according to the latter model, but, in particular, in Europe the former model has catalysed the wide scale creation of forest plantations. On the other hand, a framework of forest-based development (for example, Palo 1988) can give additional support when analyzing the terms of transition from deforestation into sustainable development.

Demographers, geographers, economists, sociologists, and scientists from some other disciplines have traditionally been engaged in research concerning the relationships of population pressure with economic development, environment, and deforestation. No interdisciplinary consensus has so far been reached (Woods 1987, Birdsall 1988, Keyfitz 1991, Tabah 1992). However, some practical conclusions have been derived concerning the adverse effects of the population pressure on economic development (World Bank 1985 and 1992a). Despite this, the Tropical Forestry Action Plan, which provides so far the largest international effort in combating the tropical deforestation, has never recognized the key rôle of population pressure in deforestation (FAO 1985a, Winterbottom 1990). This chapter contributes to clarifying the rôle of population in the most acute tropical deforestation processes.

4

International debt
and deforestation

James Kahn and Judith McDonald

4.1 Introduction

In recent years there has been considerable interest in determining the causes of the rapid deforestation of tropical rainforests. Research has focused on both microeconomic causes (Mendelsohn 1990, Repetto 1988a, Repetto & Gillis 1988) and macroeconomic causes (Von Moltke 1990, Deacon & Murphy 1992, Kahn & McDonald 1992, Capistrano & Kiker 1990, and Shilling 1992). Other arguments involving population growth, urban-to-rural migration, road-building, and high energy prices have also been postulated. All the discussions of causes of deforestation centre on factors which lead to inefficiently high rates of deforestation.

One aspect of the question which needs clarification is exactly what is meant by "inefficiently high" when describing rates of deforestation. Rates of deforestation in a forested country may be inefficiently high in a global context, as a particular country's forests provide environmental services (such as carbon sequestration and biodiversity) for the entire planet. The global marginal social benefits of preserving a tract of forest are less than the domestic marginal social benefits, so a level of deforestation develops which is inefficient in a global context. Deforestation rates may also be inefficiently high when viewed from the perspective of a forested country. Market failure and social, institutional, macroeconomic and political constraints may generate a situation where the domestic marginal social benefits of preserving a tract of forest are greater than the domestic marginal social costs, implying a level of deforestation that is inefficient from the forested country's perspective.

This distinction between inefficiency from a global perspective and inefficiency from a domestic perspective has interesting implications for policy. To address the global inefficiency issue one must equate the global and domestic benefits of preservation. This generally requires compensation for the tropical forested countries from the temperate developed countries which receive the external benefits of preservation. In order to deal with the domestic inefficiency issue, the domestic marginal social benefits must be equated with the domestic marginal social costs of preservation. This generally requires the elimination of the market failures and constraints on both individuals' and policy-makers' maximizing behaviour.

This chapter focuses on debt as a potential cause of deforestation, because debt-related policies have the potential to address simultaneously both domestic and global inefficiency in deforestation. If debt generates constraints on a country's maximizing behaviour, then reducing debt can allow the country to operate in a more optimal fashion domestically. In addition, addressing the debt issue through debt-for-nature swaps can provide the compensation necessary to raise the domestic benefits of preservation closer to the global benefits. This will generate levels of deforestation which are more optimal from the global perspective.

Focusing on any particular factor involving deforestation is a difficult exercise, as the different factors involved are linked through a variety of pathways. A subset of factors which may influence deforestation and a subset of the linkages among these factors are depicted in Figure 4.1. In this figure, the central focus is on forests, which are categorized into unprotected forests, extractive reserves (which allow certain types of production activities) and forest reserves (which exclude all types of production activities). Where these different types of forests are contiguous, there may be important boundary interactions (linkages H, I, and J).

Both macroeconomic and microeconomic conditions have an impact on unpro-

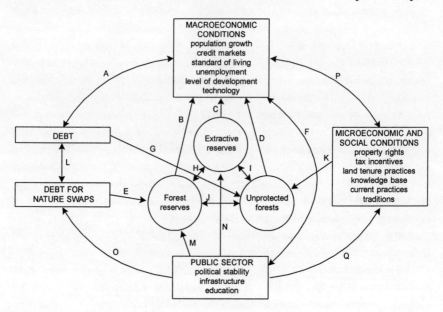

Figure 4.1 Linkages between debt, forests, macroeconomic and microeconomic conditions, and the public sector.

tected forests. Microeconomic conditions influence the incentives for deforestation (linkage K). Macroeconomic conditions constrain the behaviour of individuals, firms and governments and give further incentives for deforestation (linkage D). In addition, there are many interactions (P) between the microeconomic and macroeconomic conditions which may lead to indirect effects on forests. Forests also generate ecological services and inputs to production which have an effect on the macroeconomy and on individuals (B, C and D). It should also be noted that the public sector has a direct impact on forests through restrictions on use (M, N) and indirect effects through interactions with macroeconomic (F) and microeconomic conditions (Q).

Although many people, particularly in the media, have suggested that debt causes deforestation, it is not an easy task to establish a causal relationship between the two. The effect of debt may be primarily indirect, operating through effects on macroeconomic conditions and political stability. In order to test the hypothesis that debt influences deforestation, it is necessary to develop a behavioural model for which the optimal solution specifies a well-defined relationship.

4.2 The conceptual model

In a previous paper (Kahn & McDonald 1992), the authors developed a conceptual model of the macroeconomy of a forested country which suggests that debt may lead to myopic behaviour. The model was based on the assumption that a country's decision makers maximized the present value of the future stream of consumption of the country, and that this maximization is constrained by the necessity that consumption exceed a minimum acceptable level. In other words, there are limits to how low current consumption may be reduced in order to increase future consumption. The outcome of this model suggests that debt may exacerbate myopic behaviour by making it more difficult to meet short-term consumption needs. Deforestation accelerates to meet short-term needs, since deforestation generates income that can be used to meet the minimum acceptable consumption level. However, using deforestation in this way means that this current consumption is at the expense of future consumption. The existence of high levels of debt may cause countries to behave in a myopic fashion, with higher levels of deforestation than would occur with lower levels of debt.

Changes in other economic variables may also influence the rate of deforestation. Increases in the level of inputs such as labour and capital would tend to increase GNP and better enable current consumption needs to be met without increasing deforestation. Increases in competing uses of GNP, such as government spending, would tend to make it more difficult to meet current consumption needs and therefore increase deforestation. This is assuming that government spending is unproductive both in the sense of raising current GNP and in meeting current consumption needs. Population increases have been cited as an important factor in deforestation. In our model, population would influence deforestation through its effect on the minimum acceptable level of consumption. If the minimum acceptable level of per capita consumption is held constant, increases in population will increase the minimum acceptable level of aggregate consumption and therefore lead to more deforestation, *ceteris paribus*.

The need to service debt, pressing current consumption needs, and competing uses of gross national product (GNP) all interact to create a need for current income. Deforestation activities may increase to satisfy this current demand for income, even though future income will decline as a result of the deforestation. There also may be an interaction between the debt and micro-economic factors which are cited as causes of deforestation. For example, Repetto & Gillis (1988) show that a variety of inefficient government policies (investment incentives, credit concessions, tax provisions, agricultural pricing policies, and the nature of lease or sale of forest exploitation rights) create incentives to engage in faster deforestation. The question arises as to why these seemingly irrational policies exist. Our conceptual model develops a general framework of individual country maximizing behaviour under which governments might institute such short-sighted policies. This behaviour is one possible explanation for the existence of these policies, and this behaviour may be exacerbated by high external debt.

Capistrano & Kiker (1990) argue that debt affects deforestation in another fashion. Their hypothesis is that real currency devaluations, which are a part of IMF-prescribed economic restructuring programs as a result of high debt, lead to increased incentives for deforestation. An alternative hypothesis is offered by Deacon & Murphy (1992) who argue that debt does not cause deforestation, but that debt and deforestation are symptoms of the same myopia. Political instability is suggested as a potential source of this myopia.

As can be seen from the above discussion, it is difficult to isolate a "debt effect" as a cause of deforestation. Our research emphasis on the relationship between debt and deforestation should not be construed to mean that debt is the only, or the most important, economic factor that causes tropical deforestation. Rather, we feel that there is reason to believe that debt may cause countries to behave myopically, and we are empirically examining the degree to which debt may be a contributing factor to a country's deforestation.

Both the conceptual arguments presented in this chapter and the more formal optimal control model in Kahn & McDonald (1992) suggest that variables which contribute to the production of GNP should reduce deforestation, and variables that compete for the use of GNP should increase deforestation. The former variables include labour, the amount of forested land and the amount of non-forested land (capital is not included due to the lack of good measures for most countries). The latter variables include debt service, investment and government spending. In addition, we control for intercountry differences in economic, social, political and physical environments using regional dummy variables. These dummy variables may also control for differences in unobserved variables, such as the minimum acceptable level of consumption. It should also be noted that the regional dummy variables may capture some of the specific motivations for deforestation which vary across regions. For example, the primary activities which remove the forest tend to be fuelwood and overgrazing in Africa, logging in Southeast Asia and conversion to agricultural land in Latin America.

4.3 Data

The data used in our empirical analysis come primarily from the Food and Agriculture Organization (FAO) and World Bank sources. Our sample consists of only those tropical countries for which both deforestation and debt data are available. The countries that are included in our sample are listed in Table 4.1. The time dimension of the sample is limited by the availability of deforestation data. Deforestation is available as an average of the 1975–80 period and an average of the 1981–85 period. However, the need for lagged variables as instruments to explain right-hand side endogenous variables precludes regression analysis of the 1975–80 data.

Data on deforestation and total forested area are taken from FAO sources (1981b) and Lanly (1988) and measured in thousands of hectares. The FAO defines deforestation as the permanent conversion of forest land to other uses. Although this is the most comprehensive data set pertaining to forests and deforestation, it has several shortcomings. For example, the data are largely self-reported by the tropical countries, and therefore may be biased. The primary debt variables used in our empirical analysis are total debt service (actual payments of interest and principal on debt), relative debt service (debt service/exports) and the per cent change in public external debt. Our debt data are taken from the World Bank's *World Tables* (various years-a) and *World Debt Tables* (various years-b). We convert these data from millions of current US dollars to their real (1980) equivalent. Table 4.1 contains a listing of countries, average deforestation levels and average public debt levels for the 1981–85 period. For our labour force variable we use the 1988 per cent of population which comprises the work force (as reported by the *Human Development Report*, United Nations 1990) multiplied by total population. The government expenditure and investment variables are from the World Bank's *World Tables* (1989a). Although these variables are in current units of domestic currency, we calculate the real US dollar (1980) equivalent.

Table 4.1 Sample countries, average annual deforestation and public debt levels, 1981–85.

Country (code)	Deforestation (thousand hectares)	Public debt ($ billion)	Country (code)	Deforestation (thousand hectares)	Public debt ($ billion)
Bangladesh (13)★	8.0	5.264	Mali (104)★	36.0	1.096
Belize (16)	8.5	0.077	Mauritania (107)★	13.3	1.201
Benin (17)★	67.2	0.610	Mauritius (108)★	−0.1	0.367
Bolivia (20)★	117.2	3.306	Mexico (109)★	615.0	63.354
Botswana (21)★	20.0	0.266	Nepal (114)	84.0	0.436
Brazil (22)★	2530.0	63.856	Nicaragua (120)★	121.0	3.624
Burundi (27)★	1.1	0.321	Niger (121)★	67.1	0.729
Cameroon (29)★	110.0	1.998	Nigeria (122)★	400.0	12.135
Central African (32)★ Republic	55.0	0.255	Pakistan (125)★	9.0	10.154
			Panama (126)★	36.0	3.098
Chad (34)	80.0	0.165	Papua New Guinea (127)	23.0	0.952
Colombia (38)★	890.0	7.905			
Comoros Islands (39)	0.5	0.099	Paraguay (128)★	212.0	1.236
Congo (40)★	22.0	2.062	Peru (129)★	270.0	8.706
Costa Rica (41)★	65.0	2.991	Philippines (130)★	92.0	12.092
Dominican Rep. (47)★	4.0	2.165	Rwanda (138)★	5.2	0.266
Ecuador (48)★	340.0	5.965	Senegal (141)★	50.0	1.616
El Salvador (50)★	4.5	1.253	Sierra Leone (143)★	6.0	0.388
Ethiopia (52)★	88.0	1.403	Solomon Islands (23)	0.8	0.038
Fiji (53)	1.7	0.281	Somalia (146)	13.5	1.455
Gabon (58)★	15.0	0.860	Samoa (516)	1.6	0.061
Gambia (59)	5.2	0.169	Sri Lanka (149)★	58.2	2.441
Ghana (62)★	72.0	1.274	Sudan (153)★	504.0	5.842
Guatemala (69)★	90.0	1.614	Tanzania (159)★	130.0	2.796
Guinea-Bissau (71)	57.0	0.194	Thailand (160)★	379.0	7.847
Guyana (72)	2.5	0.706	Togo (162)★	12.1	0.813
Haiti (73)★	1.8	0.473	Trinidad-Tobago (164)★	0.8	1.112
Honduras (74)★	90.0	1.782	Uganda (168)	50.0	0.706
India (78)★	147.0	23.647	Burkina Faso (172)★	80.0	0.430
Indonesia (79)★	620.0	23.017	Venezuela (175)★	245.0	16.417
Côte d'Ivoire (85)★	510.0	5.353	Zaire (182)★	370.0	4.610
Jamaica (86)★	2.0	2.459	Zambia (183)★	70.0	2.811
Kenya (89)★	39.0	2.730	Zimbabwe (136)★	80.0	1.369
Liberia (96)★	46.0	0.793			
Madagascar (100)★	156.0	1.938			
Malawi (101)★	150.0	0.759			
Malaysia (102)★	255.0	11.863			

Note: ★ indicates that the country is included in the 1981–85 sample.
Sources: *FAO 1981b, Lanly 1988, World Bank various years-a, World Bank various years-b.*

Two compromises are required to implement our model. Both are necessitated by the high degree of collinearity between pairs of variables. Since the simple correlation coefficient between total forested area and non-forested area was −0.9996, we dropped non-forested area from our empirical analysis. Similarly, since the simple correlation coefficient between public debt and the change in public debt was 0.956, we redefined the change in public debt variable as the relative or per cent change in public debt to reduce this collinearity problem.

4.4 Empirical results

The discussion of our empirical results will focus on the 1981–85 period. This is when both deforestation rates and external debt accumulation increased dramatically, leading to the speculation that they were linked. It should be cautioned that since our paper is based on cross-sectional data from 1981–85, our results may not necessarily be applicable in the 1990s.

As discussed earlier, our empirical analysis is based on our optimizing model. These regressions are estimated using the two-stage least-squares procedure of the personal computer version of Limdep, with investment and per cent change in debt treated as right-hand side endogenous variables. Additionally, since government spending may be an endogenous variable (it is treated as endogenous in some studies and exogenous in others), it is treated as an endogenous variable in column (1) and an exogenous variable in column (2) of Tables 4.2 and 4.3.

Since the sum of the direct and indirect effects (which may be greater than or less than the direct effects alone) is of primary interest, we have presented our most important regression results in Tables 4.2 and 4.3, with primary emphasis on column (1) which treats government spending as an endogenous variable. It should be emphasized that since column (1) in each table measures both direct and indirect effects and treats government spending as endogenous, it represents the most meaningful regression in each table. All the regressions in Tables 4.2 and 4.3 are scaled to account for size. The reason for doing this is to eliminate the generation of a relationship between deforestation and debt that is merely based on size. In other words, large countries could tend to have large debt, large forests and large deforestation, while small countries could tend to have small debt, small forests and small deforestation. An unscaled regression could establish a relationship between debt and deforestation that is completely driven by country size and has nothing to do with economic relationships. Two procedures were used to adjust for size. In the first, all relevant variables (that is, all variables except dummy and proportional variables) were divided by population. These regressions are contained in Table 4.2. The second procedure scales by real US dollar GNP, with the regression results presented in Table 4.3.

The lack of significance and the negative tendencies of the government spending variable may be due to several factors. First, government spending may have a sig-

Table 4.2 Two-stage least squares regressions explaining deforestation (1981–85) using total debt service relative to exports (t-statistics in parentheses). Dependent variable: deforestation (1000 ha) per million people.

Independent variables:	Direct & indirect effects (Investment and change in debt variables excluded)	
	(1)	(2)
Intercept (South America)	26.26	25.55
	(2.37)	(2.34)
Africa dummy	−13.82	−12.67
	(−2.66)	(−2.49)
Central America & Caribbean dummy	-13.71	−14.66
	(−2.49)	(−2.72)
Asia dummy	−19.63	−18.83
	(−3.26)	(−3.19)
Forested land area (10^3 ha/10^6 people)	0.00054	0.00049
	(1.92)	(1.80)
Labour force (% of population)	−0.27	−0.27
	(−1.07)	(−1.12)
Government spending (real US\$/10^6 people)	-0.12×10^{-7} [a]	0.10×10^{-8} [b]
	(−0.88)	(0.09)
Relative debt service (total debt service/exports (in real US\$))	67.80	61.96
	(2.31)	(2.15)
R^2:	0.292	0.316
Number of observations:	55	55

Notes: a) government spending is treated as an endogenous variable; b) government spending is treated as an exogenous variable.

nificant consumption component, which would contribute to the meeting of consumption needs. Second, current government spending may be related to past government spending and actually proxying for an infrastructure variable. Since both of these effects are likely to be negative, they may counteract the hypothesized positive effect.

Table 4.3 contains regressions that are scaled by real US dollar GNP. This scaling eliminates much of the difference among regions (as evidenced by the insignificance of all but one of the regional dummy variables) and leads to the forested land area variable becoming more highly statistically significant, with more forests leading to more deforestation, *ceteris paribus*. The labour force variable (labour force relative to GNP) is statistically significant, with a greater labour force (relative to GNP) leading to more deforestation. Our model suggested that the direct effects of expanding the labour force (unscaled) should be to reduce deforestation. This pre-

Table 4.3 Two stage least squares regressions explaining deforestation (1981–85) using total debt service relative to exports (t-statistics in parentheses). Dependent variable: deforestation (1000 ha) per million real US$ GNP.

Independent variables:	Direct & indirect effects (Investment and change in debt variables excluded)	
	(1)	(2)
Intercept (South America)	−0.022	−0.025
	(−1.28)	(−1.50)
Africa dummy	0.0024	−0.0014
	(0.18)	(−0.11)
Central America & Caribbean dummy	−0.011	−0.015
	(−0.82)	(−1.16)
Asia dummy	−0.013	−0.016
	(−0.96)	(−1.14)
Forested land area (10^3 ha/10^6 of US$ real GNP)	0.0012	0.0011
	(3.10)	(2.98)
Labour force (per 10^6 of US$ real GNP)	0.000012	0.000012
	(2.88)	(2.91)
Government spending (real US$/$10^6$ of US$ real GNP)	-0.34×10^{-7} [a]	0.33×10^{-7} [b]
	(−0.37)	(0.46)
Relative debt service (total debt service/exports (in real US$))	0.26	0.24
	(3.03)	(2.94)
R^2:	0.304	0.332
Number of observations:	55	55

Notes: a) government spending is treated as an endogenous variable; b) government spending is treated as an exogenous variable.

dicted negative impact was observed (although insignificant) in Table 4.2. However, the labour force variable in Table 4.3 (the labour force divided by real US dollar GNP) has a positive effect on deforestation. This labour force variable may be serving as a proxy for development or technology, with lower values of the variable implying higher states of development or more sophisticated technologies. This interpretation means that countries that have more developed economies tend to deforest less, *ceteris paribus*.

One way of measuring the burden of the debt is to measure debt relative to the ability to repay it, which would be related to foreign exchange earnings. The debt variable used in the regressions of Tables 4.2 and 4.3 is designed to capture this, with the debt variable constructed as total debt service divided by total exports. In Table 4.2, the regressions are scaled by population. In particular, the relative debt service variable is highly significant, with the t-statistic equal to 2.31 in column (1).

This debt variable (total debt service/total exports) also worked well when the regressions were scaled by real US dollar GNP. These results are contained in Table 4.3.

As noted in the discussion of data, the regional dummy variables are included to control for differences in production technologies and differences in non-debt related motivations for deforestation (and other factors that may vary by region). After controlling for these region-specific reasons for deforestation, there exists an additional effect due to debt which is strongly significant (see column (1) of Tables 4.2 and 4.3).

One interesting result of our work is that in this cross-sectional analysis, population did not appear as a significant variable. In both unscaled regressions and the regressions scaled by real US dollar GNP, the coefficients on population and population growth variables were statistically insignificant. These regressions, as well as regressions which looked at alternative specifications of variables and functional forms, are not reported here.

In summary, our empirical results suggest that there is a strong statistical relationship between debt and deforestation. An examination of the first column of each table, representing the most meaningful regression models, shows this relationship, which is robust across scaling procedures and measures of debt burden. Consequently, it may be reasonable to reject the null hypothesis that debt and deforestation are unrelated and accept the alternative hypothesis that there is a positive relationship between debt and deforestation.

4.5 Conclusions

The economic factors which cause inefficiently high levels of deforestation are part of a complex system of economic and ecological relationships. Consequently, it is extremely difficult to isolate individual factors as "causes" of deforestation. Nonetheless, we have developed a model of economic behaviour which suggests that high levels of debt may generate myopic behaviour which leads to inefficiently high levels of deforestation. This deforestation is inefficiently high from the perspectives of both the forested country and other countries throughout the world. Our empirical results tend to suggest that debt may be an important factor in the deforestation of tropical countries. It should be emphasized that we are not suggesting that there are not important microeconomic (or other macroeconomic) causes of deforestation, which may vary from country to country.

The link between debt and deforestation which is suggested in this paper implies that debt-for-nature swaps may have a dual effect on deforestation. The contractual part of the agreement generates preservation, and makes a globally-potential Pareto improvement into an actual Pareto improvement. The reduction of debt may also relieve internal pressures for deforestation. However, our empirical results indicate that this indirect effect is likely to be small. Our research tends to indicate that there is evidence that reducing debt may reduce deforestation, which may be an argument to utilize more fully debt-for-nature swaps as a tool for preserving environmental quality. However, the global benefits to be realized through debt-for-nature swaps can be better understood with the conduct of more research in this area. In particular, more research is required to examine the interactions among economic phenomena which are linked to deforestation, on intercountry differences in the relationship between debt and deforestation, and on the intercountry differences in the costs and benefits of these swaps.

5

Tropical forest depletion and the changing macroeconomy, 1967–85

Ana Doris Capistrano

5.1 Introduction

The large scale depletion of the world's tropical forests is among the most serious environmental problems of our time. The ecological consequences of forest depletion represent substantial costs in terms of lost productive potential which developing countries, in particular, can ill afford. While increasing population has been identified as a major underlying factor behind forest depletion in tropical regions, the rôle played by government policies and market forces is now well recognized (Winpenny 1990, Panayotou 1991, Shaw 1992). The influences of the changing macroeconomic and international trade context on forest depletion and overall environmental degradation are also increasingly being explored (Hansen 1990, Killick 1990). This chapter examines the macroeconomic factors which contributed to forest depletion in 45 tropical countries from 1967 to 1985, a period characterized by turbulence in the world's economic and financial systems.

Forest depletion occurs in varying degrees. Forests may be qualitatively impoverished through marginal or pronounced disturbance without complete removal of forest cover such as during selective logging. Or, they may be deforested, i.e., forest cover may be completely and permanently removed as land-use shifts from forestry to non-forestry use (Myers 1986c). While forest conversion to pasture has been the common pattern of deforestation in Latin America, in most other regions, commercial logging is often the first step to deforestation (Fearnside 1987, Walker 1990). Even selective logging damages at least half of the remaining forest stock beyond recovery and leaves behind access roads and trails which tend to invite landless peasants, ranchers and speculators into the forest (Jacobs 1988, Mahar 1989a, Repetto 1990). A tendency towards urban-biased development coupled with widespread landlessness would appear to have been important contributors to tropical deforestation. As relative land rents increase, the poorest and least productive agricultural producers tend to be pushed into the forest by their inability to compete for cultivable lands. Many governments, particularly in countries where land and political power are highly unequally distributed, also tend to favour forest settlement and conversion to agriculture as means of achieving better population dispersal and generating much needed foreign exchange (Shresta 1987, Geores & Bilsborrow 1991).

Over 90 per cent of tropical forests are public forests and, in many countries, they are a significant source of government revenues (Repetto 1990). One could argue that political and macroeconomic factors and financial pressures contribute to forest depletion by making governments more likely to allow logging and forest land conversion to what are perceived to be more profitable types of land-use. This chapter examines the hypothesis that tropical forest depletion during the period 1967 to 1985 was largely a response to externally-mediated macroeconomic forces and financial pressures on developing countries. It analyzes the influence of variables including population, income, export prices, international debt, currency devaluation, and food security on tropical forest depletion during the study period. The hypothesis is statistically tested using the extent of logging of tropical broadleaved forests as an approximation of deforestation in 45 countries in Asia, Africa, Central America and the Caribbean, and South America (see Table 5.1). The countries included oil exporters, low income oil importers, middle income oil importers, highly indebted countries, and non-highly indebted countries. The approximation of tropical deforestation using industrial logging of closed broadleaved forests is justified by the fact that closed broadleaved forests make up the majority of tropical forests and provide the bulk of commercial tropical timber harvest. Of the estimated 1195 million hectares of tropical closed forests standing in 1980, 97 per cent were closed broadleaved forests (FAO 1981a). Available data also indicate a high degree of correlation between industrial logging and deforestation of closed broadleaved forests in the countries included in the study. Data on average annual deforestation were compared with the average area of forest logged annually during 1976–80 for the 45 countries in the study. The Pearson correlation coefficient and the rank order correlation coefficients between the two measures of forest depletion were 0.79 and 0.81, respectively.

5.2 The study period

The study focuses on the years 1967 to 1985. The 19-year study period witnessed the collapse of the global system of fixed exchange rates, two major oil crises, and the early years of the debt crisis. It could be roughly divided into four sub-periods: 1967–71, 1972–75, 1976–80, and 1981–85.

Table 5.1 Classification of countries by region, income group and international credit status

	Highly indebted/ major borrower	Low income oil importer	Middle income oil importer	Oil exporter
Central America				
	Costa Rica		Costa Rica	Trinidad & Tobago
	Jamaica		Dominican Republic	
	Mexico		El Salvador	
			Guatemala	
			Haiti	
			Honduras	
			Jamaica	
			Mexico	
			Nicaragua	
			Panama	
South America				
	Bolivia		Bolivia	Venezuela
	Brazil		Brazil	
	Colombia		Colombia	
	Ecuador		Ecuador	
	Peru		Paraguay	
	Venezuela		Peru	
Asia				
	India	Burma	Malaysia	Indonesia
	Indonesia	India	Philippines	
	Malaysia	Sri Lanka	Thailand	
	Philippines			
Africa				
	Côte d'Ivoire	Burundi	Côte d'Ivoire	Cameroon
		Central African Rep.	Liberia	Gabon
		Ethiopia		Nigeria
		Gambia		
		Ghana		
		Kenya		
		Madagascar		
		Nigeria		
		Rwanda		
		Sierra Leone		
		Somalia		
		Sudan		
		Tanzania		
		Togo		
		Zaire		
		Zambia		

The period 1967–71 was a period of relative stability in the international markets. It marked the end of a fixed currency exchange system with the devaluation of the pound sterling in November 1967 and the suspension of the gold convertibility of the US dollar in August 1971 (de Vries 1986). Subsequent currency adjustments worldwide ushered in an era of greater volatility in international prices and greater uncertainty in international markets.

The period 1972–75 was characterized by shortages and rising commodity prices. It started with worldwide grain shortages caused by poor harvests in the major grain producing areas and by the low levels of US grain reserves. Speculation on primary commodities sent prices in 1972–73 to levels not experienced since the Korean War in the early 1950s. The high prices proved devastating for grain-importing countries and heightened concerns about food security and self-sufficiency in the developing world. Many countries adopted policy measures aimed at increasing food security by expanding domestic food production. For some, this meant expansion of agriculture into previously forested lands. Triggered by the outbreak of the Arab–Israeli war and precipitated, in part, by disputes over sharing of oil revenues following the devaluation of the US dollar, the first oil crisis hit in 1973. Oil prices increased sharply and created trade deficits for oil importing countries and surpluses for oil exporters, reduced real income and growth in oil importing countries, and fuelled inflationary pressures worldwide. In industrialized countries, consumers switched to coal, natural gas and other energy sources. In non-oil-producing developing countries, consumers increasingly turned to wood-based fuels which already supplied the major portion of energy needs. By mid-decade, the world economy was faced with stagflation.

The period 1976–80 was characterized by a commodity price boom, increased international lending to developing countries, and greater influence of creditor countries over debtor country policies. Agricultural commodity prices rose to peak levels in 1976 until 1977, making governments more favourably disposed towards lucrative export crops. World prices of pulp and paper also rose to unprecedented levels. Since developing countries were net importers of pulp and paper products, many governments also found investment in pulpwood plantations very attractive (Bruenig 1987, Kuusela 1987, Bourke 1988). The need to recycle surplus oil revenues realized by oil exporting countries encouraged increased international lending to newly industrializing developing countries. The funds financed mostly large-scale development projects such as multipurpose dams, monocultural plantations, agro-industrial complexes, and transmigration projects, many of which were associated with significant social and environmental disruptions. For example, road construction and transmigration projects destroyed vast areas of the Amazon forest (Hecht 1985, Mahar 1989a). In some projects, natural forests were replaced by plantations of fast-growing trees grown for pulp and paper or fuel.

Concerned about mounting developing country debt, the World Bank and IMF, starting in 1977, made further lending conditional on structural adjustments primarily through exchange rate devaluation to be undertaken by a borrowing country. Real devaluations were supposed to reduce real income and spending by increasing the price of tradable goods, and to correct trade imbalance by making exports and import-substitutes more profitable and imports more expensive. To reduce fiscal deficits that contribute to aggregate demand and domestic credit creation, debtor countries were asked to cut back public outlays, eliminate subsidies, increase taxes and interest rates, contain nominal wage levels, and increase prices for public services (Fishlow 1985). These prescriptions have been controversial and were suspected to have intensified rates of exploitation of forests and other environmental resources in debtor countries (George 1988, Killick 1990). The second round of oil price increases in 1979 following the Iranian Revolution further increased demand for firewood and charcoal in developing countries as consumers sought cheaper substitutes for petroleum-based fuels. The increased demand coupled with declining supply of fuelwood caused wood fuel prices to rise, creating what has been called "the other energy crisis" for more than a third of the world's population (Eckholm 1980, Leach et al. 1986, Pereira et al. 1987).

The period 1981–85 was a period of shrinking export markets, high interest rates, and reduced inflow of new lending to developing countries. The recession of 1981–82 was the world's deepest and most prolonged in 50 years. The unfavourable market conditions magnified the burden of debt service among debtor countries. The crisis precipitated by Mexico's default in August 1982 forced debtor countries to follow emergency stabilization programmes geared towards creating large trade surpluses in very short periods of time (Dornbusch 1987, Lessard & Williamson 1987, Edwards 1989). In highly indebted countries, adjustment was complicated by fiscal deficits resulting from shrinking tax revenue base and persistent exchange rate overvaluation which contributed to capital flight (Cuddington 1987). In theory, since both the agriculture and forestry sectors produce tradable goods, they stand to benefit from devaluation. However, the reality of depressed international commodity prices during the first half of the 1980s prevented developing countries from reaping the supposed benefits of devaluation. Currency devaluations may also have encouraged tropical forest depletion in two ways: directly, by increasing the competitiveness of wood exports, and indirectly, because of the need to produce food domestically, by increasing demand for land for agricultural production.

5.3 Data

The data used in this study were obtained from various international statistical sources. The International Monetary Fund's *International Financial Statistics* was the main source of data on national income, aggregate prices and international exchange rates. Data on international debt came from the World Bank's *World Debt Tables*. Agricultural and forestry data were obtained from publications of the Food and Agricultural Organization (FAO). Land-use, population and cereal production data came from the *FAO Production Yearbook*. Data on volume and value of exports and imports of cereals and selected agricultural commodities were obtained from the *FAO Trade Yearbook*. Forestry production and trade data came from the *Forest Products Yearbook*. Data on forest productivity and average rates of deforestation for 1976–80 were obtained from the Food and Agriculture Organization and United Nations Environment Programme's (FAO/UNEP) *Tropical Forest Resources Assessment Project Reports*.

The quality of the published data left much to be desired. There was a problem with consistency of the data reported by different agencies, each differing in their assumptions and definitions. From the point of view of this research, perhaps the most important weakness of the data lay in the underestimation and under-reporting of industrial wood extraction from tropical closed forests. Illegal logging, smuggling and bureaucratic corruption contributed significantly to this under-reporting. To make the data series consistent across countries for the years under investigation, data were all expressed in the same units of measure, base period (1980), and currency (US dollars). Missing data were individually estimated using, whenever possible, highly correlated predictors from the same data source. When data on the same variable were available from two different sources, missing values were estimated using figures from the other source as predictors. Information from other sources was also used to adjust inconsistent figures. Despite these adjust-ments, the final data set still reflects some of the underlying problems with quality of the raw data particularly when alternative sources of information for their cor-rection were unavailable. Hence, the results of this study must be interpreted with caution. The trends and relative magnitudes of the parameters estimated using the data are, perhaps, more instructive than the absolute values of the estimates them-selves. Nevertheless, as the results indicate, the data provide valuable insights into the linkages between economic factors and tropical forest depletion.

5.4 Variables

Extent of deforestation

The area of closed broadleaved forest industrially logged, expressed in 1000 hectare units, was used as the dependent variable. It was calculated by dividing the reported total volume (in 1000 cubic metres of roundwood equivalent) of wood removed from natural forests for industrial use by the volume actually commercialized (VAC).

The VAC is the average volume of marketable wood, in cubic metres of roundwood equivalent, obtained per hectare of undisturbed closed broadleaved forest. It is an average measure of commercial productivity across all site classes of closed broadleaved forests and is expressed in cubic metres per hectare.

Price indices

Tropical deforestation was expected to increase with increasing export value of tropical wood products and also with increasing prices of major agricultural export crops that could be grown on forest lands. In this study, the unit export value of tropical broadleaved saw logs and veneer logs, expressed in US dollars per cubic metre roundwood equivalents, was used as the indicator of the price of tropical wood in the international market. It was converted into 1980 dollars using the GDP deflator and expressed as an index with 1980 as base year.

Unit export prices of six major agricultural crops – banana, coffee, cacao, cereals, rubber and tea – were used to construct an agricultural export price index. These crops are the most common cash crops grown on deforested lands and are important foreign exchange earners in many countries in the study. Each crop's price was weighted by the crop's share in the total export value of all six crops combined. The resulting weighted prices were deflated by the GDP deflator and expressed as an index with 1980 as the base year.

Income

Per capita real gross national product, expressed in 1980 US dollars, was used as an indicator of the country's level of economic development and effective demand in the domestic market. Since higher income implies greater demand for both wood and agricultural products which, in turn, increases the opportunity cost of keeping the forest unexploited, a positive association between deforestation and income could be hypothesized. However, if the pristine quality of the forest is a normal good whose demand increases with income, deforestation could then be viewed as a manifestation of environmentally destructive poverty (Mellor 1988, FAO 1985a). Thus, it could also be hypothesized that deforestation and per capita income would be negatively related.

Obviously, any relationship between income level and forest depletion would be far from direct and simple. Higher levels of per capita income could be expected to have both positive and negative influences on forest depletion, to encourage increased forest exploitation while at the same time creating demand for forest preservation. The net effect of higher income on the deforestation, therefore, is an empirical question. Nevertheless, it was expected that higher income among the countries included in the study would be associated with a net increase in forest

74

depletion. Since per capita incomes in many countries were close to the subsistence level, additional income was more likely to have been used to satisfy unmet needs which, in turn, may have encouraged greater forest exploitation or forest land conversion to agriculture.

The extent of forest depletion was also expected to vary among countries belonging to different income groups. To capture the effect of income grouping, the countries included in the study were divided into three groups according to the World Bank's classification: low income oil-importing countries (Income Group 1); middle income oil-importing countries (Income Group 2); and oil-exporting countries (Income Group 3). Oil-importing countries with per capita GNP less than $400 were classified as low income and those with per capita GNP of $401 or more were classified as middle income countries.

Debt burden

It was hypothesized that the greater the country's debt burden, the greater the pressure to liquidate forest resources and, hence, the greater the extent of forest depletion. In this study, the debt service ratio was used as a measure of debt burden. The debt service ratio was calculated as the per cent ratio of total amortization and interest payments on public and publicly guaranteed private debt, with maturity of one year or longer, to total export earnings. The debt service ratio is a short-term measure which considers the availability of foreign exchange as the major constraint in meeting external obligations. The higher the debt service ratio, the greater the debt burden on governments. It was also hypothesized that forest depletion in highly indebted countries would be higher than in non-highly indebted countries. In this study, countries were classified into two groups based on the World Bank's assessment of their international credit standing as of year-end 1985. Included in the second group are the 15 highly-indebted and major borrower countries with disbursed and outstanding long-term debt of more than $17 billion as of year-end 1985. The rest of the countries were included in the first group.

Real exchange rate changes

Tropical forest depletion was hypothesized to increase with real currency exchange rate devaluation. Changing relative incentives signalled by changing real currency exchange rates influence economy-wide allocation of resources, including allocation of land between agriculture and forestry. When the exchange rate is overvalued such as happens when domestic inflation exceeds world inflation, resources tend to move away from export-producing sectors. Currency overvaluation lowers incentives for exportable goods production while encouraging imports, leading to deficits and declining international reserves. Since agriculture is a tradable goods producing sector, currency overvaluation hurts both exportable and import-competing agricultural products. The products of the forestry sector, also a tradable goods producing sector, are similarly hurt by currency overvaluation.

A devaluation is supposed to favour agricultural exports and increase demand for non-tradable agricultural products which the devaluation would make more competitive. The consequent increase in the demand for cultivable lands would raise land rent on existing arable lands and encourage agricultural expansion into marginal lands. Real exchange rate devaluations which encourage agricultural expansion could lead to deforestation if the expansion occurs on forest lands. Exchange rate changes which increase the competitiveness of agriculture increase the opportunity cost of keeping land under forest cover. The increased competitiveness of forest products following devaluations is also likely to encourage increased wood harvesting and, thus, contribute to deforestation.

In this study, the change in the real exchange rate was operationally defined as the change in the bilateral real exchange rate between the country's domestic currency and the US dollar. The use of the US dollar is justified by the fact that during the study period, it was the dominant medium of international transactions as well as the major currency in developing countries' reserves (Gandolfo 1987). The real exchange rate was calculated by multiplying the nominal exchange rate, expressed in units of domestic currency per US dollar, with the ratio of the US wholesale price index to the domestic consumer price index. The domestic and US price indices reflect domestic and US inflation, respectively. They also serve as indices of the prices of non-traded and traded goods, respectively.

Food self-sufficiency

The instability in international food trade in the 1970s created a greater degree of insecurity in developing countries' food consumption. For most countries, the solution to the problem of food security became synonymous with food self-sufficiency through greater domestic production. Because of the importance of cereals in most developing countries' food needs, food self-sufficiency is often defined in terms of cereal self-sufficiency (Paulino 1986, Cheng 1989). Where agricultural lands are limited, cereal self-sufficiency could mean agricultural expansion into forest lands. Thus, the extent of forest depletion was expected to increase with increasing cereal self-sufficiency. In this study, self-sufficiency in cereal was measured by the proportion, as a percentage, of domestically produced cereal out of total cereal consumption. Cereal consumption was calculated as the sum of domestic production and imports less exports.

Land and population

Forest depletion in tropical countries was hypothesized to increase with both increasing population and increasing rural landlessness. The ratio of arable land to agricultural population, in hectares per capita, was used as the indicator of land availability for agriculture. The smaller the ratio, presumably, the greater the constraint on cultivable land, and, the greater the likelihood that forests would be cleared for agriculture. Population, in million individuals, was included as an explanatory variable and also served as a scaling variable. Larger countries with larger populations were also expected to have greater deforestation.

5.5 Statistical estimation

Tropical deforestation, as indicated by the extent of logging of tropical closed broadleaved forest, was hypothesized to be a linear function of income, agricultural and forestry products export prices, currency exchange rate changes, international debt service, cereal self-sufficiency, rural landlessness, and population. The average annual values of all variables were calculated for each of the four periods. These average annual values were then used in the regressions. Thus, there were 45 observations for each period, and 180 total observations for all periods pooled together.

Two types of linear models were specified and estimated using least squares regression. Model 1 assumed common intercept and slope coefficients for all countries regardless of geographic region, income group, and conditions of international indebtedness. Model 2 retained the assumption of common slope coefficients but allowed the intercept term to vary differentially by region, income group, indebtedness, and, for regressions using pooled data, by period. This was accomplished by the introduction of dummy variables as additional regressors into the second model. The Breusch–Pagan test was used to test statistically the null hypothesis of homoscedasticity of the residuals from each model in each of the four periods. Test results rejected the null hypothesis with 99 per cent level of confidence in all instances, indicating heteroscedasticity in the residual variances from each model in all periods. Homogeneity of the residual variances across the four periods was a precondition for pooling the data. To determine whether this condition was satisfied, the Bartlett test was used. Calculated χ^2 with 3 degrees of freedom were 111.12 and 115.00 for Model 1 and Model 2, respectively. Again, the null hypothesis of homogeneous residual variances was rejected at the 0.01 significance level in both models. The results of the Bartlett test were verified by Chow's test using the F statistic. To eliminate the problem of heteroscedasticity and make the residual variances homogenous, the observations in each period were divided by the square root of the mean square error obtained from the respective individual period regressions. The resulting homoscedastic data were then used in the regressions.

Condition numbers calculated from individual period and pooled regressions using the heteroscedasticity-corrected data indicated mild to moderate multicollinearity in the explanatory variables. Examination of partial correlation coefficients among explanatory variables and Glauber–Farrar tests for the presence and location of multicollinearity confirmed the existence of multicollinearity and pointed to the export value index of tropical wood as the variable that contributed the most to the problem. The export value index of tropical wood tended to move together with per capita income and, not unexpectedly, to some extent, with the export price of agricultural commodities. This tendency was most pronounced during the first period. The removal of the unit export price index as explanatory variable in Models 1 and 2, creating corresponding Models 1a and 2a, significantly reduced the linear association among the explanatory variables. However, it did not drastically change the regression results from both models.

Table 5.2 Coefficient estimates from Model 1 for individual periods and for all periods pooled (standard errors).

Variable	Period 1 1967–71	Period 2 1972–75	Period 3 1976–80	Period 4 1981–85	Pooled 1967–85
Constant	-1.80^a	-2.93^b	-2.48^a	-2.11	-0.10
	(0.84)	(0.67)	(0.93)	(1.05)	(0.19)
Log export value index	1.42^b	-0.57	-8.08	19.99	0.93^b
	(0.38)	(0.66)	(4.76)	(10.13)	(0.29)
Agricultural export	-0.33	1.44	-1.39	-3.67	4.03E-03
price index	(0.39)	(0.79)	(3.46)	(8.68)	(0.41)
Per capita income	0.02	0.34^b	0.40^a	-0.34	0.07
	(0.07)	(0.05)	(0.17)	(0.34)	(0.04)
Debt service ratio	-11.26	-32.23^a	-25.83	-8.52	-2.95
	(11.59)	(14.72)	(22.05)	(30.77)	(8.97)
Real devaluation rate	8.99	-0.02	84.12^b	15.08^a	6.16^a
	(7.34)	(2.51)	(22.57)	(5.84)	(2.45)
Cereal self-sufficiency	4.53	14.61^b	21.73^a	7.42	-1.00
ratio	(2.54)	(3.43)	(8.67)	(11.36)	(1.22)
Arable land per	67.06	8.35	868.01	2304.66^b	21.66
agricultural capita	(63.56)	(146.80)	(599.21)	(704.09)	(70.56)
Population	9.48E-04	$1.88\text{E-}03^a$	2.14E-03	2.91E-03	$1.39\text{E-}03^a$
	(7.13E-04)	(8.52E-04)	(1.52E-03)	(2.01E-03)	(5.84E-04)
R-square	0.85	0.70	0.58	0.47	0.48
Adjusted R-square	0.82	0.64	0.48	0.35	0.46
F-ratio	25.72^b	10.62^b	6.10^b	3.91^b	19.93^b

Notes: a) significant at 0.05 level; b) significant at 0.01 level.

5.6 Results and discussion

Regression results showed significant influence of certain macroeconomic factors at certain periods during the years under study (Tables 5.2 to 5.6). F-tests indicated that the over all regression of Models 1 and 1a using pooled and individual period data were all statistically significant at 0.01 level of significance (Tables 5.2 and 5.3). The over all regression of Models 2 and 2a using pooled and individual period data were also statistically significant at the 0.01 level except during the fourth period (Tables 5.4 and 5.5). These results were unchanged by the removal of the export value index of tropical wood as explanatory variable. The explanatory power of the models was highest in Period 1 and lowest in Period 4. Apparently, with major changes in the global economy and in developing countries' economies along with them, the driving forces for tropical forest depletion had become more complicated than could be represented by the linear functional relationships specified in this study. During the relative stability of the first period, all models (1, 1a, 2 and 2a), explained over 80 per cent of the variation in the dependent variable. By the fourth period, the same models were able to explain only about 40–50 per cent of the variation.

Table 5.3 Coefficient estimates from Model 1a for individual periods and for all periods pooled (standard errors).

Variable	Period 1 1967–71	Period 2 1972–75	Period 3 1976–80	Period 4 1981–85	Pooled 1967–85
Constant	−3.14[b]	−2.97[b]	−3.13[b]	−1.14	−0.22
	(0.67)	(0.67)	(0.83)	(0.94)	(0.19)
Agricultural export price index	0.07	1.04	−2.18	−2.63	0.57
	(0.44)	(0.63)	(3.51)	(9.00)	(0.40)
Per capita income	0.23[b]	0.32[b]	0.34	−0.20	0.17[b]
	(0.04)	(0.04)	(0.17)	(0.34)	(0.03)
Debt service ratio	−18.80	−31.07[a]	−20.32	2.91	−3.38
	(13.26)	(14.60)	(22.36)	(31.38)	(9.33)
Real devaluation rate	9.90	−0.02	60.18[b]	10.35	6.27[a]
	(8.52)	(2.50)	(18.08)	(5.53)	(2.41)
Cereal self-sufficiency ratio	10.00[b]	14.29[b]	23.83[a]	6.15	−0.38
	(2.41)	(3.40)	(8.80)	(11.78)	(1.31)
Arable land per agricultural capita	27.97	47.70	991.48	2056.64[b]	−38.10
	(72.85)	(139.01)	(609.79)	(719.32)	(74.03)
Population	1.56E-03	1.86E-03[a]	1.96E-03	3.20E-03	1.61E-03[b]
	(8.07E-04)	(8.49E-04)	(1.56E-03)	(2.08E-03)	(6.07E-04)
R-square	0.79	0.70	0.54	0.41	0.43
Adjusted R-square	0.75	0.64	0.45	0.30	0.41
F-ratio	20.29[b]	12.12[b]	6.24[b]	3.63[b]	18.51[b]

Notes: a) significant at 0.05 level; b) significant at 0.01 level .

Table 5.4 Coefficient estimates from Model 2 for individual periods (standard errors).

Variable	Period 1 1967–71	Period 2 1972–75	Period 3 1976–80	Period 4 1981–85
Constant	−2.29[a]	−2.82[b]	−1.90	−2.48
	(1.06)	(0.86)	(1.17)	(1.31)
Region 2 (South America)	−0.51	0.07	0.02	−0.07
	(0.63)	(0.62)	(0.66)	0.68
Region 3 (Asia)	0.19	0.66	−0.77	0.39
	(0.67)	(0.62)	(0.69)	(0.72)
Region 4 (Africa)	0.13	−0.22	−0.49	0.34
	(0.55)	(0.55)	(0.57)	(0.61)
Income group 2 (middle income oil importing countries)	0.58	0.12	−0.13	0.40
	(0.53)	(0.54)	(0.57)	(0.60)
Income group 3 (oil exporting countries)	0.87	−0.34	−0.45	−5.02E-03
	(0.58)	(0.61)	(0.68)	(0.81)
Debt group 2 (highly indebted countries)	0.24	−0.03	0.60	−0.13
	(0.58)	(0.50)	(0.56)	(0.69)
Log export value index	1.54[b]	−0.71	−12.24[a]	22.40
	(0.47)	(0.73)	(5.77)	(11.90)
Agricultural export price index	−0.13	1.86[a]	−1.54	−5.09
	(0.45)	(0.87)	(4.01)	(10.71)
Per capita income	−5.59E-03	0.35[b]	0.47[a]	−0.34
	(0.08)	(0.07)	(0.20)	(0.47)
Debt service ratio	−7.17	−41.98[a]	−45.96	−0.47
	(12.60)	(16.91)	(27.07)	(39.67)
Real devaluation rate	8.59	0.51	95.99[b]	15.32[a]
	(8.95)	(2.66)	(26.48)	(6.56)
Cereal self-sufficiency ratio	4.40	13.62[b]	25.86[a]	3.71
	(2.70)	(3.78)	(9.85)	(15.38)
Arable land per agricultural capita	87.27	92.99	680.13	2487.76[b]
	(71.40)	(168.35)	(684.65)	(862.24)
Population	8.82E-04	1.63E-03	1.88E-03	3.42E-03
	(8.57E-04)	(1.04E-03)	(1.96E-03)	(2.79E-03)
R-square	0.87	0.73	0.62	0.48
Adjusted R-square	0.80	0.61	0.45	0.24
F-ratio	13.77[b]	5.91[b]	3.52[b]	2.00

Notes: a) significant at 0.05 level; b) significant at 0.01 level.

The declining predictive power of the model is consistent with the results of Chow's test for structural change over the four periods. Calculated F statistics were

Table 5.5 Coefficient estimates from Model 2a for individual periods (standard errors).

Variable	Period 1 1967–71	Period 2 972–75	Period 3 1976–80	Period 4 1981–85
Constant	−3.71[b]	−2.96[b]	−3.28[b]	−1.15
	(0.85)	(0.85)	(0.97)	(1.12)
Region 2 (South America)	0.07	−0.03	−0.26	0.10
	(0.60)	(0.61)	(0.67)	0.68
Region 3 (Asia)	0.69	0.57	−0.32	0.25
	(0.63)	(0.61)	(0.67)	(0.71)
Region 4 (Africa)	0.43	−0.25	−0.07	0.06
	(0.53)	(0.55)	(0.55)	(0.59)
Income group 2 (middle income oil importing countries)	0.62	0.21	0.02	0.36
	(0.52)	(0.54)	(0.58)	(0.60)
Income group 3 (oil exporting countries)	0.49	−0.19	−0.41	0.15
	(0.56)	(0.59)	(0.70)	(0.80)
Debt group 2 (highly indebted countries)	−0.83	−0.01	0.52	−0.13
	(0.53)	(0.50)	(0.58)	(0.69)
Agricultural export price index	0.16	1.44	−2.49	−1.57
	(0.51)	(0.75)	(4.20)	(10.97)
Per capita income	0.24[b]	0.32[b]	0.39	−0.29
	(0.04)	(0.05)	(0.21)	(0.49)
Debt service ratio	−11.18	−39.93[a]	−30.62	1.00
	(14.40)	(16.76)	(27.53)	(41.25)
Real devaluation rate	12.80	0.63	59.98[b]	10.54
	(10.17)	(2.66)	(21.44)	(6.29)
Cereal self-sufficiency ratio	9.24[b]	13.54[b]	26.99[a]	2.80
	(2.60)	(3.77)	(10.37)	(15.99)
Arable land per agricultural capita	64.46	140.48	1017.00	2060.55[a]
	(81.57)	(160.85)	(702.59)	(865.35)
Population	1.67E-03	1.64E-03	1.59E-03	3.37E-03
	(9.45E-04)	(1.04E-03)	(2.06E-03)	(2.90E-03)
R-square	0.82	0.73	0.57	0.42
Adjusted R-square	0.74	0.61	0.38	0.18
F-ratio	10.62[b]	6.31[b]	3.10[b]	1.74

Notes: a) significant at 0.05 level; b) significant at 0.01 level.

$F(27,171) = 5.49$ and $F(42,162) = 2.41$ for Model 1 and Model 2 and were significant at the 0.01 and 0.05 levels, respectively. Structural change is defined as a switch in a regression equation and is manifested by statistically significant change in one or more of the regression coefficients (Johnston 1984, Hsiao 1986,

Table 5.6 Coefficient estimates from Model 2 and Model 2a for all periods pooled (standard errors).

Variable	Model 2	Model 2a
Constant	−3.98[b]	−3.72[a]
	(0.72)	(0.61)
Region 2 (South America)	0.51	0.49
	(0.33)	(0.31)
Region 3 (Asia)	0.26	0.26
	(0.33)	(0.32)
Region 4 (Africa)	0.18	0.18
	(0.29)	(0.27)
Income group 2 (middle income oil importing countries)	0.39	0.36
	(0.28)	(0.27)
Income group 3 (oil exporting countries)	0.06	−9.19E−04
	(0.32)	(0.30)
Debt group 2 (highly indebted countries)	−0.01	−0.02
	(0.27)	(0.26)
Dummy variable for period 2	1.52[b]	1.09[b]
	(0.37)	(0.30)
Dummy variable for period 3	2.83[b]	2.52[b]
	(0.55)	(0.45)
Dummy variable for period 4	3.18[b]	2.92[b]
	(0.61)	(0.51)
Log export value index	0.24	−
	(0.31)	−
Agricultural export price index	0.24	0.40
	(0.40)	(0.39)
Per capita income	0.19[b]	0.22[b]
	(0.05)	(0.03)
Debt service ratio	−9.92	−9.63
	(8.77)	(8.96)
Real devaluation rate	6.23[b]	5.99[b]
	(2.35)	(2.27)
Cereal self-sufficiency ratio	8.80[b]	9.49[b]
	(2.13)	(2.08)
Arable land per agricultural capita	81.06	81.67
	(68.31)	(72.58)
Population	1.55E−03[a]	1.61E−03[a]
	(6.08E−04)	(6.24E−04)
R-square	0.59	0.56
Adjusted R-square	0.54	0.52
F-Ratio	13.55[b]	13.16[b]

Notes: a) significant at 0.05 level; b) significant at 0.01 level.

Broemeling & Tsurumi 1987). In the context of this study, structural change indicates that the nature of the relationship between tropical forest depletion and the explanatory variables changed with changing global economic conditions.

In all models, the intercept terms were consistently negative, suggesting that the explanatory variables had to be larger than some minimum level before the associated forest depletion became observable. The intercept terms were statistically significant during the first three periods in Models 1, 1a and 2a. In Model 2, the intercept was significant only during the first two periods. Results from the pooled regressions showed the intercept terms to be highly significant in Models 2 and 2a but to be non-significant in Models 1 and 1a.

Period 1

Results from Models 1 and 2 indicate that during Period 1, the export value of tropical wood was the only variable statistically associated with forest depletion. The significantly positive coefficient indicates that logging and forest depletion increased with increasing value of tropical wood in the export market. When this variable was excluded in Models 1a and 2a, per capita income and cereal self-sufficiency ratio became statistically significant at the one per cent level in both models. That per capita income would become highly significant in Models 1a and 2a was to be expected given its high correlation with the tropical wood export value index.

Period 2

During Period 2, results from Models 1 and 1a showed forest depletion to be highly statistically associated with rising per capita income and increasing cereal self-sufficiency. Population and debt service ratio were also statistically significant explanatory variables. While forest depletion increased with population, it apparently decreased with increasing debt service. The negative relation between forest depletion and debt service was counter to prior expectations, suggesting that in the early 1970s, less logging and forest depletion occurred in countries with higher debt service obligations. This could conceivably be a manifestation of the effect of the sudden rise in fuel prices which not only prompted many countries to borrow to finance oil imports but also raised the cost of logging operations, leading to reduced logging and forest depletion.

When dummy variables for region, income group and debt category were introduced in Models 2 and 2a, the coefficients of per capita income and cereal self-sufficiency remained highly significantly positive and that of debt service remained significantly negative. However, the coefficient of population was no longer statistically significant. The coefficient of the agricultural export price index was significantly positive in Model 2 but was not significant in Model 2a.

Period 3

During the third period, when exchange rate adjustments were required of borrower countries, the rate of real currency devaluation was the most highly significant explanatory variable in all models. The results suggest that logging and depletion of tropical forests during this period increased with higher rates of domestic currency devaluation. The cereal self-sufficiency ratio continued to be significantly positively associated with tropical forest depletion in all models. Per capita income also continued to be positively and significantly associated with forest depletion in Models 1, 2 and 2a but not in Model 1a. The export value of tropical wood was statistically significant with a 95 per cent confidence level in Model 2. In Model 1, the export value of tropical wood could be considered statistically significant if one were to settle for only 90 per cent level of confidence in the estimate. However, in both models, it had a negative coefficient, suggesting that wood harvesting increased even while the value of tropical wood exports declined. This result implies that countries tended to increase the volume of wood harvested to make up for declining value in order to maintain revenue.

Period 4

Coefficient estimates for the fourth period indicated a highly significant association between forest depletion and arable land per agricultural capita. The ratio of arable land to agricultural population was significantly positive in all models and was the only significant explanatory variable in Models 1a, 2 and 2a. In Model 1, the real devaluation rate also had a statistically significant positive coefficient indicating that logging and depletion of tropical forests increased as the domestic currency falls in value relative to the US dollar.

All periods pooled

Results of Model 1 regression using pooled data indicated that forest depletion during the whole study period was most significantly associated with the export value of tropical wood followed by population and real exchange rate changes. All three variables had significantly positive coefficients suggesting greater logging and forest depletion with higher wood export value, increasing population, and falling value of the domestic currency. With the removal of wood export value in Model 1a, population and real exchange rate devaluation became more highly significant. Per capita income which was not significant in Model 1, also became highly and positively significant. Consistent with the results from Models 1 and 1a, per capita income, real devaluation rate, and population all had significantly positive coefficients in Models 2 and 2a (Table 5.6). However, the cereal self-sufficiency ratio, which was not significant in Models 1 and 1a, was shown to be highly positively associated with forest removal in Models 2 and 2a both. The dummy variables for period were all highly statistically significant. Their relative magnitudes suggest that the extent of logging and forest depletion has increased over the four periods. None of the other dummy variables was statistically significant.

5.7 Conclusion

Over the 19-year period covered by the study, there appears to have been a secular increase in the magnitude of forest depletion in tropical developing countries. Presumably, this was in response to the turbulence in the global macroeconomy experienced during the study period which provided the context of countries' policies and resource allocation decisions. Viewed longitudinally, tropical forest depletion seems to have increased with increasing population and income, increasing domestic cereal self-sufficiency, and with downward adjustments in the value of the domestic currency relative to the US dollar. However, these variables exerted varying levels of influence on tropical forest depletion at different times within the study period.

Viewed over a series of shorter time intervals, tropical forest depletion appeared to be a response to what, at the time, were the most dominant incentives and disincentives for forest preservation as opposed to logging and forest land conversion to agriculture. During 1967–71, a period of relative global market stability, logging and depletion of tropical forests was most strongly related to the value of tropical wood in the export market. With the grain shortages, increasing oil prices and heightened concerns for food security during the early 1970s, tropical forest depletion was most strongly associated with countries' per capita income and degree of self-sufficiency in cereals. During the latter half of the 1970s, with currency exchange rate devaluations as the centrepiece of most countries' macroeconomic structural adjustment programmes, forest depletion was most strongly linked to the rate of devaluation of the domestic currency relative to the US dollar. With the debt crisis and reduced capital and trade flows to developing countries in the 1980s, tropical forest depletion appears to have been associated with the availability of arable land for agriculture.

6

Macroeconomic causes of deforestation: barking up the wrong tree?

Nemat Shafik

6.1 Introduction

Traditional explanations of the causes of deforestation have focused on the clearing of land for agriculture or livestock and on the harvesting of forest products. However, more recently, studies of deforestation have often linked the problem to macroeconomic variables such as income, investment, exchange rates, trade policy, and indebtedness. These conclusions are often derived from localized studies of the causes of deforestation that draw associations between the macroeconomic situation and the state of forest resources in a particular country. Rarely have these associations between macroeconomic conditions and deforestation been evaluated in a systematic manner across a large number of countries.

This chapter analyzes the causes of deforestation econometrically for a large sample of countries. The following sections will review the issues and describe some of the methodological and empirical issues that constrain such an analysis. Thereafter, empirical evidence on the relationship between deforestation and per capita income as well as a large number of policy variables is presented. The final section draws some conclusions about the limits of aggregate analysis and the apparent absence of generalized macroeconomic explanations for deforestation.

6.2 The issues: macroeconomics and deforestation

The direct causes of deforestation are fairly obvious – forests are cleared either to harvest timber or other forest products or they are cleared to use the land for agriculture or livestock. In the case of tropical deforestation, some estimates of the relative rôle of these causes have been made. A study by Johnson suggests that 64 per cent of tropical forest depletion can be attributed to the conversion of land for agriculture, 18 per cent to commercial logging, 10 per cent to collection of fuelwood, and 8 per cent to ranching (Johnson 1991). Estimates by the World Bank indicate that the bulk of tropical deforestation (about 60 per cent of the area cleared each year) is lost to agricultural settlement, with the remainder split roughly between logging and other uses such as roads, urbanization and fuelwood (World Bank 1991a).

The indirect causes of deforestation are more complex and controversial. Clearly macroeconomic variables such as exchange rates, indebtedness and investment can alter the incentives to clear land for agriculture and livestock or to harvest forest resources. A number of studies have tried to identify such links directly. The effects of international trade on the environment, including deforestation, have been assessed to respond to environmentalist criticisms that free trade forces natural resource endowed developing countries to exploit forests excessively (Low 1992, Shrybman 1990). Studies have also tried to link developing country debt and tropical deforestation (Kahn & McDonald 1992, Bigman 1990, Barbier 1989b). Some authors have drawn associations between deforestation and high rates of inflation (Thampapillai 1992). With few exceptions, those studies that have linked deforestation to macroeconomic variables have been based on single country studies or a small number of selected countries. Because of methodological and empirical difficulties, few have attempted to test whether such apparent links between excessive deforestation, and policies such as free trade, competitive exchange rates, and indebtedness stand up to the scrutiny of systematic statistical analysis.

6.3 Methodological issues

The first obstacle to econometric analysis is the availability of data. Data on land-use and specifically on forest area are compiled by the Food and Agriculture Organization of the United Nations based on national governments' responses to annual questionnaires (FAO, various years). Forest and woodland is defined as land under natural or planted trees, regardless of whether it is intended for eventual harvesting. Data are available for a large sample of countries on a consistent basis only since the 1960s, which is long after many countries had deforested most intensively. Moreover, the quality of these data is often disputed, in part because national governments often have incentives to misrepresent actual forest area. Because of the availability of satellite imaging techniques, there are now ways to confirm reported forested areas. This is expected to improve the quality of available data, although disputes are likely to remain. Sources for all other data are the World Bank, unless otherwise specified.

The second obstacle is how deforestation should be measured. The annual variation in total forest area is deceptive since countries that depleted their forests in the distant past and have slowed down more recently (such as many of the now high income countries) would appear to be doing better than countries with substantial forest resources that have only begun to draw down timber stocks. Looking at a longer time period does not resolve the problem since data are not available sufficiently far in the past to include times when some countries were cutting down forests most intensively (in some cases, during the Middle Ages). Given the limitations of the data, two different measures of deforestation are used in the econometric work below. The first is the yearly change in forest area between 1962 and 1986, which was available for 66 countries and will be referred to as annual deforestation. The second was the total change in forest area between the earliest date for which substantial data are available, 1961, and the latest date, 1986, which was available for 77 countries. (This variable was defined as: $\log [(F_{61} - F_{86}) \times 100)/ F_{61}]$ where F = forest area).

6.4 Empirical evidence

Deforestation and income

It is often argued that poverty forces individuals with limited access to resources to encroach on forests, especially where property rights over forest resources are poorly defined. There is also a view often associated with the South that forest protection is a luxury that only rich countries can afford; meanwhile poor countries must exploit whatever natural endowment they have in order to improve living standards. Thus the relationship between per capita income and the rate of deforestation is perceived to be a central issue.

In order to understand the relationship between income and deforestation, panel regressions were used to test three models – log linear, quadratic and cubic – to explore the shape of the relationship between per capita income (Y) and the annual and total rates of deforestation (D):

$$D = a_1 + a_2 \log Y + a_3 \text{time} \tag{1}$$

$$D = a_1 + a_2 \log Y + a_3 \log Y^2 + a_4 \text{time} \tag{2}$$

$$D = a_1 + a_2 \log Y + a_3 \log Y^2 + a_4 \log Y^3 + a_5 \text{time}. \tag{3}$$

Per capita income was defined as real per capita gross domestic product in purchasing power parity terms. This is variable RGDPCH in the Penn World Table

Mark 5 (Summers & Heston 1991). The chain based method of indexing was used to take into account the changing production bundle over the period. Estimates using conventional measures of per capita income were also made, but in general the purchasing power parity estimates performed relatively better. The constant term varied for each country or city to capture country-specific fixed effects. A time trend was added in the case of annual deforestation to proxy improvements in technology, either autonomous or policy-induced, over time. The results are reported in the first three columns of Tables 6.1 and 6.2.

The results for both the annual and total measures of deforestation indicate that per capita income has an insignificant effect, regardless of the functional form. The best fit, relatively speaking, is the quadratic form, particularly with the total deforestation variable and is depicted in Figure 6.1. This implies that deforestation, like some measures of local air pollution, follows a "bell-shaped curve" – tending to worsen with rising per capita income and then improve after some "turning point." Nevertheless, statistically speaking, per capita income has virtually no explanatory power in every case, as evidenced by the R^2 statistics equal to zero. This is in stark contrast to most other environmental indicators where per capita income is by far the most important variable for explaining changes in environmental quality (Shafik 1993).

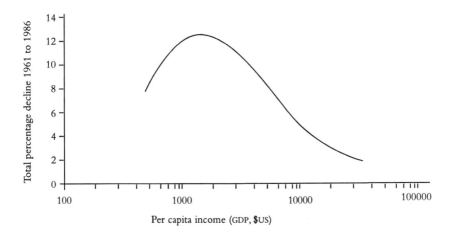

Figure 6.1 Total deforestation.

In order to explore the importance of policy, a number of indicators of policy stance were added sequentially controlling for the albeit small effect of per capita income. The results for the quadratic functional form, which was the best fit, are the only ones reported in Tables 6.1 and 6.2. All policy variables were tested with all functional forms and the results were not substantially different.

Table 6.1 Annual rate of deforestation, 1962–86.

Intercept	3.49 (0.79)	0.64 (0.05)	15.57 (0.64)	-2.12 (-0.15)	3.36 (0.25)	155.85 (0.20)	-24.62 (-1.69)	45.81 (2.44)	25.18 (0.53)	9.58 (0.31)	-0.12 (-0.34)	-1.21 (-0.4)
Income	-0.02 (0.77)	0.65 (0.85)	-5.34 (-0.86)	0.68 (0.88)	0.14 (0.18)	-7.58 (-1.57)	4.56 (4.78)	1.53 (1.24)	-3.97 (-2.41)	1.31 (1.01)	0.82 (0.71)	1.61 (0.92)
Income squared	-0.04 (-0.88)	0.76 (0.70)	-0.05 (-0.91)	-0.02 (-0.37)	0.54 (1.61)	-0.30 (-4.59)	-0.11 (-1.30)	0.29 (2.56)	-0.08 (-0.93)	-0.05 (-0.64)	-0.06 (-0.82)	
Income cubed			-0.04 (-0.74)									
Time trend	0.00 (0.83)	0.00 (0.81)	0.00 (0.23)	0.00 (0.43)	0.00 (0.15)	-0.06 (-0.16)	0.01 (0.84)	-0.02 (-2.52)	-0.01 (-0.54)	-0.00 (-0.29)	0.01 (0.42)	0.00 (0.07)
Income growth				0.49 (0.87)								
Investment share					0.33 (4.22)							
Electricity tariff						-0.70 (-2.91)						
Trade share in GDP							-1.06 (-11.81)					
Parallel market premium								-0.06 (-1.92)				
Dollar index									3.13 (0.49)			
Debt per capita										0.06 (0.63)		
Political rights											0.07 (1.81)	
Civil liberties												0.13 (2.60)
Adjusted R^2	0.00	0.00	0.00	0.00	0.01	0.11	0.10	0.01	0.02	0.00	0.00	0.00
Number of observations	1511	1511	1511	1508	1509	71	1248	895	657	892	762	762

Note: t-statistics are shown in parentheses.

Table 6.2 Total change in forest area, 1961-86

	(1)	(2)	(3)	(4)	(5)	(6)	(7)	(8)	(9)	(10)	(11)	(12)
Intercept	2.99 (0.04)	-9.74 (0.96)	-41.42 (0.56)	-9.99 (-0.96)	-9.41 (-0.91)	-22.27 (-1.21)	-21.94 (-1.83)	-16.78 (-1.19)	-11.63 (-0.59)	-15.25 (-0.82)	-14.31 (-1.23)	-15.53 (-1.34)
Income	-0.87 (0.66)	3.33 (1.22)	16.04 (0.54)	3.40 (1.22)	3.31 (1.20)	7.00 (1.37)	6.58 (2.06)	5.26 (1.37)	0.60 (0.17)	4.89 (1.07)	5.16 (1.58)	5.49 (1.79)
Income squared		-0.23 (-1.25)	-1.91 (-0.49)	-0.23 (-1.26)	-0.22 (-1.21)	-0.48 (-1.36)	-0.45 (-2.11)	-0.36 (-1.39)	-0.06 (-0.23)	-0.34 (-1.07)	-0.38 (-1.84)	0.40 (-1.95)
Income cubed			0.07 (0.43)									
Income growth				0.65 (0.22)								
Investment share					-0.22 (-0.81)							
Electricity tariff						-0.30 (-1.14)						
Trade share in GDP							-0.32 (-0.89)					
Parallel market premium								0.05 (0.62)				
Dollar index									0.13 (0.78)			
Debt per capita										-0.26 (-1.01)		
Political rights											-0.16 (-1.53)	
Civil liberties												-0.18 (-1.35)
Adjusted R^2	-0.01	0.00	-0.02	-0.02	-0.01	0.00	0.05	0.02	-0.04	-0.02	0.07	0.06
Number of observations	58	58	58	58	58	36	47	41	60	54	56	56

Notes: t-statistics are shown in parentheses.

Deforestation, economic growth and investment

Economies experiencing rapid economic growth and investment may have lower environmental quality relative to the average for their income level if regulations are slow to respond to changing circumstances. However high investment rates may imply relatively higher environmental quality to the extent that cleaner technologies are embodied in newer machines. In the case of forests, the rate of investment and growth could imply more pressure on natural resources (to the extent that forests are an input or complementary to physical capital in the production process) or less pressure on forests (where new investments in areas such as agricultural intensification serve to substitute for forest inputs).

In the case of income growth, defined as the growth rate of real per capita GDP in purchasing power parity terms, there is no significant effect on either annual deforestation (Table 6.1) or total deforestation (Table 6.2). The investment rate, defined as total capital formation as a share of GDP, has no effect on total deforestation, but has a significantly positive effect on the annual rate of deforestation. This seems to imply that forest resources tend to be complementary to physical investment, rather than a substitute, so that high investment economies tend to deforest more than low investment economies.

Deforestation and energy pricing

Subsidized energy tends to be associated with greater environmental degradation, but in the case of forests, the relationship is complex and depends upon the cross-elasticities between biomass and other sources of energy. Subsidies to domestic energy consumption may reduce the incentives to rely on forests as a fuel source, especially if biomass fuels are considered inferior goods.

Because energy subsidies affect a number of different energy types – such as kerosene, coal, oil and nuclear – it is difficult to quantify their magnitude. Consistent cross-country data on electricity prices were available only for about 70 countries for the late 1980s. These electricity prices were added to the core quadratic model in Tables 6.1 and 6.2. In the case of annual deforestation, electricity tariffs have a significantly negative sign, implying that higher energy prices are associated with a lower rate of annual deforestation. This result is somewhat counter-intuitive since higher electricity tariffs would be expected to encourage the use of virtually free energy sources such as forests. A possible explanation may be that electricity prices are acting as a proxy for some other factor, such as price distortions or market orientation, which may be significant for explaining deforestation.

Trade policy

The relationship between trade policy and deforestation is complex and operates through a number of channels. More open economies are characterized by greater specialization and, where countries have comparative advantage in timber production, a liberal trade regime may be associated with substantial timber exports. This is why timber bans have been advocated in countries such as Thailand where there was extensive deforestation for export purposes. But open trade policies are also associated with greater efficiency, improved technologies, and often more rapid growth − all of which may work in favour of the environment.

Evaluating the impact of trade policy on deforestation is complicated by the lack of adequate indicators of trade policy. Many of the indicators in use are not correlated with each other and generate conflicting results (Pritchett 1991). Because of this lack of consensus, three different measures of trade orientation were tested empirically: total imports and exports as a share of GDP, Dollar's index of trade orientation based on a weighted average of mean price distortion in an economy relative to other countries of similar endowment, and the parallel market premium which serves as a measure of the degree to which tradable goods prices are distorted in an economy. The Dollar index was available for 87−90 countries between 1973 and 1985 and was computed as the weighted average of mean price distortion in the period 1973−85 and its standard deviation. The price distortion was calculated as the residual of a regression of the relative price of consumption goods on urbanization, GDP per capita, and an interactive term. All variables were entered in logs, so that the residual could be interpreted as a percentage deviation from an appropriate level as determined by the countries' endowment (Dollar 1991). Although each of these measures of trade policy has its drawbacks, they are all available for a relatively large number of countries and are all empirically-based, rather than some more subjective indicators in use.

The results in Tables 6.1 and 6.2 indicate no effect of trade policy on total deforestation, but some significant effects in the case of the annual rate of deforestation. Countries that have a larger trade share in GDP tend to put less pressure on forest resources − implying that more open policies reduce deforestation. In contrast, countries characterized by greater distortions (as measured by an overvalued official exchange rate and a large premium in the parallel market) seem to deforest less on an annual basis. These results are somewhat contradictory and inconclusive and may reflect the measurement problems on both the left and right hand side of the regressions.

· Debt

It is often argued that countries with heavy debt burdens are forced to exploit their natural resources, such as forests, suboptimally in order to meet their debt servicing obligations. Debt per capita was included in the regressions as an indicator of the severity of a country's debt burden. Debt was defined as disbursed amounts of short and long term external liabilities outstanding and IMF credit (World Bank, 1992c). Because of methodological problems with defining the debt of industrial countries, essentially because of the complexity of multinational borrowing, the sample was truncated so that debts of industrial countries were set to equal zero. The econometric evidence indicates the absence of any significant relationship between indebtedness and deforestation.

Political and civil liberties

There is a perception that more open and democratic societies will have better environmental quality because of the public-good character of many natural resources. However, more democratic societies may be subject to greater interest group pressure that could undermine environmental protection. To test such hypotheses, two indices were added to the regressions. One is an index of political liberties that measures rights such as free elections, the existence of multiple parties, and the decentralization of power. Countries' political rights are evaluated on a scale of one to seven. A high ranking country must have a fully operating electoral procedure, usually including a significant opposition vote. It is likely to have had a recent change of government from one party to another, an absence of foreign domination, and decentralized political power (Gastil 1989). The civil liberties index measures freedom to express opinions without fear of reprisal. This index reflects rights to organize and demonstrate as well as freedom of religion, education, travel and other personal rights; more weight was given to those liberties that are more directly related to the expression of political rights (Gastil 1989).

Both political rights and civil liberties seem to have an insignificant effect on total deforestation, but a significantly positive effect on the annual rate of deforestation. Countries that are more democratic are likely to experience a more rapid loss of forest area, perhaps because they are more subject to local pressures and are reluctant to enforce forest protection.

6.5 Conclusions

All of the results presented above should be treated with caution given the caveats about the data described at the beginning of the paper. Future refinements in data and methodology should make continuously improved estimates of the causes of deforestation possible. It may be too early to conduct such econometric analysis, given the inadequacy of the data, and it is certainly too early to draw any conclusive results.

Nevertheless, the conclusions from this paper do seem to indicate that there are very few macroeconomic causes of deforestation at the aggregate level. Many of the variables often cited in the literature – poverty, trade and indebtedness – are consistently insignificant. The only variables that seemed to matter at all – the investment ratio and political and civil liberties – are not ones that have featured prominently in the debate. In the case of investment, the evidence points to the complementarity of capital formation and deforestation in many countries. In the case of political and civil liberties, more democratic regimes seem to deforest more. None of these results was very robust, as evidenced by the absence of any variable that could explain both the annual rate and the total change in forest area over the period 1961–86 and the remarkably low R^2 of the regressions.

The absence of any conclusive results on the causes of deforestation at the macroeconomic level would seem to point to the preferability of microeconomic and case study work. But these results must also serve as a warning to more microeconomic research. Drawing aggregate conclusions from local associations can be very deceptive – the coexistence of debt and deforestation in Brazil does not imply causality, neither in the case of Brazil nor in other countries. The evidence in this paper indicates that while macroeconomic conditions may exacerbate the incentives to deforest, the causes of such deforestation are more likely to lie in microeconomic conditions. Research on issues such as the impact of investment and tax incentives, subsidized credit and property rights over forest areas would seem to be more promising avenues, particularly since effective policies to reduce excessive deforestation are more likely to be microeconomic than macroeconomic.

7

Population, development and tropical deforestation: a cross-national study

Thomas Rudel

7.1 Introduction

Because tropical deforestation has human rather than natural causes, the search for reasons why it has accelerated in the late twentieth century and why it varies in extent from place to place leads directly to phenomena familiar to social scientists. Changes in rural populations, their social structures, and their ties to the larger world system offer a plausible starting point in the search for causes of variable rates of deforestation.

As with other attempts to interpret change in Third World settings (London 1987), the initial attempts to explain tropical deforestation have taken either a human–ecological or a political–economic form. The ecological perspective identifies growing populations of peasant or subsistence cultivators as the chief cause of tropical deforestation (Myers 1984, World Resources Institute 1985). In this argument, small farmers, compelled by the Malthusian necessity of a growing population, leave their lands fallow less and less. The shortened fallow periods prevent forest regrowth, and a region gradually becomes deforested. A Peruvian report (Dourojeanni 1979, cited in Myers 1984: 150) describes a variant of this process in graphic terms:

> The population overflowing from the Andes down to the Amazon plains do not settle there. They advance like a slow burning fire, concentrating along a narrow margin between the land they are destroying and are about to leave behind, and the forests lying ahead of them.

In other words, deforestation results from a growing population's *sustenance-related endeavours* (Poston et al. 1984: 116).

The second perspective on the causes of tropical deforestation is the political–economic one. This perspective emphasizes the rôle of public and private capital in raising rates of deforestation (Hecht 1985, Shane 1986). For example, government funds for colonization open up rainforest regions for settlement and thereby accelerate the rate of deforestation, while government loans to farmers speed up the rate at which farmers convert forests into fields (Hecht 1985). Private investors convert large areas of forest into plantations for the cultivation of export crops; spontaneous colonists follow roads constructed by other investors in pursuit of oil, minerals, or timber. Taken together, these considerations suggest that rapid deforestation coincides with the incorporation of rainforest regions into an expanding national and world economy. In other words, deforestation occurs as part of a process in which capitalism penetrates the countryside. In some instances, such as the deforestation of Indonesia's outer islands, foreign capital plays a rôle in the deforestation process (Peluso 1983). While the eventual result of this process may be regional underdevelopment (Bunker 1984), rapid deforestation initially occurs as part of an accelerated expansion in the larger economy (Hecht 1985). Extending this line of argument, some analysts might argue that rapid deforestation has its origins in the classical forms of dependency which tie peripheral nations to the core states in the world system (development sociologists frequently contrast classical and new industrial forms of dependency (Evans 1979), but as Hecht (1985: 673) has pointed out, the newer forms of dependency almost always involve investments by multinational companies in industrial plants in urban areas. Given this pattern of investment, there is little reason to expect a direct relationship between deforestation and the newer forms of dependency). In this interpretation, rapid deforestation is a by-product of a pattern of trade in which peripheral countries export agricultural commodities and raw materials like timber to the core nations.

In addition to its inconclusiveness on issues of causation, the deforestation literature has been uneven in its coverage of the world's rainforest regions. Latin America has been well studied, while Africa has received almost no attention. In this context, a cross-national study which includes African, Asian, and Latin American countries promises to add breadth to the depth of understanding available in the case studies. If the same study can provide partial tests of the adequacy of the population-growth and capital-availability explanations for deforestation, it should enhance our understanding of the causes of tropical deforestation.

7.2 Data and methods

Data quality considerations influenced the choice of a data set and the selection of a sample for study. Cross-national data on tropical deforestation are found in two data sets, Food and Agriculture Organization (FAO) production data and data from an FAO-United Nations Environmental Program (UNEP) study (FAO 1981a). The FAO-UNEP data set seemed more useful for several reasons. First, the FAO-UNEP study used a restrictive definition of deforestation. It measured deforestation as a decline in closed tropical forest areas, while the FAO production data lump together declines in all types of forests: arid and alpine, as well as tropical. Closed forests are defined as having unbroken canopies, whereas in open forests tree cover is broken, the trees growing wider apart. Secondly, the FAO–UNEP study distinguishes between countries in terms of data quality, while FAO's production reports do not. For these reasons, this study uses FAO–UNEP data. Of the 60 countries containing tropical forests, 36 had data on deforestation which the study's directors regarded as either satisfactory, good, or very good (Lanly 1983: 297). This paper analyzes data from these countries. Since the data on deforestation from the remaining 24 countries were poor in quality, these countries were excluded from the study. The methods of data collection varied among the 36 study countries. Satellite imagery, airborne radar, and aerial photography provided estimates for 18, 3, and 4 countries, respectively. The data for the remaining countries (11) came from reliable on-ground surveys of land-use conversion in and around forested areas (Lanly 1983: 296).

The countries in the sample are spread across Africa, Asia, and Latin America in roughly proportional numbers and contain approximately 77 per cent of the world's tropical forests.[1] On several measures (the extent of urbanization, rates of economic growth, and rural population growth rates), the included and excluded countries fo not differ significantly. The included countries have somewhat larger forested areas and higher GNPs per capita ($792 to $532) than the excluded countries. The sample does have countries with small tropical forests (Burundi, Costa Rica, Guinea-Bissau, Haiti, Jamaica, Rwanda, and Sri Lanka) and low per capita GNPs (Burundi, Guinea-Bissau, Haiti, India, Nepal, Togo, Zaire), so it contains the full range of variation among the variables and can be used to assess their effects on deforestation. In sum, the excluded and included countries are not identical, but the differences between them do not appear to bias the analysis in a serious way.

The analyses reported below include the following variables[2]

Deforested area 1976–80

This variable measures the average annual decline in hectares of a country's tropical forests during the 1976–80 period. Prior to computing this figure, FAO–UNEP personnel established a uniform system for categorizing forests by their humidity and the density of their canopies (FAO 1981a). The uniformity established by these definitions makes it possible to compare deforested areas cross-nationally.

Data quality

Each equation contains a variable which measures the quality of a country's data (1 = very good; 2 = good; 3 = satisfactory), as judged by the United Nations' personnel who compiled the data. The inclusion of this variable in each equation provides a control for variations among countries in the quality of the data.

Closed-forest area 1975

There is a necessary relation between the area deforested annually and the extent of tropical forests in a country. Only countries with large or medium-sized tropical forests will experience the deforestation of large areas each year. To prevent this relationship from confounding other relationships in the analysis, closed-forest area has been included in the equations.

Population growth 1960–75

This variable measures the impact of urban and rural population growth on tropical deforestation. In this line of reasoning, high rates of urban and rural population growth generate strong demand for agricultural and wood products which, in turn, promotes deforestation.

Rural population growth 1960–70

Demographic explanations often attribute declines in forest area to growth in the rural populations living near forests (Myers 1984, World Resources Institute 1985). While additional pressure to expand the cultivated area by clearing forests occurs with each birth, the largest declines in forest area occur when a child reaches adolescence. At this point, a family usually claims and clears new lands in order to provide an economic base for a male child; alternatively, a young man may begin taking out contracts to log nearby forests. In both cases, the effect of local population growth on forest clearing lags about 15 years. To capture this effect, the study measures population growth between 1960 and 1970.

GNP per capita 1975

The numerous ways in which the availability of capital promotes rapid deforestation makes it difficult to select a single measure for this effect and suggests that a general measure of the level of economic activity in a country may provide the best measure for this effect. GNP indirectly measures the wealth which provides local capital for activities which spur deforestation, such as logging, mining, and plantation agriculture. GNP measures the economic output which generates public revenues for roads and concessional loans to farmers which accelerate deforestation.

In an attempt to specify the relationship between economic development and deforestation, two additional variables were added to equations containing demographic and economic-development variables. These variables, export trade in for-est products and export trade in agricultural products, provide a preliminary test of the classical dependency explanation for deforestation.

Value, wood exports 1975

Plantations produce only a small fraction of the tropical hardwoods exported to developed countries; most of the exported wood comes from forests which are logged without replanting (FAO 1981a). For this reason, one would expect a posi-tive relationship between deforestation and the value of foreign trade in tropical hardwoods.

Export agriculture 1975

In a number of well known instances expansion in export agriculture has spurred tropical deforestation. The expansion of Central American cattle ranches to meet a growing demand for imported beef in the US market entailed the widespread destruction of tropical forests (Shane 1986). If this argument is correct, countries which experienced rapid deforestation during the 1960s and early 1970s should have had larger-than-average agricultural export sectors by 1975. This variable measures the value of all agricultural exports, including wood, as a proportion of gross-domestic product. The analysis uses a ratio variable in this instance in order to avoid the problems of collinearity which occur with national accounts data when large countries have high scores and small countries have low scores on all measures.

The dependent variable, hectares deforested, and several independent variables (closed-forest area, population growth, GNP per capita, and wood exports) have skewed distributions. Because the arguments presented above postulate linear rela-tionships between these variables and deforestation, the skewed distributions could mask the existence of a relationship in the data. To counter this tendency, these variables have been logged (Cohen & Cohen 1975: 244–5). To avoid problems of simultaneity bias (Greenwood 1975), all of the independent variables with the exception of the one lagged variable (rural population growth) are taken from 1975 data. The dependent variable concerns the 1976–80 period.

The analyses of the 36 countries presented below use both weighted and unweighted samples. Because this study attempts to answer questions about global patterns rather than intercountry differences in deforestation, most of the analyses weight cases by the size of a country's closed-forest area in effect, weighting coun-tries with large tropical forests heavily, and countries with small tropical forests lightly. This procedure makes equal units of forest area, belonging in varying pro-portions to different nations, the unit of analysis.

7.3 Findings

Table 7.1 presents the correlation matrices; Table 7.2 reports the results from the regression analyses. In diagnostic tests for multicollinearity, the highest condition index achieved by any of the equations is 3.4. Because this score is well below the level at which the effects of collinearity begin to be observed (Belsley et al. 1980: 128), it indicates that the equations do not suffer from serious problems of collinearity.[3] Equation 1 (column 1) in Table 7.2 indicates that in an unweighted sample, population growth (not GNP per capita) explains substantially the variations in tropical deforestation. Allen & Barnes (1985) found a similar association between population growth and deforestation in a study of arid, alpine, and tropical environments. An analysis of the residuals found one influential case: Guinea-Bissau. The removal of this case from the analysis in Equation 2 improves the explanatory power of the population growth variable, but otherwise leave the equation unchanged.

A comparison of the findings from the weighted and unweighted samples of all 36 countries clarifies the relationship between level of development (GNP) and

Table 7.1 Means, standard deviations, and zero-order correlations.

Variables	(1)	(2)	(3)	(4)	(5)	(6)	(7)	(8)
(1) Average deforested area 1976–80 (thousand hectares)								
(2) Data quality	.265							
(3) Forest area (thousand hectares)	.872	.241						
(4) GNP per capita, 1975	.293	−.049	.425					
(5) Population growth, 1960–75 (in thousands)	.529	.130	.411	−.151				
(6) Rural population growth, 1960–75 (in thousands)	.475	.202	.352	−.312	.928			
(7) Value, wood exports, 1975 (thousands, US$)	.544	−.013	.655	.214	.250	.301		
(8) Export average, 1975	−.157	−.260	−.260	−.144	−.182	−.161	.174	
Mean*	153.1	2.3	25,964	648	10,184	3,577	47,777	.133
SD	286.6	.8	62,726	758	29,817	13,482	107,294	.109

Note: * the means and standard deviations reported here come from an unweighted sample, and they have not been transformed to correct for skewness.

Table 7.2 Regression analyses on tropical deforestation.

Independent variables	Unweighted analyses		Weighted analyses			
	(1)	(2)	(3)	(4)	(5)	(6)
Data quality	.133	.049	−.354	−.460	.357	−.260
	(.223)	(.212)	(.244)	(.245)	(.261)	(.254)
Forest area	.667[c]	.595[c]	.418[c]	.486[c]	.411[b]	.450[c]
	(.093)	(.092)	(.105)	(.113)	(.121)	(.113)
GNP	−.014	.226	520[b]	.617[b]	.517[b]	.555[b]
	(.235)	(.241)	(.172)	(.214)	(.176)	(.179)
Population growth	.180[a]	.344[b]	.355[c]	−	.356[c]	.369[c]
	(.086)	(.106)	(.069)		(.072)	(.071)
Rural population growth	−	−	−	.370[c]	−	−
				(.093)		
Wood exports	−	−	−	−	.005	−
					(.083)	
Agricultural exports	−	−	−	−	−	1.54
						(1.92)
R^2 (adjusted =	.772	.808	.830	.790	.823	.828
N =	36	35	36	36	36	36

Notes: the numbers in parentheses are standard errors of the B coefficients. a) $p < .05$; b) $p < .01$; c) $p < .001$.

deforestation.[4] GNP explains a substantial amount of variation in the weighted analysis in Equation 3, but it fails to explain much variation in the unweighted analysis in Equation 1. Because the weighted analysis magnifies the importance of countries with large rainforests, this pattern of findings suggests an interaction between the size of a country's forests and the effects of capital availability on deforestation. Unweighted zero-order correlations between GNP and deforested area for countries with large and small rainforests provide further evidence for this interaction. For the eight countries with the largest forests, the correlation between GNP per capita and the area deforested is 0.575; for the remaining 28 countries with smaller forests, the correlation is -0.011.

These findings suggest the importance of large, capital-intensive projects in sustaining high rates of deforestation in countries with large forests. Without expensive projects like Brazil's trans-Amazon highway to open up distant markets and bridge natural barriers, such as fast-running rivers, rates of deforestation would be much lower in these places. In countries with small, scattered rainforests, encroachment by growing rural populations, with and without capital, leads to deforestation.

Equation 4 examines an alternative line of influence of population growth on tropical deforestation. It substitutes a country's rural population growth for total population growth in the equation. The success of this variable along with total population growth indicates that population growth contributes to deforestation both directly (by increasing the population which clears the land) and indirectly (by increasing the demand for wood products in a country). The foreign economic-influence variables in Equations 5 and 6 fail to explain much variation in the extent of deforestation. Regional differences in the pattern of foreign influence may explain the negative findings.

While exports of wood have accelerated deforestation in Southeast Asia, most African and Latin American countries have experienced rapid deforestation without exporting significant amounts of tropical hardwoods. Similarly, export agriculture may have contributed to extensive deforestation in Central America, but this pattern does not characterize most African or Amazon Basin countries. Taken together, these findings suggest caution in attributing rapid deforestation to relations of dependency between peripheral and core nations in the world system[5].

7.4 Discussion

The analyses presented in Table 7.2 allow us to assess in a preliminary way the accuracy of the two arguments about the causes of deforestation presented at the beginning of this chapter. The analyses provide empirical support for the Malthusian idea that population growth contributed to high rates of deforestation. The significance of these demographic variables in both the weighted and unweighted analyses suggests that all of the various types of countries with high rates of deforestation have recently experienced rapid population growth. The analyses also indicate that political-economic factors contribute to deforestation in a more limited set of circumstances. In countries like Burundi, Rwanda, and Haiti which have small rainforests, growing peasant populations are largely responsible for deforestation, but in countries like Brazil which have large forests, capital investments in frontier regions make possible rapid rates of deforestation.

The last finding raises questions about the extent to which capital expenditures and local population growth interact in the exploitation of forested regions. In some places, deforestation is an extractive process prompted by increasing global demands for timber, and population growth plays a minimal rôle in the process. In other places, population growth alone causes deforestation. In still other places, capital expenditures and population growth interact in creating conditions which accelerate deforestation. Governments or private investors spend money on infrastructure, such as roads which open up regions for settlement and deforestation. Questions about the form and prevalence of these different deforestation processes can only be answered through comprehensive, comparative studies which focus on regional and subregional dimensions of deforestation both within and across nations.

The finding about capital investment and deforestation suggests that the efficacy of policies designed to preserve tropical forests will vary with the size of the forest. Political pressures on lending organizations like the World Bank which provide capital for dams, mines, and penetration roads should slow deforestation in countries with large rainforests because the forested areas cannot be opened up for development and deforestation without major expenditures of capital. Conversely, in countries with small forests, encroachment by the populations surrounding the remaining islands of forest appears to be sufficient to generate high rates of deforestation. To preserve these forests, conservation groups or the state may have to purchase the land, either outright or through debt-for-nature swaps. By denying poor, rural populations access to land, these policies raise important issues of equity (Fortmann & Bruce 1988: 273) which policy makers and researchers will want to acknowledge in evaluating the policies.

Notes

1. The sample included 16 African countries (Benin, Burundi, Cameroon, Gabon, Gambia, Ghana, Guinea-Bissau, Côte d'Ivoire, Kenya, Liberia, Nigeria, Rwanda, Sierra Leone, Tanzania, Togo, Zaire), 15 Latin American countries (Bolivia, Brazil, Colombia, Costa Rica, Ecuador, Haiti, Jamaica, Mexico, Nicaragua, Panama, Paraguay, Peru, Trinidad, Tobago, and Venezuela), and 7 Asian countries (India, Malaysia, Nepal, Papua New Guinea, Philippines, Sri Lanka, and Thailand). The sample excluded 9 African countries (Angola, Central African Republic, Congo, Equatorial Guinea, Guinea, Madagascar, Mozambique, Sudan, and Zambia), 8 Latin American countries (Belize, Cuba, Dominican Republic, El Salvador, Guatemala, Guyana, Honduras, and Surinam), and 7 Asian countries (Bangladesh, Bhutan, Burma, Indonesia, Kampuchea, Laos, and Vietnam).

2. Data Sources: deforested area – FAO/UNEP, Tropical Forest Resources Assessment Project (FAO 1981a); closed-forest area – FAO/UNEP, Tropical Forest Resources Assessment Project (1982); rural population growth – UN, Estimates and Projections of Urban, Rural, and City Populations, 1950–2025 (1985); population growth, 1960–1975 – World Bank, World Tables, Vol. 2 (1983–1984); GNP per capita, 1975 – World Bank, World Tables, Vol. 2 (1983–1984); wood exports, 1975 – FAO, Yearbook of Forest Products (1976); agricultural exports, 1975 – World Bank, World Tables, Vol. 2 (1983–1984); data quality – Lanly (1983: 308).

3. Scores on condition indices have to be above 15 before the effects of multicollinearity can be observed in an equation. The index is calculated from "the eigenvalues of the matrix $X'X$, where X is the data matrix, divided into the largest eigenvalue". For further details on the method, see Belsley et al. (1980).

Simpler but less precise measures of collinearity also indicate that it is not a problem in these equations. One way to assess the degree of collinearity among the independent variables is to regress each independent variable on the other independent variables. Econometricians usually do not worry about collinearity if the r^2 from these equations does not exceed the r^2 in the original analysis. In this study, the highest r^2 involving just the independent variables is 0.58, more than 0.20 below the r^2 in the regressions on deforestation. Accordingly, collinearity would not appear to be a problem.

4. As mentioned in the text, there is a close relationship between closed-forest area and the area deforested. Only countries with large rainforests will have large areas deforested each year. Several reviewers expressed concern that the closeness of this relationship could distort the other, more substantive relationships in the analysis. In response to these concerns, I re-estimated the equations using another measure of the extent of forests (the percentage of a country's land area covered by closed forests); this alternate measure does not create the problem noted above of high correlations between large numbers. The equations presented below are re-estimations of Equations 1 and 3 in Table 7.2, using the new measure of forest area.

Unweighted analysis

area deforested = −3.85 + 0.311data + 0.046forest[b]
(2.30)   (.313)quality   (.014)extent

+ 0.537population[c] + 0.216GNP[c]
(0.103)change   (0.352)

$r = 0.775$, $r^2 = 0.601$, $n = 36$.

Weighted analysis

area deforested = −4.63 + 0.353data + 0.020forest
(2.02)   (0.266)quality   (0.011)extent
+ 0.601population[c] + 0.756GNP[c]
(0.076)change   (0.186)

$r = 0.892$, $r^2 = 0.796$, $n = 36$.
a) $p < 0.05$, b) $p < 0.01$, c) $p < 0.001$.

These re-estimations of Equations 1 and 3 suggest that the results of this analysis are quite robust. Despite the change in the measure of forest extent, the patterns in the re-estimated equations do not differ significantly from the patterns in Table 7.2. Of the two measures of forest extent, I chose to present analyses using forest area rather than the percentage of a country in forests because the latter variable is a ratio variable. Given the controversy surrounding the use of ratio variables in regression analyses, it seemed advisable to avoid the use of ratio variables wherever possible.

5. I do not present the results from the inclusive model because the inclusion of all of the variables in a single equation would violate the principle that there should be substantially more cases than degrees of freedom in a model. The rule of thumb in multivariate analyses calls for no more than one variable for every eight-to-ten cases (London 1987: 35).

8

Population, land-use and the environment in developing countries: what can we learn from cross-national data?

Richard Bilsborrow and Martha Geores

8.1 Introduction

In recent years there has occurred an explosion of concern about the growing environmental degradation in the world. Moreover, many of these problems are apparently continuing to get worse. For example, estimates of the net annual rate of deforestation in the world (almost all in the developing countries) have recently been revised from the figure of 11 million hectares per year used in the 1980s to 17 million (FAO 1990a) and now 20 million hectares per year (World Bank 1991a). On the surface the litany of environmental problems appears necessarily related to the size and characteristics (including location and concentration) of population as well as its practices (technology of resource extraction, resource use and resource disposal). Yet existing knowledge of these linkages is almost entirely of a descriptive, *ad hoc* nature. The purpose of this chapter is to attempt to advance this limited knowledge base, first, by reviewing the ways in which population change can influence land-use patterns (and thereby the environment), and secondly, by examining what available cross-country data can tell us about these linkages.

The need for a much better understanding of population–environment linkages, and therefore for more intensive research, is highlighted in Chapter 5 of Agenda 21 of the United Nations Conference on Environment and Development held in Rio de Janeiro in 1992. The issue is important because of the continuation of high rates of population growth in many developing countries; because of the rapid pace of colonization of so-called frontier areas, especially tropical moist forests; because of the continuation of high rates of absolute poverty and hunger, especially in rural areas (viz. World Bank 1990b); and because of increasing reservations about the sustainability of existing agricultural practices based on high inputs of chemicals, water (especially through irrigation), and capital-intensive machinery.

The organization of this chapter is as follows. In the next section we summarize some of the salient literature on the interrelationships between demographic processes, patterns of land-use, and environmental changes in rural areas during the course of socio-economic development. A figure is presented to illustrate the complexity of the processes, as well as where policy factors enter. The next section presents and discusses the data available on a cross-national basis, and the following one explores the extent to which these data shed light on the processes described in the theoretical section. In our conclusion we try to do more than simply throw up our hands in consternation at the poor quality of the data by indicating particular, remedial gaps in the data, as well as promising directions for future research, at not only the cross-country level but also the within-country level.

8.2 Conceptual linkages between population, land-use and the environment

The conceptual relationships between demographic processes, land-use practices and the environment in rural areas of developing countries are very complex. Since many aspects of these relationships have been previously discussed in the literature (Pingali & Binswanger 1987, Bilsborrow 1987, Bilsborrow & Geores 1992, Davis & Bernstam 1991), we provide here only a quick summary.

For purposes of discussion, we focus on the relationships from population to land-use to the environment. Although the relations between population and agriculture have long attracted the attention of scholars, extending these linkages to the environment has been largely (but see below) of recent origin. The former can be traced back at least to Malthus, who postulated (in 1798) a tendency for human populations to grow geometrically (whenever economic conditions temporarily improve) while the "means of subsistence" grows only arithmetically, the former thus tending to outstrip the latter over time. Of course, history appears to have proved Malthus wrong, as the "Malthusian crisis" has not occurred; but this has been due to the effects of not only technological advances but also massive international out-migration movements from the area of the world (Europe) undergoing the most rapid natural population growth during the century or so following Malthus.

More recent scholars, with the benefit of hindsight, have further postulated specific types of responses to population growth, looking at the effects of population growth and hence (it is presumed) increased population density in rural areas. This is thought to create pressures upon living standards which, following the historian Arnold Toynbee, stimulate a response at the household level. Thus Kingsley Davis developed his "theory of the multiphasic response," postulating that families respond by altering their demographic behaviour: postponing marriage, reducing fertility within marriage by whatever means are available at the time, and/or out-migration (Davis 1963). Davis viewed the responses as "multi-phasic" in the sense that several could occur simultaneously. But he did not consider the possibility of non-demographic responses, nor consider the full implications of out-migration. Ester Boserup (1965) filled in the first gap, noting that as population grows relative to land, there is a tendency to use land more intensively, by reducing fallow times and increasing labour per unit of land. The term "intensification of agriculture" has thus come to be associated with her name.

But there is a third possibility as well, which should be explicitly recognized in the same context as the above and which has been, through most of human history since the development of sedentary agriculture, the major means by which agricultural production has increased. Moreover, it is hinted at by Davis and can actually be traced back to Parson Malthus himself. This is the "extensification of agriculture", or the process whereby agricultural production grows through an extension of the land area, usually associated with population mobility and appropriation of new lands. Internal migration, the neglected demographic variable in recent decades, is then the key link or mechanism between population pressures and agricultural expansion.

The environmental consequences of these processes depend upon the extent of the population pressures; land-use practices and the extent of out-migration in areas of origin; and the ecological characteristics (the so-called "carrying capacity") relative to the population influx in places of destination. The rate of out-migration to new lands is determined by the availability of unused, potentially productive (or thought to be productive) lands. Where such lands have been available extensification has been a major response, even in recent decades. This has occurred particularly where access has been facilitated by government road-building and other infrastructure. In recent years, the extent to which this occurs depends on a wide range of *other* policies and contextual factors, precisely because of the advance of technology which makes more alternative responses possible. Thus the degree of agricultural intensification occurring depends on a variety of policies related to agriculture, including relative prices, direct and indirect subsidies and taxes, exchange rates for agricultural exports and imported inputs such as fertilizer and agricultural machinery, credit availability and cost, and agricultural research and dissemination through extension services. On the other hand, the extent to which another alternative (demographic) response, fertility decline, occurs depends on (a) a host of factors that influence desired family sizes (including the expansion of education systems, female labour force participation and urbanization) and (b) couples' ability to realise those desires (mainly dependent upon public and private provision of fertility regulation methods – availability and cost). Because the responses of land-use to a growing population are multi-phasic, the extent to which any *one* of the three categories of responses occurs – demographic, land intensification or land extensification – depends on the other two, and hence on all the factors influencing those other two responses.

The severity of environmental degradation resulting from these processes depends on the density of human habitation, the ecological conditions of the land (for example, the so-called "carrying capacity", see Higgins et al. 1982), and the land-use practices engaged in by the human population. This is evidently true of areas of origin as well as areas of destination, though the overwhelming focus of the recent literature has been on the latter, because these areas are increasingly tropical rainforests which have an enormous wealth of biodiversity (Myers 1990). In the case of origin areas, farmers may respond to having larger surviving families by reducing fallow times and clearing (deforesting) more of their own land, reducing the vegetative cover that retains moisture and protects the soil. Increased soil erosion is a likely consequence of the above. But modern forms of agricultural intensification are also possible responses in origin areas, at least in those developing countries at what Higgins et al. describe as an intermediate level of technology, in the form of increases in the use of fertilizer and irrigation. In some cases fertilizer is applied excessively, there is chemical runoff into nearby bodies of surface water, seepage into ground water supplies, and/or chemical poisoning of the soil. In other cases where more water is required for irrigation than available supply allows, salts build up in the soils (salinization occurs).

Figure 8.1 illustrates these processes. Natural population growth, on the left of the diagram, combined with the local availability of land and the prevailing system of land tenure and land distribution, determines rural population density. Over time this growth eventually creates pressures to adapt. The first and most immediate response, which does not require significant stress or changes in behaviour ("cognitive dissonance"), is to either extend the area cultivated *in situ*, if additional lands are available nearby, or reduce fallow time. The former requires clearing more land, and thereby deforestation, if it has trees, thus increasing the risk of erosion. Similarly, the reduction in fallow time leads to a decline in soil fertility.

Under the usual conditions prevailing in Third World countries, compensating changes in technology and land management usually do not occur, so that a time comes when accumulating demographic pressures stimulate additional responses. In the absence of technological change, families have to abandon their plots and migrate elsewhere. Where this is to other rural areas and new farm plots are established by clearing land, this is land extensification. As the sites with good soils tend to have been settled first, this increasingly requires appropriation of marginal lands – in lowland rainforests, steep slopes or semi-arid lands. While the environmental effects vary with the density of human habitation, land practices and ecosystem, varying degrees of deforestation, desertification and soil erosion and degradation tend to occur, sometimes quite dramatically. For example, in the case of lowland rainforests, the land typically has low quality/fertility soils that support crops for

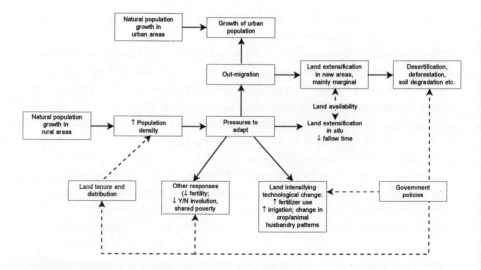

Figure 8.1 Illustration of possible land-use responses to population growth in rural areas of developing countries.

only a few years (for example, in the Brazilian Amazon, according to Martine 1988, Schmink & Wood 1987, and others). Therefore, plots soon become barren or are converted to pasture (see Joly 1989, on Panama). There is evidence that all three forms of environmental destruction are occurring on a large scale in the tropical rainforests – not only deforestation and its direct consequence of declining soil fertility, but also something akin to desertification in patches of the rainforest, for example, in Brazil and Guatemala (Leonard 1987). In highland areas, settlement increasingly further up the slopes involves a tree-clearing process that leads to increases in soil erosion, flooding and downstream siltation of rivers and dams, as has been widely reported in the Indian subcontinent (Ives & Messevli 1989) and many other areas (Leonard 1987). Links between the expansion of the human and animal populations and desiccation in Africa are also widely reported in the literature (for example, Ibrahim 1984).

The extent to which the out-migration occurs to other rural areas (see upper-right lines in Figure 8.1) depends on, first, the availability of untapped lands *perceived* as potentially productive and unclaimed; second, on these lands being made accessible by roads; third, on other public policies that stimulate land colonization of new areas (as the governments of all the countries bordering the Amazon basin have had); and fourth, on lack of alternative attractive destinations. Given the widespread urban bias in development policies, however, along with the inherent perceived attractions of urban areas, many persons and families migrate from rural areas to urban areas, accounting for nearly half of the overall growth of urban areas in developing countries in recent decades. This has implications for resource use,

110

since these migrants generally attain higher incomes than those of rural non-migrants, and therefore tend to have higher claims upon natural resources from the demand side.

But, as noted earlier, other forms of response to the demographic pressure are also possible, as indicated by the two lower-left solid lines in Figure 8.1. And both are increasingly possible in developing countries today compared to a few decades ago. Land-intensifying technological change – in the form of increased applications of labour per unit of land, accompanying changes in crops grown and/or increased use of hybrid seeds, fertilizer and irrigation – may be facilitated by agricultural research and extension policies, but a critical issue from a demographic point is the extent to which poor, small farmers have real access to it and the complementary inputs it requires. The rôle of policies here is indicated by the dotted lines, which also, with the advent of modern fertility-regulation technology, extend to the possible response of fertility decline as well. That is, to the extent policies achieve either a land-intensifying increase in labour absorption and agricultural output or a reduction in fertility, the inherent pressures for out-migration are mollified. The converse is also true, following the multi-phasic argument.

For at least two reasons linkages between demographic processes and environmental deterioration in rural areas of developing countries have rarely been convincingly demonstrated for countries or other significant geographic areas. The first reason is the inherent complexity of the process, requiring an examination of the intermediate variables of land-use, and therefore of all the *other* factors affecting land-use changes. The effects of such factors, moreover, vary with the ecological context, and also include many contextual and institutional factors that are difficult to measure. The second reason is the lack of adequate and linkable data sets. For example, it is becoming recognized that a major problem is that demographic (and other socio-economic) data are available for countries by political or administrative units (such as provinces and districts) while environmental data are often available only for areas defined by their ecological characteristics, such as forests and watersheds (Clarke & Rhind 1991). This makes the task of the remainder of this chapter exceedingly difficult.

8.3 The data, such as they are

The three types of information of interest here are demographic data, land-use data and environmental data for rural areas of developing countries. The quality of the information available declines precipitously from the first to the second to the third, and also varies greatly across countries.

While vast advances have occurred in the quantity and quality of demographic data over the past four decades, there are still great differences across countries and also important gaps. Regarding the former, the second significant digit is usually unreliable, even for population size (even the first digit has been in doubt until recently for the most populous country of Africa, Nigeria) and certainly for the total fertility rates obtained from the plethora of fertility surveys in the past 20 years. Data on rural–rural migration – so vital for the present investigation – are not available on a cross-country basis from any standard UN or other source[1]. Data on rural population density are also not available. Definitions of "urban" and "rural" vary greatly across countries, as is painfully evident in perusing the foot-notes to the UN Demographic Yearbooks, and rural–urban migration rates are also affected by changes in the classification of areas over time (e.g., of rural areas near cities as urban), which vary both across and within countries.

Land-use data, such as the proportion of a country's land area in agricultural use (land in arable and permanent crops, or A&P land), use of fertilizer, forest cover, and land distribution, are usually available from agricultural censuses. But such censuses are carried out less frequently than population censuses, and often omitted altogether when budget crises occur. Thus very few countries in Latin America have carried out an agricultural census since the 1970s, and many countries else-where have never had one. In such situations, estimates of land-use are prepared by countries or the FAO based on much more limited data from agricultural surveys or even only production data.

Finally, data on environmental conditions at the country level are the least reli-able of all, and are virtually non-existent for two of the three major forms of envi-ronmental degradation relevant to this paper – desertification, or "desiccation", and soil erosion. For the third measure, deforestation, separate estimates exist from the FAO for the loss of closed (-canopy) forests and open forests by country. But differing values are reported for some countries by others, including Ledec et al. (1985) and the World Bank. For example, the latest estimates of the FAO and the World Bank on the annual rate of deforestation in the 1980s in the country with the third-highest annual loss in the world differ by a factor of three.

With these caveats in mind, let us briefly look at some broad trends in the data, as presented in Tables 8.1 and 8.2, before examining possible interrelationships.

Tables 8.1 and 8.2 present the main data used in this paper. The population data show that rural population growth was less than overall population growth in all countries between the 1960s and the 1980s. Sub-Saharan Africa led both popula-tion growth categories, while the Middle East had the lowest rural growth and Asia had the lowest overall growth rate. Using medians as the measure of central ten-dency, because of the skewed nature of the data, yields the following average annual rural population growth rates for the different regions: Asia, 1.9; Middle East, 1.4; Latin America, 1.6; and sub-Saharan Africa, 2.2. The countries experiencing nega-tive rural population growth (eight of 85) were the Republic of Korea, Cuba, Brazil, the three Southern Cone countries (Argentina, Chile, and Uruguay), Gabon and Liberia. The overall median growth rates in 1965–85 were: Asia, 2.2; Middle East, 2.5; Latin America, 2.4; and sub-Saharan Africa, 2.8. Population density, measured by total persons/land area, is given in column (3), but it should be remembered that most developing countries have significant portions of their land area which are not habitable, or at least not usable for agriculture.

Columns (4), (5) and (6) of Table 8.1 report measures of land-use, whose change over time reflect land extensification: the percentage of the total land area devoted to agricultural uses in 1965 and 1985, and the amount of land per agricultural worker in 1965 and 1985. Median changes in the percentages of agricultural land were as follows across regions: Asia, 0.7; the Middle East, 0.4; Latin America, 1.6; and sub-Saharan Africa, 0.6. The much larger change in Latin American is noteworthy. Land per agricultural worker is measured in hectares of arable and permanent crop (A&P) land per economically active person engaged in agriculture. As one would expect from population growth, the median dropped in Asia and sub-Saharan Africa: from 0.38 ha/person in 1965 to 0.30 ha/person in 1985 in Asia, and from 1.44 to 1.35 ha/person in sub-Saharan Africa. But because of an increase in agricultural land in some countries and net rural out-migration in others, in the Middle East and Latin America, the medians actually rose: from 1.17 ha/person in 1965 to 1.39 ha/person in 1985 in the Middle East, and from 2.3 to 2.7 in Latin America.

Table 8.2, column (1) shows the 1987 per capita Gross National Product for each country in 1987 US dollars. Column (2) shows the average annual change in real GNP during 1977–87: Asia has the highest regional median growth of 5.0 per cent, followed by the Middle East with 3.8 per cent, sub-Saharan Africa with 1.6 per cent, and Latin America with 1.1 per cent.

The change in average number of kilograms of fertilizer used per hectare of crop–land between 1975/77 and 1985/87 (column 3) is taken here as a measure of agricultural intensification. Again, the regional data are highly skewed, reflecting differing economic and technological abilities to produce or import fertilizer. The highest median change is in Asia, which rose 39 kg/ha; the Middle East's median is an increase of 18 kg/ha; Latin America has a median increase of 9 kg/ha, and sub-Saharan Africa has a median increase of only 1 kg/ha. The greater increase in Asia may be due to greater population pressures, faster economic growth in the 1980s, or other factors.

The deforestation measures (columns 4 and 5) are based on WRI data and are particularly problematic, as noted above. Both report the average annual loss of forests (in thousand hectares) during the 1980s. Column (4) measures the loss of closed-canopy forests, i.e., forests where sunlight does not reach the forest floor and no grasses grow. Closed forests occur naturally in tropical moist areas. Column (5) measures the loss of open-canopy forests, i.e., those in which trees are interspersed with grasslands. Open forests occur naturally in savannah regions, such as the Sahel

Table 8.1 Population and land data.

	Rural population growth 1960s–80s[a]	Population growth rate 1965–85[b]	Population density 1985c	Change in A&P land 1965–87[d]	Land per agricultural worker 1965[e]	Land per agricultural worker 1985[f]
Asia						
Cambodia	–	0.8	–	0.7	0.60	0.60
China	1.6	1.7	108	−0.6	0.18	0.13
Indonesia	1.9	2.2	89	2.0	0.24	0.25
Korea, DPR	–	2.4	–	3.4	0.28	0.30
Korea, Rep.	−1.7	1.8	411	−1.2	0.15	0.18
Laos	–	1.9	–	0.3	0.38	0.29
Malaysia	2.3	2.5	40	2.0	0.68	0.80
Mynamar	–	2.2	57	−0.2	0.66	0.54
Philippines	1.9	2.8	178	3.2	0.37	0.29
Thailand	2.3	2.6	100	14.2	0.50	0.60
Vietnam	–	2.3	–	2.5	0.19	0.17
Afghanistan	2.4	0.9	022	0.2	0.94	0.84
Bangladesh	–	2.8	676	0.7	0.18	0.12
India	1.9	2.2	253	3.2	0.47	0.34
Iran	1.8	3.3	27	−0.3	0.85	1.09
Nepal	2.2	2.5	122	3.6	0.18	0.15
Pakistan	–	3.0	109	1.7	0.53	0.37
Sri Lanka	1.8	1.9	241	0.0	0.30	0.22
Latin America						
Costa Rica	2.1	2.9	47	0.7	2.3	2.1
El Salvador	–	2.3	230	3.5	1.2	1.2
Guatemala	2.1	2.8	073	3.0	1.6	1.5
Honduras	2.8	3.3	39	2.5	1.8	2.2
Mexico	1.5	2.9	41	0.7	4.0	2.7
Nicaragua	2.0	3.1	23	0.7	2.9	2.9
Panama	2.2	2.5	28	0.3	2.9	2.7
Cuba	−0.6	1.3	91	13.6	2.1	3.9
Dominican Rep.	1.6	2.6	112	8.8	1.3	1.8
Haiti	0.9	1.8	188	−2.1	1.1	0.8
Jamaica	–	1.8	204	−2.1	1.1	0.8
Trinidad	–	1.3	220	4.5	2.4	3.1
Bolivia	–	2.5	5	1.6	1.8	3.8
Brazil	−0.2	2.4	16	3.1	2.4	5.7
Colombia	–	2.4	28	0.3	2.0	1.9
Ecuador	2.3	3.0	33	0.5	3.0	2.7
Guayana	0.3	1.0	4	0.7	5.9	5.9
Paraguay	2.0	3.0	8	3.2	2.4	3.5
Peru	2.0	2.6	15	0.8	1.4	1.6
Venezuela	0.8	3.3	19	0.5	6.4	5.0
Argentina	−0.7	1.6	11	2.4	17.9	28.9
Chile	−0.5	1.8	16	1.6	5.8	9.4
Uruguay	−0.4	0.6	17	0.2	9.1	8.5
Middle East						
Iraq	0.9	3.5	27	1.3	1.23	1.39
Jordan	–	2.8	30	0.7	0.48	1.51

Lebanon	–	1.4	–	0.4	0.47	0.96
Syrian Arab Rep.	2.6	3.4	56	−5.3	2.37	1.94
Turkey	1.3	2.4	58	1.8	1.17	1.13
Algeria	0.8	3.0	8	0.0	–	–
Egypt	2.0	2.3	46	−0.2	–	–
Libya	2.9	4.3	–	0.1	–	–
Morocco	1.4	2.5	46	2.7	–	–
Tunisia	1.1	2.2	45	3.8	–	–
Sub-Saharan Africa						
Botswana	2.4	3.4	2	0.4	4.33	4.52
Burundi	1.5	1.9	184	6.7	0.74	0.78
Ethiopia	3.1	2.3	39	0.7	1.44	1.35
Kenya	2.9	3.7	34	0.5	0.55	0.47
Lesotho	–	2.3	40	−2.4	0.71	0.30
Madagascar	1.9	2.7	–	1.2	0.76	0.84
Malawi	–	3.1	70	3.0	1.12	1.01
Mozambique	–	2.5	15	0.2	1.19	1.21
Rwanda	2.6	3.3	192	11.5	0.40	0.44
Somalia	–	3.4	–	0.1	1.13	0.75
Sudan	1.1	2.9	8	0.3	3.20	2.84
Tanzania	2.4	3.4	25	0.5	0.98	0.87
Uganda	–	3.4	–	3.7	1.41	1.30
Zambia	1.4	3.4	7	0.3	4.21	3.66
Zimbabwe	–	3.1	20	0.6	1.98	1.78
Angola	–	2.6	–	0.0	3.52	3.25
Cameroon	2.4	2.2	20	2.2	2.13	2.19
Central African Rep.	2.2	2.3	4	0.2	1.96	1.80
Rep. of Congo	–	2.4	–	0.2	3.46	3.72
Gabon	−1.2	3.4	–	0.8	1.27	2.27
Zaire	1.2	2.8	12	0.2	1.96	1.80
Burkina Faso	–	2.3	–	1.3	0.86	0.86
Chad	2.7	2.1	3	0.2	2.31	2.21
Mali	3.0	2.4	7	0.3	0.64	0.63
Mauritania	0.4	2.4	–	−0.1	0.84	0.47
Niger	–	3.0	–	0.6	2.18	2.31
Benin	0.4	2.5	30	2.2	2.45	2.41
Ghana	2.5	2.5	–	0.2	1.40	1.26
Guinea	–	1.8	–	0.0	1.01	0.09
Liberia	−1.3	3.2	17	0.0	0.90	0.74
Nigeria	–	3.2	–	0.9	2.12	1.96
Senegal	–	2.9	26	2.9	3.21	3.00
Sierra Leone	–	2.1	–	4.1	1.95	2.09
Togo	2.1	3.1	–	0.9	2.18	1.94

Notes: – means data are not available. a) Average annual rural population growth rate based on available population census data from each country for a year in the 1960s and in latest year in the 1980s. The definition of rural depended on each country's own definition, which varied greatly; also, not all countries reported separate rural and urban populations. b) Average annual overall population growth rate, based on 1965 and 1985 UN estimates. c) Persons per km^2, using 1985 population data and total land area. d) The change in the percentage of total land in the country which was in cropland (arable) and pasture, between 1965 and 1987. e) & f) Hectares of A&P land per economically-active person engaged in agriculture, 1965 and 1985.

Sources: col. 1, UN 1988; cols 2 & 3, UN 1990; cols. 4, 5, 6, FAO 1984, 1988a.

Table 8.2 Income, agricultural technology and deforestation data.

Country	Per capita GNP 1987[a]	Growth in GNP 1977–87[b]	Change in fertilizer use 1976–86[c]	Loss of closed forests 1980s[d]	Loss of open forests 1980s[e]	Productivity index 1985/65[f]	Labour prod. index 1985/65[g]
Asia							
Cambodia	–	–	0	2,264	510	0.89	0.72
China	292	9.3	121	–	–	2.28	1.17
Indonesia	444	5.1	73	10,7116	2,100	1.92	1.81
Korea, DPR	910	–	84	–	–	1.94	2.10
Korea, Rep.	2,689	6.9	61	–	–	1.59	1.67
Laos	166	–	2	10,092	3,129	2.10	1.97
Malaysia	1,820	5.2	86	25,195	–	2.13	1.99
Mynamar	212	4.8	13	67,076	–	2.21	1.90
Philippines	589	1.2	16	14,265	–	1.63	1.24
Thailand	850	5.7	12	–	–	1.26	1.32
Vietnam	200	–	3	17,540	–	1.71	1.49
Afghanistan	220	–	3	–	–	–	–
Bangladesh	164	3.7	39	834	–	1.36	1.15
India	311	4.8	30	149,814	–	1.63	1.22
Iran	1,756	0.1	40	–	–	2.14	1.60
Nepal	161	3.7	14	8,346	–	1.07	0.95
Pakistan	353	6.8	49	655	206	2.12	1.30
Sri Lanka	406	5.0	57	5,806	–	1.60	1.07
Latin America							
Costa Rica	1,608	0.9	20	12,448	–	1.61	1.44
El Salvador	842	–2.3	22	451	–	1.30	1.32
Guatemala	947	0.1	18	8,884	–	1.45	1.14
Honduras	808	1.9	1	8,733	–	1.40	0.90
Mexico	1,825	2.5	24	60,125	2100	1.45	1.08
Nicaragua	829	–2.3	34	12,139	–	1.08	0.79
Panama	2239	4.0	–2	3,748	–	1.72	1.60
Cuba	–	–	73	291	–	0.81	1.50
Dominican Rep.	734	2.0	–17	377	–	1.07	1.21
Haiti	363	1.1	2	182	–	1.10	1.25
Jamaica	940	–1.2	–26	201	–	1.30	0.80
Trinidad	4,149	–0.3	–1	83	–	0.51	0.92
Bolivia	496	–2.0	1	8,802	2275	0.86	1.39
Brazil	2,021	3.2	10	786,456	109900	1.36	0.19
Colombia	1,238	2.9	27	83,520	6890	1.67	1.42
Ecuador	1,044	1.7	15	34,200	–	1.51	1.44
Guayana	389	–4.8	–9	0	44	0.87	0.89
Paraguay	995	4.0	4	19,129	1564	0.92	1.26
Peru	1,467	1.8	–13	27,872	–	0.91	0.90
Venezuela	3,226	–0.9	88	12,748	12000	1.59	1.88
Argentina	2,394	–0.9	2	–	–	1.13	1.73
Chile	1,358	1.5	20	–	–	1.11	1.70
Uruguay	2,198	–0.5	9	–	–	1.23	1.31
Middle East							
Iraq	2,400	–	28	–	–	1.25	1.39
Jordan	1,560	5.4	18	–	–	0.80	3.87
Lebanon	–	–	7	–	–	–	–
		2.6	27	–			

	(a)	(b)	(c)	(d)	(e)	(f)	(g)
Arab Rep.							
Turkey	1,213	3.8	19	–	–	1.56	1.50
Algeria	2,629	10.2	18	–	–	1.96	2.54
Egypt	678	5.9	159	–	–	1.89	1.38
Libya	5,453	–4.2	3	–	–	2.56	3.75
Morocco	615	2.9	13	–	–	1.19	1.10
Tunisia	1,182	4.2	11	–	–	1.81	1.94
Sub-Saharan Africa							
Botswana	1,059	11.9	–2	–	3,256	0.93	1.14
Burundi	241	2.7	1	70	41	1.35	0.97
Ethiopia	126	1.8	2	870	9,120	1.23	0.78
Kenya	330	3.8	24	1,878	2,008	1.46	0.80
Lesotho	355	2.3	7	–	–	1.26	0.79
Madagascar	207	–0.7	1	15,450	580	1.18	1.20
Malawi	164	2.3	7	–	15,114	1.42	1.07
Mozambique	146	–5.7	–2	1,028	11,600	1.07	0.42
Rwanda	301	4.7	1	312	198	1.37	1.11
Somalia	290	0.3	–1	308	751	1.42	0.66
Sudan	331	–1.4	–2	390	52,415	1.56	1.08
Tanzania	180	1.3	3	994	4,200	1.38	0.72
Uganda	260	–0.1	0	1,008	12,180	0.97	0.61
Zambia	248	–0.8	4	3,913	2,650	1.33	0.93
Zimbabwe	585	3.8	8	–	7,848	–	–
Angola	–	2.8	0	4,350	5,070	0.82	0.29
Cameroon	966	9.5	4	9,900	8,840	1.27	1.32
C. African Rep.	334	0.6	0	359	6,460	1.34	1.27
Rep. of Congo	873	7.6	1	2,134	–	1.31	0.61
Gabon	2,733	–0.8	3	2,050	–	0.78	0.88
Zaire	153	0.2	–1	21,150	21,552	1.43	1.32
Burkina Faso	191	5.1	2	298	7,588	1.38	0.98
Chad	139	–	0	–	7,800	1.12	1.04
Mali	200	2.5	10	–	3,375	1.23	1.62
Mauritania	439	1.9	0	70	1,260	1.53	0.87
Niger	258	–0.1	0	250	6,370	0.85	0.40
Benin	305	3.9	5	382	6,494	1.80	0.99
Ghana	393	0.2	–6	2,233	4,882	1.38	1.03
Guinea	316	1.6	–1	3,690	5,160	1.21	0.80
Liberia	451	–2.2	–10	4,600	–	1.70	1.11
Nigeria	368	–0.9	8	29,750	9,680	1.72	0.96
Senegal	510	2.4	–5	–	5,412	1.36	0.96
Sierra Leone	249	1.5	–1	592	–	1.13	1.05
Togo	286	0.2	5	213	910	0.98	0.64

Notes: – means data are not available. a) Per capita gross national product in 1987 US dollars. b) Average annual percentage change in real GNP between 1977 and 1987. c) Change in average kilograms of fertilizer used per hectare of cropland, based on three-year average figures for 1975–1977 and 1985–1987. d) Average annual loss of closed-canopy forests during the 1980s, in thousands of hectares. e) Average annual loss of open-canopy forests during the 1980s, in thousands of hectares. f) Relative index of land productivity in 1985 divided by land productivity in 1965, where land productivity in 1965 is the index of FAOgricultural production for 1966/67 divided by the amount of A&P land in 1965 (similarly for the 1985 index). g) Relative index of labour productivity in 1985 divided by labour productivity in 1965, where labour productivity in 1965 is the index of FAO agricultural production for 1966/67 divided by the number of persons employed in agriculture in 1965 (similarly for the 1985 index). *Sources: cols. 1–5, WRI 1990; 988a.*

and northeast Brazil, and are often found as the result of partial deforestation in areas where closed forests are natural but had been depleted earlier. No data are available on deforestation for the Middle East. Asia has the highest median loss of closed forests (10,092 thousand hectares), followed by Latin America (8,884) and sub-Saharan Africa (1,008). The regional order for the loss of open forests is reversed, with sub-Saharan Africa having the highest median loss (6,370), followed by Latin America (2,275), and Asia (510). These figures, however, are reflective of the country size, the limited samples of countries reporting, and their initial stocks of forests. In any case, the overall loss of forests is highest for Latin America.

Columns (6) and (7) contain the land and labour productivity indices, showing the change in productivity as a ratio of the respective country-level productivities for 1985 compared to 1965 (see table note for fuller definitions). Asia has the highest median increase in the relative land productivity index at 1.71, followed by the Middle East at 1.56, sub-Saharan Africa with 1.32, and Latin America with 1.22. Regarding changes in the labour productivity index, the Middle East had the highest median change of 1.94, followed by Asia with 1.32, Latin America at 1.25, and sub-Saharan Africa with 0.92 (or a net decline).

In addition to the data reported in this chapter, we examined several other variables which did not prove fruitful in graphs or correlations. These include: Gini coefficients for land distribution and income equality; Higgins et al.'s (1982) potential population-supporting capacity of the land; pesticide and tractor usage; food production indices; and total fertility rates. In addition, we sought data for new road construction, to relate it to deforestation and agricultural extensification. However, data were not available for sufficient countries or appeared completely unreliable.

8.4 What do the data show?

The relationships between demographic, land-use and environmental variables can be conceptualized at either a static or dynamic level. The static level is of less interest precisely because we are interested in the dynamic interrelationships between the three phenomena, but it is a useful starting place. Simply put, we expect there to be a strong relation between, for example, population density and land-use, as measured by the proportion of a country's land in agricultural use, even though many other factors also influence land-use, including land quality and the general level of non-agricultural development (which reduces dependence on land-based agriculture). In fact, we observe a high correlation across countries between overall population density (not rural, which is not available for most countries) and the percentage of A&P land. The correlation coefficient for the 68 countries with the available data in 1965 is 0.59, significant at the 0.0001 level. Similarly, the correlation between population density and mean hectares of land per agricultural worker is −0.24, significant at the 0.05 level. There is also a strong negative correlation between overall density and the proportion of a country's land covered by forest. However, it is interesting that there is *no* relationship between density and the *rate* of depletion or loss of forests. This is to be expected; the static relationships reflect the cumulative outcome of a long historical process of human habitation and land-use in a country, while more recent changes (such as deforestation) ought to be related to other factors which have also been changing recently.

Neither existing theory nor existing data permit a truly dynamic analysis, but we can examine the relationships between *changes* over time in demographic variables and *changes* in land-use and the environment by means of a comparative statics analysis. This is far more relevant to the question at hand than the static analysis. A series of correlations, figures graphing two-way relationships (particularly useful for identifying "outliers"), and further correlations were prepared, both for different regions and for testing the effects of outliers. This was done for various alternative measures of the "independent" (population) and "dependent" variables, given the uncertainties regarding their reliability and differences in the self-selected sample and therefore composition of countries available for the different measures. Definitions of variables and problems in measurement are discussed in Section 8.3 and the notes to Tables 8.1 and 8.2. In the discussion below a "very significant" (simple) correlation refers to one significant at better than the 0.01 level (two-tailed test), "significant" at better than the 0.05 level, and "weakly significant" at better than the 0.1 level. The simple correlation coefficients referred to are found in Table 8.3.

The demographic variables looked at in the comparative static analysis include rural population growth in the 1980s; total observed population growth 1965–85, 1965–80 (to explore possible lagged effects on changes in the dependent variables in the 1980s), and 1985–90 (to focus on the most recent period); and total fertility rates, 1965 and 1985–90. For each of the "dependent" variables of interest, we report the relationships only for those demographic variables that seem the most logical or relevant *a priori*. Following Figure 8.1, but ignoring the lower left arrow responses (changes in fertility and involution), we examine first the interrelationships between demographic processes and changes in land-use, before considering environmental consequences.

We begin by assessing whether there is any relationship between population changes and changes in the amount (or proportion) of a country's land-used for agricultural purposes (A&P land) and in hectares of (A&P) land per agricultural worker. The former provides a direct measure of the land extensification hypothesis viewed across countries, so a positive relationship is anticipated. What we appear to observe, however, is apparently no relationship between the percentage change in A&P land and any of the demographic variables.

Table 8.3 Correlation coefficient matrix.

	RURPOP	PGR	APX	APHX	HEAPX	FERX	IRRY	LOSSC	LOSSO	LANINDX	INDX	GNP77	CAPITA87	UNDER
RURPOP Rural Population Growth rate 1960s–1980s	1.000 .000 57													
PGR Population growth rate 1965–85	.334 .011 57	1.000 .000 85												
APX Change in A&P as % land 1965–1985	-.043 .752 57	-.099 .367 85	1.000 .000 85											
APHX Change in A&P land 1965–85 (000ha)	-.193 .151 57	-.035 .748 85	.247 .022 85	1.000 .000 85										
HEAPX Change in ha/agricultural worker 1975/77–1985/7	-.386 .005 52	-.202 .072 80	.167 .139 80	.406 .000 80	1.000 .000 80									
FERX Change in fertilizer use kg/ha 77/87	-.049 .718 57	-.049 .654 85	-.013 .906 85	-.073 .506 85	-.056 .623 80	1.000 .000 85								
IRRY Change in irrigated land 77/87 (000ha)	.011 .936 55	-.076 .499 81	.075 .506 81	.378 .000 81	-.005 .967 76	.239 .031 81	1.000 .000 81							

Correlation matrix (each cell: correlation / significance / N). Columns 1–7 correspond to earlier variables (labels not shown on this page); columns 8–14 correspond to the seven variables listed at left.

Variable	(1)	(2)	(3)	(4)	(5)	(6)	(7)	LOSSC	LOSSO	LANDINDX	INDX	GNP77	CAPITA87	UNDER
LOSSC Average annual loss closed forest in 1980s (000ha)	-.238 / .150 / 38	-.041 / .757 / 59	.036 / .784 / 59	.971 / .000 / 59	.582 / .000 / 59	.080 / .549 / 59	.330 / .013 / 56	1.000 / .000 / 59						
LOSSO Average annual loss open forest in 1980s (000ha)	-.480 / .018 / 24	.016 / .920 / 40	.040 / .808 / 40	.869 / .000 / 40	.587 / .000 / 40	-.016 / .923 / 40	.342 / .036 / 38	.865 / .000 / 34	1.000 / .000 / 40					
LANDINDX Land productivity index 1986/1966	.194 / .151 / 56	.239 / .030 / 82	-.300 / .006 / 82	-.088 / .434 / 82	-.214 / .062 / 77	.418 / .000 / 82	.186 / .104 / 78	.101 / .447 / 59	.067 / .684 / 39	1.000 / .000 / 82				
INDX Labor productivity index 1986/1966	.053 / .700 / 56	.118 / .291 / 82	-.064 / .563 / 82	-.150 / .178 / 82	.124 / .282 / 77	.272 / .014 / 82	-.019 / .871 / 78	-.193 / .143 / 59	-.289 / .074 / 39	.496 / .000 / 82	1.000 / .000 / 82			
GNP77 Average annual change in GNP 1977–1987	.174 / .213 / 53	-.050 / .667 / 76	.093 / .426 / 76	.029 / .803 / 76	-.078 / .518 / 71	.360 / .001 / 76	.223 / .060 / 72	.142 / .301 / 55	-.046 / .788 / 37	.234 / .043 / 75	.158 / .177 / 75	1.000 / .000 / 76		
CAPITA87 Per capita GNP 1987	-.287 / .032 / 56	.122 / .279 / 81	-.086 / .445 / 81	.114 / .309 / 81	.267 / .020 / 76	.121 / .280 / 81	-.067 / .559 / 78	.210 / .119 / 56	.347 / .033 / 38	.126 / .267 / 79	.493 / .000 / 79	-.070 / .551 / 75	1.000 / .000 / 81	
UNDER % agricultural holdings under 5 ha	.168 / .266 / 46	-.031 / .814 / 60	-.003 / .982 / 60	-.216 / .097 / 60	-.410 / .002 / 55	.061 / .642 / 60	.056 / .680 / 57	-.419 / .009 / 38	-.322 / .144 / 22	-.125 / .351 / 58	-.359 / .006 / 58	.151 / .265 / 56	-.472 / .000 / 59	1.000 / .000 / 60

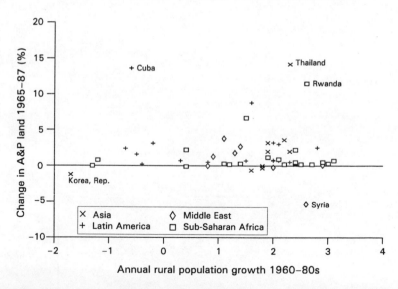

Figure 8.2 Relationship between change in A&P land and rural population growth. (Note, A&P land is land under arable or permanent crops.)

Figure 8.2 plots the relationship against the theoretically preferred measure of the demographic variable (among those available), and helps us see that the lack of any significant correlation is due to the values for two countries, which may be considered outliers, Cuba in the upper left and Syria on the bottom right. Syria, despite rapid rural population growth, reported a substantial *decline* in A&P land. Whether this actually occurred, and, if so, why, we do not know. And Cuba, if the data are correct, which we are less likely to second-guess, experienced a massive increase in its agricultural land, perhaps to increased production of sugar cane and other agricultural exports to earn sorely needed foreign exchange, even though out-migration from rural areas continued (which is expected at Cuba's advanced level of social development). The Cuba case indicates the potential power of government policies, particularly in directed economies, to swamp the "natural" forces of demographic pressures. Otherwise we would have expected the A&P change value for Cuba to have been near the values for the other four Latin American countries which experienced negative rural population growth during the period (the three Southern Cone countries and Brazil). The gist of all this is that the expected positive relationship clearly obtains if these two "outliers" can be excluded.

But it is also worth considering whether the expected positive relationship is plausible for *negative* rates of rural population growth. It would imply that land should be taken out of production as the rural population declined. To the extent this decline results from out-migration to urban areas rather than from negative natural growth in rural areas, there is no reason to expect such a decline because the growing urban population also contributes to a growing demand for agricultural products. We thus computed correlations for only those countries that had experienced positive rates of rural population growth. But unless Syria is omitted, no significant positive relation exists for the apocopated sample.

Our conclusion of this exercise is that a positive relationship *does* appear to exist, but it is weaker than expected and depends on the exclusion of two outliers. The same is true of the regional breakdowns. Positive relationships for Asia, Latin America and Africa depend respectively on the inclusion of both Republic of Korea and Thailand, the exclusion of Cuba, and the inclusion of Rwanda. Finally, both to examine the relationships between changes in A&P land and *other* measurable factors thought to influence it over time, and to compare the relative magnitudes of the relationships, we also correlated and plotted it against measures of total GNP growth (representing demand side effects) and measures of land distribution (both the proportion of holdings with less than five hectares and the proportion of all agricultural land in holdings under five ha). This tests the hypothesis that demographic pressures to out-migrate to new areas and clear land for agriculture depend directly on the concentration of existing landholdings in potential places of origin (evidence for this hypothesis has been found for Guatemala and several other countries: see Bilsborrow & Stupp (1988), Bilsborrow & DeLargy (1991), and Bilsborrow & Geores (1992), among others). No statistical relationship was found.

The next land variable examined is the change in hectares of agricultural land per economically active person in agriculture (HEAPX). This reflects the net combined

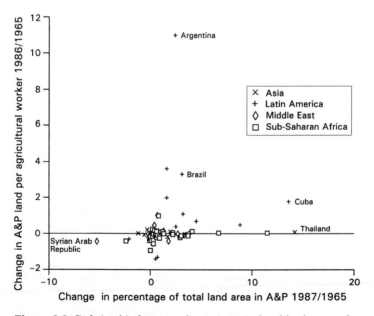

Figure 8.3 Relationship between change in agricultural land per worker and change in A&P land.

effects of both the change (usually an increase) in land and the change (also usually an increase) in the rural labour force, which directly reflects the change in the rural population. The former increases HEAPX through the numerator (the simple correlation is only 0.14, however – see Figure 8.3 – due to the two outliers indicated) and the latter mathematically reduces it through the denominator. But we also expect the growth in the rural population to stimulate an increase in land, but not by enough to compensate for itself, so that HEAPX should decline over time in most developing countries. With some striking exceptions (e.g., Brazil, Bolivia, Cuba, Paraguay) it does, and this decline is very significantly related to rural population growth as well as the overall rate of population growth (1965–80, 1985–90 and 1965–85). However, this relationship disappears if the seven countries with negative rates of rural population growth are excluded. HEAPX is also strongly negatively correlated with the proportion of land in farms under five hectares (land distribution inequality). This may suggest that greater land inequality is associated with greater out-migration to new lands and creation of new, large farms.

Figure 8.1 argues that land intensification should also be positively related to population pressures. In the absence of data on changes over time in average fallow times (some limited data are collected together in Bilsborrow 1987), we use changes in fertilizer use (FERX) and changes in the proportion of agricultural land irrigated as measures of land intensification. While a positive relation is observed between FERX and rural population growth, it is weak overall (Figure 8.4). For

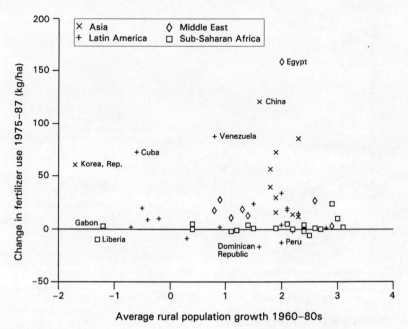

Figure 8.4 Relationship between change in fertilizer use and rural population growth.

Sub-Saharan Africa it is strongly significant, perhaps reflecting the greater relative strength of the relation at lower levels of development, closer to subsistence. Changes in fertilizer use are also strongly related to concurrent growth in GNP in the overall sample, but separately only for Asia. Finally, they are not related to either population density or land distribution inequality.

The other measure of intensification, change in proportion of land irrigated, is not related to population growth, however measured, and only slightly to GNP growth. It is also weakly related to density and not at all to land concentration. To the extent increases in fertilizer use and increases in irrigation are alternative responses to population growth (and not in general complementary, which is sometimes the case and sometimes not), then the multi-phasic theory would also imply that they should be negatively related to each other. In fact, this is observed overall, though not within any region.

Given the multiphasic theory postulated in Figure 8.1, we expect a negative relationship between measures of land intensification and land extensification. We test this hypothesis by correlating and plotting the changes in A&P land against the increases in fertilizer use and irrigation. In fact neither of the latter appears significantly related to land extensification, as seen in Figure 8.5 for fertilizer, though there is clearly a negative, non-linear (hyperbolic) relation if the outliers are accepted as valid observations.

With such scintillating results for the relationships between demographic changes and changes in land-use, we now move on to examine the relationships between both of these and environmental degradation, as measured by deforestation. Two measures of the latter are available, the annual loss of closed forests in the

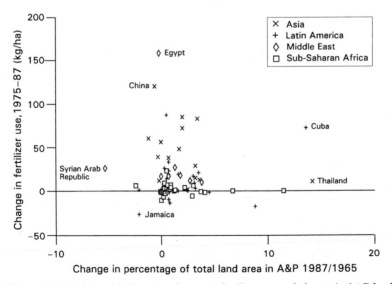

Figure 8.5 Relationship between change in fertilizer use and change in A&P land.

1980s, the loss of open forests, and the total loss, constructed by adding the two. Since a number of studies in different countries have suggested that clearing land for agriculture is the principal direct (but not underlying) cause of deforestation of tropical rainforests (well ahead of logging, fires caused by lightning, roads and development projects), deforestation should be positively related to the change in A&P land. Figure 8.6 plots the relationships, evidently dominated (as will be every subsequent figure on deforestation) by the single, striking outlier, Brazil. To the extent that what happened in Brazil was a reflection more of two events or factors completely exogenous to the process we are investigating – a combination of a particular government policy of internal expansionism via colonization of the Amazon interior and a World Bank policy of providing billions to finance the means to achieve this (mainly via roads) – one could again make a case for excluding Brazil as an outlier. But we are reluctant to do this because of the enormous proportion of forest loss accounted for by Brazil – it would be akin to "throwing out the baby with the bath water". In fact, Figure 8.6 does not look much different if Brazil is excluded. The fact is that many countries reported *no* loss of closed forests in the 1980s, whether because they experienced none, had no more left to lose, or government officials had no basis (or time) to prepare separate estimates at two points in time and hence simply reported the same proportion of the country's area covered by forests at both time points. Given the scarcity of agricultural censuses mentioned above, the latter is not implausible for some countries. The number in

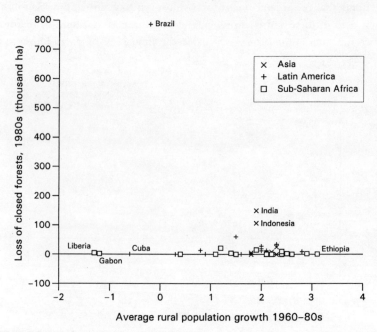

Figure 8.6 Relationship between loss of closed forests and rural population growth.

126

Africa reporting zero loss of closed forests is certainly suspicious, including countries at each extreme (Liberia, Gabon, and Ethiopia).

In any case, if Brazil is *included*, there is a weak, *inverse* correlation between rural
population growth and loss of either (or both) closed and open canopy forests;
however, if excluded, the relationship is positive but insignificant *unless* the countries reporting zero loss of closed forests are also excluded. No relationship is
observed with population growth either, overall or for any region.

Based on Figure 8.1, we also expected a positive relation between deforestation
and change in A&P land. This is a clear case where the importance of the units of
measurement can be demonstrated: while *no* relationship is observed if the change
in the percentage of land in agricultural use is used, as shown in Figure 8.7 (such a
change being only of the order of 3 in Brazil, but huge in several other countries
such as Cuba and Thailand – see Figure 8.2). However, a strong positive relationship is indeed observed when the change in A&P land is expressed in the same
units (hectares) as the loss of forests: but if Brazil is included, the slope of the apparent regression line is about 1/3 while if excluded it is about 1/6. If all *three* of the
countries with the largest amounts of deforestation and increase in agricultural land
are dropped (i.e., also excluding India and Indonesia), we are left with an amor-
phous blob.

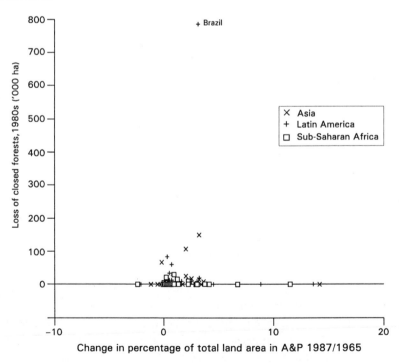

Figure 8.7 Relationship between loss of closed forests and change in A&P land.

We also examined the relationships between deforestation and a number of other variables, starting with change in land per worker (HEAPX). As with change in A&P land, there is a strong positive relation, for each of the three measures of deforestation, but again due largely to Brazil. A positive relationship to growth in GNP is also expected, but only a weak one with GNP growth in 1967–77 (and none with 1977–87) is observed for both open and closed forests and a bit stronger one ($r = 0.011$) for total forests, for both the overall sample and Latin America. It should be noted that the number of cases for regional breakdowns is small for all except Latin America, and particularly limited for loss of open forests (Table 8.2). Looking also at the level of GNP per capita across countries, little relation is found except for a weak positive one for total forests. Finally, we expect a positive relation between land concentration and deforestation for the same reason as we expected it for the change in agricultural land and deforestation, but we observe instead a significant negative relationship overall for deforestation of closed forests and a weak one for open forests; but both are entirely due to the outlier, Brazil (a weak positive relationship appears to exist if it is excluded).

It is also of interest to look at the overall changes in land and labour productivity over the past several decades, by country (see Table 8.2 and above), and see to what degree these changes have been related across countries to the same aggregate demographic and economic trends investigated above. While data are not available to measure productivity differentials directly across countries or over time except for individual crops, the *relative* productivity of land (and labour) in a country can be compared from one time period to another by comparing changes in the index of (the value of) agricultural production with changes in the agricultural land area (this evidently abstracts from substantial changes in either/both the composition or basket of agricultural commodities comprising the index in the country or in their relative prices). Thus the index in one year may be divided by the land area for that year and compared to the same ratio in another year to estimate the proportionate change in (the) average land productivity (index). The *numerator* of this ratio should increase over time to the extent that there is successful land intensification, i.e., through increased inputs of labour or other inputs such as fertilizer per unit of land. Thus Boserup would anticipate a positive association with population growth. On the other hand, the ratio would tend to fall to the extent that extensification onto inferior or more marginal agricultural land occurs. Whether this ratio rises or falls in a country thus may be considered a simple, albeit crude, reflection of the net forces of agricultural intensification versus land extensification.

What do we observe? In almost all cases the land productivity index is above one. Moreover, Figure 8.8 shows that there seems to be a statistically significant positive relation between the proportionate change in land productivity between 1965–66 and 1985–86 and rural population growth. While the slope seems quite modest, that is an artifact of the scale on the vertical axis. But again this relation-

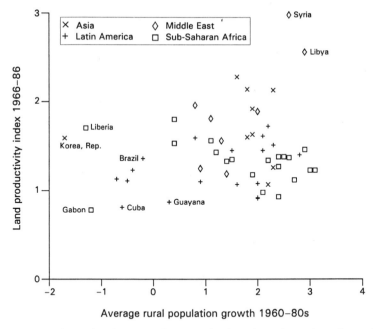

Figure 8.8 Relationship between change in land productivity and rural population growth.

ship is positively affected by the Syria and Libya outliers, and negatively by Republic of Korea and Liberia. Which of these four observations are correctly measured, we cannot say. The overall correlation is .15, and is stronger for both the Middle East and Latin American regions. With respect to other variables, a strong negative relation exists with the change in A&P land, showing the effects of inferior quality land coming into production with extensification, for both the overall sample and the same two sub-regions. A weaker negative correlation exists with HEAPX, as expected given its complexity. More interestingly, a strong positive association exists with the change in fertilizer use, as expected, both for the whole sample and for Asia. A weak positive relation exists between the proportionate change in land productivity and the change in irrigation, and none with the change in tractors, population density, concentration of landholdings, or GNP growth.

To complete the picture, we also computed the proportionate change in *labour* productivity over time for each country and compared it to the various available demographic and other factors. In the absence of intensification we would expect it to fall with population growth due to diminishing returns (as population has grown faster than agricultural land for most countries). In fact, no such decline is evident, there being no association with any of the measures of population change. The relative change in labour productivity was also not related to the change in agricultural land, growth in GNP, increase in irrigation or increase in tractors, but it

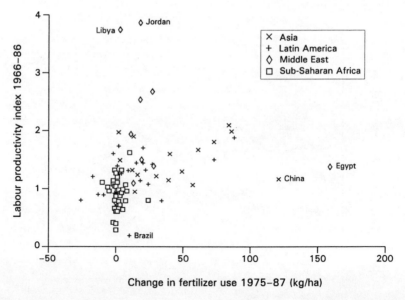

Figure 8.9 Relationship between change in labour productivity and change in fertilizer use.

was to fertilizer increase, both for the overall sample and for Latin America (see Figure 8.9). This suggests that increases in fertilizer use have played particularly important roles in increasing both land and labour productivity in recent decades.[2]

8.5 Some conclusions and recommendations

The results of the exercises above – from examining the relationships between demographic factors, land-use and the environment on the basis of simple correlation and graphical analyses – may be summarized as follows. There does indeed appear to be a positive relationship across countries between population growth (especially rural growth) and the increase in agricultural land (land extensification), land intensification in the form of increases in use of (chemical) fertilizers, and deforestation. However, in each case the relationship is weak, and dependent on the inclusion or exclusion of particular "outlier" countries. This analysis surely must lead one to question the findings of certain studies which have purported to find strong relationships between population growth and deforestation based upon cross-country data, such as those of Allen & Barnes (1985) and Rudel (Ch. 7). Perhaps some solace comes from the fact that the relationships between the changes in land-use and deforestation with the other non-demographic (primarily economic) variables are in general no more significant. The same problems that frustrate the overall analysis also, in general, vitiated observing particularly interesting or different regional patterns. Nevertheless, one positive finding is that increases in fertilizer use have occurred widely across developing countries in recent decades and have contributed to generalized increases in land productivity.[3]

This type of analysis is evidently extremely frustrating, with suspicious or extremely poor quality data and missing data omnipresent. Problems also exist in determining how to measure some of the concepts, and what measures of what variable (ratios, percentage changes, absolute changes) should be compared with what measures of another. And what is appropriate for one comparison is often not for another. Particularly serious data shortcomings include the following:

(a) Perhaps surprisingly, inadequacies exist even with available cross-country *demographic* data, including the lack of data on rural–rural migration (rarely in country census publications and never – to date – in UN compendia such as the *Demographic Yearbook*). These data are sorely needed to investigate hypotheses relating land extensification to demographic (and other) factors. In addition, data are not available on *rural* population density, with HEAP the closest, generally available, albeit indirect, (inverse) indicator, in lieu of using simply the overall density in the country. Indeed, the *change in rural population density* is the conceptually preferred variable rather than rural population growth, since the latter does not always represent a measure of pressure effectively stimulating a response until some density threshold is reached or approached; this will differ greatly by country according to soil quality, etc. (or "carrying capacity").

(b) Problems also exist with some of the land-use and agricultural variables. Land-use data on A&P land do not distinguish land in use from fallow land, or marginal land from good quality land. Data on fertilizer use do not include use of non-chemical fertilizers, and do not control for *excess* use. Available irrigation data for countries rarely allow for *reductions* or losses in irrigated land due to salinization, inadequate water, or failure to maintain irrigation canals (a major problem in northern Mexico, northern India, Egypt, and many other areas).

(c) Available data on environmental degradation across countries are in a pristine state (to make a poor pun). There is not enough data on soil erosion or desertification to even try to investigate them (though hopefully this will change within a few years), and deforestation data are evidently unreliable. Data on the loss of open forests are often missing for countries and are surely (even) less reliable than existing cross-country data on the loss of closed forests. Data from satellite (e.g., Landsat) imagery need to be converted into reliable estimates and provided at little or no cost to the countries themselves and to international agencies and investigators in usable, understandable form. This will permit much better measures of the extent and pace of deforestation, which will facilitate better analyses of its causes on an international scale (see also Clarke & Rhind 1991.)

Despite the problems pointed out here with cross-country analyses, they can still be useful for indicating particular data needs if communicated directly to the data-producing agencies, by identifying not only variables needed for potentially important cross-country research but also outliers which beg for re-examination – are they real numbers or artifacts?

Finally, this type of cross-country examination can help identify which relationships are the key ones to investigate at a *lower* level of aggregation, that is at the country or sub-country level. The conceptual framework in Figure 8.1 also indicates the rather substantial advantages of the latter. Cross-country data are too aggregated for a proper investigation of the relationships studied here. What is important, for example, is not overall population density, or even density relative to arable land, but *where* the population is concentrated, and whether thresholds of density (plots too small to be economically viable or human habitation too large for the sustainability of the ecosystem) are being surpassed in certain areas as populations continue to grow. Sub-national data for provinces or even districts are better suited for this. They would also greatly facilitate "ecological" or cross-area analyses that measure and therefore permit controlling for a number of *additional* economic, institutional and contextual factors available within particular countries which are not measurable in a comparable way nor even available across countries. Country-level analyses can profitably use data from population and agricultural censuses to investigate relationships either across areas within the country (the more micro the area the better) or over time (and ideally, both). Examples of country-level studies that carry the analysis well beyond the present cross-country examination include those of Bilsborrow & Stupp (1988), Bilsborrow & DeLargy (1991), and Mendez (1988) for Guatemala; Stonich (1989) for Honduras, Panayotou et al. (1989, 1991) for Thailand; and Cruz et al. (1988, 1992) for the Philippines.

Nevertheless, there are real advantages in conducting investigations into the interrelationships between demographic factors, land-use practices and environmental degradation at an even more disaggregated level, such as at the *household or farm* level. This is the level at which resource use decisions are made, and households and farm units are the immediate actors whose behaviour must be much better understood to address the rural environmental problems of developing countries. This is also the level at which real theorizing exists and formal hypotheses can be elaborated and tested. And it is at this level that fully multidisciplinary collaboration in *both* the data collection in the field and the analysis – involving ecologists and social scientists working together – can be most useful, and is most needed. Several investigations at this level are currently underway in developing countries, including a survey of migrant colonist households in the Ecuadorian Amazon in which one author, Bilsborrow, is involved in collaboration with Francisco Pichon and various Ecuadorian institutions (see Pichon & Bilsborrow 1992). While such studies should materially enhance our understanding of the processes involved and help point to specific policy recommendations, the importance of the particular socio-economic, institutional, and natural resource context to the nature of the relationships observed should not be ignored. Therefore, many more such studies are needed before we can begin to try to delineate more general relationships, or at least what the relationships are likely to be in certain types of contexts. The field is virgin. The environmental destruction is not.

Notes

1. Data on rural–rural migration (as well as rural–urban, urban–urban, and urban–rural migration) for nine countries from the 1970s and 1980s rounds of censuses are reported in the United Nations (1991). The data vary from being based on place of birth to place of previous residence to place of residence five years earlier. The largest migration flows were rural–rural in three countries (India, Malaysia, Thailand), rural–urban in two (Philippines, Republic of Korea), urban– urban in three (Egypt, Pakistan, Brazil), and equally urban–urban and rural–rural in one (Honduras). Unfortunately, census data are not usually reported on these four categories of mobility.
2. We also observe a strong negative relation with land concentration (both overall and for the Middle East), suggesting perhaps that extremely concentrated landholdings are a barrier to disseminating/adopting new forms of technology. However, this relationship was highly dependent on a single country outlier (Libya).
3. We intended to carry out multivariate analyses to estimate the "causes" or deter- minants of changes in land-use, incorporating demographic and other variables simultaneously, to assess the relative strengths of the relationships; but the domi- nating effects of outliers evident from the scatter plots has persuaded us that it is more fruitful to focus on re-examining the outliers themselves.

9

Tropical deforestation and agricultural development in Latin America

Douglas Southgate

9.1 Introduction

Tropical deforestation arouses widespread concern. Available evidence suggests that global climate is being affected (Detwiler & Hall 1988). In addition, biological diversity is threatened because tropical forests, which cover less than 10 per cent of the Earth's land surface, harbour half the world's plant and animal species (Myers 1984, Wilson 1988). In many countries, deforestation is the result of excessive timber extraction. As Repetto & Gillis (1988) emphasize, the royalties loggers pay for access to publicly owned primary forests in Southeast Asia fall far short of stumpage values. Responding to opportunities to capture sizable rents, they are inclined to "cut and run". Although destructive logging also takes place in Latin America, deforestation in the region is primarily an agricultural phenomenon. Brazil and a few other countries have implemented projects to relocate farmers to tree-covered hinterlands. More frequently, conversion of forests into crop land and pasture is "spontaneous', being driven by various economic forces.

For the most part, the existing literature addressing those forces consists of case studies. Moran's (1983) analysis of migration to Altamira, a settlement on Brazil's Transamazon Highway, is representative. The geographic focus of this paper, by contrast, is very broad. Regression analysis is used to explain farmers' and ranchers' encroachment on tropical forests and other natural environments throughout Latin America. In particular, the possibility that frontier expansion in the region is symptomatic of agricultural underdevelopment is explored.

The model and database used to study agriculture's geographic expansion are described in the following section. Next, the results of regression analysis are presented. Land clearing is shown to be inversely related to trends in crop and livestock yields. This finding prompts a brief discussion of the factors influencing agricultural productivity and leads to suggestions about how to conserve natural environments in the developing world.

9.2 A model of agricultural frontier expansion

Simple Malthusian explanations of tropical deforestation, which are widely circulated, leave one with the sense that "surplus people" are heading for the developing world's agricultural frontiers in droves. This is indeed happening in some places, including parts of Latin America. For the most part, however, cities bear the burden of mounting demographic pressure in the Western Hemisphere. Even under the most miserable circumstances, urban dwellers rarely move to the Amazon Basin or the Caribbean lowlands of Central America. In addition, emigrants from the countryside, where fertility far outstrips mortality, usually go to cities and towns, not the agricultural frontier. As indicated in Table 9.1, urbanization is a more pronounced phenomenon in the region than population growth per se.

If there is a relationship between population growth and frontier expansion, then, it is primarily an indirect one. Domestic demand for agricultural commodities is rising in most countries primarily because the number of consumers is growing. In turn, increased demand for food enhances derived demand for land inputs to crop and livestock production. Another potential source of demand growth is external. Pursuing development strategies that emphasized import substitution and industrialization, Latin American governments long discouraged exports by levying taxes and over-valuing domestic currencies (Valdes 1986). In recent years, however, these distortions have been reduced in a number of countries. As a result, specialization has increased in the production and export of agricultural commodities in which the region holds a comparative advantage.

All else remaining the same, increased domestic or international demand for agricultural commodities leads to an outward shift in the sector's extensive margin.

Table 9.1 Population growth and urbanization, selected Latin American countries.

Country	Total population in 1988 (million)	Annual growth 1980–88 (%)	Urban population in 1988(million)	Annual growth 1980–88 (%)
Brazil	144	2.2	108	3.6
Colombia	32	2.1	22	3.0
Costa Rica	3	2.3	1	1.9
Ecuador	10	2.7	6	4.7
Guatemala	9	2.9	3	2.9
Honduras	5	3.6	2	5.6
Mexico	84	2.2	60	3.1
Paraguay	4	3.2	2	4.5
Peru	21	2.2	14	3.1

Source: World Bank 1990c.

But the magnitude of that shift depends on two "supply side" factors. The first is a "land constraint". The second is the supply of "non-land" inputs in the agricultural sector (e.g., human capital and managerial talent).

The land constraint on settlers' behaviour largely reflects property arrangements. Where all land, agricultural and non-agricultural, is privately owned, frontier expansion is influenced by some of the opportunity costs of land clearing. In particular, agents of deforestation are forced to take into account the income associated with timber production. Along Latin America's agricultural frontiers, however, all opportunity costs of creating new cropland and pasture are, from a settler's perspective, external costs. Because destruction of natural vegetation is a prerequisite for formal or informal property rights (Mahar 1989a, Southgate et al. 1991), no one is in a good position to internalize forestry rents. In addition, a settler who is slow about clearing land runs the risk that somebody else will "jump" their claim. Accordingly, colonists deforest immediately whenever agricultural rents can be captured by doing so (Southgate 1990).

Given the nature of frontier tenurial regimes in Latin America, the land constraint on colonists' behaviour is important only if virtually all soils that are both accessible and suitable for crop or livestock production have been occupied by farmers and ranchers. In a non-statistical analysis of deforestation in Brazil, Schneider et al. (1990) point out that soils featuring both those characteristics are becoming increasingly scarce as evidenced by the declining gap between frontier and infra-marginal land prices. Unfortunately, sub-national data on land values are not accurate enough in most Latin American countries to allow price differentials to be used as a barometer of land scarcity.

In this study, current agricultural land-use was compared with land-use capabilities in order to determine whether or not agricultural colonization is seriously constrained. As indicated in Table 9.2, natural conditions do not favour continued frontier expansion in two Andean countries: Bolivia and Peru. In addition, the frontier

Table 9.2 Current versus potential agricultural land-use in selected Latin American countries with widespread nutritional deficits.

Country	1987 Agricultural land[a] (million hectares)	Potential agricultural land[b] (million hectares)
Bolivia	30.149	30.031
Colombia	17.480	43.973
Ecuador	7.646	12.532
El Salvador	1.343	1.320
Haiti	1.399	0.645
Honduras	4.315	3.267
Peru	30.845	33.565

Sources: a) FAO 1989a; b) OAS 1974.

is all but closed in Uruguay and five Central American countries: Costa Rica, Nicaragua, Honduras, El Salvador, and Guatemala. In Haiti, agriculture's extensive margin has advanced well beyond what natural conditions warrant. The prospects for frontier expansion are also limited in the Dominican Republic and Jamaica.

Agricultural colonization is also affected by the availability of non-land assets for crop and livestock production. As those assets are formed, yields increase and substitution away from land takes place. Consequently, the pressure to create new cropland and pasture is eased.

Principal factors affecting agriculture's geographic expansion having been identified, let us turn to specification of the dependent variable as well as the regression model itself. Since property arrangements oblige agricultural colonists to ignore the value of tree-covered land, it makes little sense to use the ratio of cleared area to remaining forests as a dependent variable. Instead, growth in the area used to produce crops and livestock (AGLNDGRO) appears on the left-hand side of this chapter's regression model of the causes of frontier expansion in Latin America:

$$AGLNDGRO = B_0 + B_1 \, POPGRO + B_2 \, EXPGRO + B_3 YLDGRO + B_4 \, NOLAND. \quad (1)$$

The coefficients of population growth (POPGRO) and agricultural export growth (EXPGRO), which both tend to stimulate frontier expansion, are expected to be positive. By contrast, the coefficient of yield growth (YLDGRO), which is associated with the formation of non-land assets in the agricultural sector, is probably negative. Finally, NOLAND is a dummy variable indicating that closure of the agricultural frontier has occurred or is imminent. Its coefficient is expected to be negative.

9.3 Data

The twenty-four countries listed in Table 9.3 comprise the sample used in this study. Data on agricultural land-use, population growth, exports, and agricultural yields for each country were obtained from annual publications of the Food and Agriculture Organization of the United Nations (FAO) as well as the International Bank for Reconstruction and Development (the World Bank). For 21 of the countries, data on crop land and pasture (FAO 1989a) were applied to the following logarithmic formula in order to calculate the regression model's dependent variable:

$$AGLNDGRO = \frac{100\left[\log(1987 \text{ agricultural land}) - \log(1982 \text{ agricultural land})\right]}{5} \quad (2)$$

This approach was not appropriate, however, for determining dependent vari-

Table 9.3 Data used in the regression analysis (% p.a.).

Country	Frontier expansion[a] (AGLNDGRO)	Population growth[b] (POPGRO)	Export growth[c] (EXPGRO)	Yield growth[d] (YLDGRO)
Argentina	−0.1	1.4	−8.5	−0.5
Belize	1.2	2.4	2.2	–
Bolivia	0.4	2.7	0.0	−1.4
Brazil	0.6	2.2	−3.5	3.2
Chile	0.1	1.7	17.5	3.6
Colombia	0.7	2.1	0.0	1.0
Costa Rica	1.1	2.3	5.2	1.5
Cuba	0.7	1.1	−4.3	−0.9
Dominican Republic	0.1	2.4	−7.4	0.1
Ecuador	2.0	2.7	11.4	−0.2
El Salvador	0.1	1.3	−8.5	−3.6
Guatemala	0.8	2.9	−2.5	−2.0
Guyana	0.1	−4.4	0.0	−3.9
Haiti	0.0	1.8	−8.0	1.0
Honduras	0.4	3.6	3.4	0.8
Jamaica	−0.3	1.5	0.0	3.5
Mexico	0.6	2.2	14.0	1.7
Nicaragua	0.8	3.4	−14.1	−4.8
Panama	0.7	2.2	−6.0	1.7
Paraguay	1.0	3.2	0.0	3.1
Peru	0.1	2.2	0.0	1.8
Surinam	3.2	1.1	0.0	−9.3
Uruguay	−0.1	0.6	−8.2	0.5
Venezuela	0.3	2.8	0.0	4.1

Sources: a) FAO 1989a; b)World Bank 1990c; c) FAO 1989b; d) FAO 1989a.

able values for Bolivia, Mexico, and Paraguay because land-use data for those three countries are very questionable.

Remote sensing studies conducted by the FAO suggest that annual deforestation currently amounts to 117,000 hectares (ha) in Bolivia and 615,000 ha in Mexico (World Resources Institute 1990). By contrast, FAO (1989a) reported that crop land expanded by just 24,000 ha and that pasture declined by 250,000 ha between 1982 and 1987 in the former country. According to the same source, Mexico had exactly 74,499,000 ha of pasture in 1972, 1977, 1982, and 1987. In Paraguay, estimated deforestation is 212,000 ha *per annum* (World Resources Institute 1990). However, the area planted to crops was supposed to have risen by 210,000 ha between 1982 and 1987 and pastures were supposed to have increased by 3,460,000 ha in the same period (FAO 1989a). Instead of reflecting an actual shift in the agricultural frontier, the latter change is probably indicative of range lands being reclassified as pasture.

Because of these incongruities between deforestation and agricultural land-use data, AGLNDGRO values were calculated for Bolivia, Mexico, and Paraguay by dividing estimated deforestation (World Resources Institute 1990) by 1987 agricultural land (FAO 1989a). This substitute procedure probably understates actual frontier expansion since forests are not the only natural environment being penetrated by farmers and ranchers. With respect to the regression model's first independent variable, POPGRO, the World Bank's (1990c) estimates of annual population growth during the period, 1980 through 1988, were used.

Estimates of annual growth in agricultural exports were obtained by applying trade data (FAO 1989b) for each of the twenty-four countries in the sample to the following regression:

$$[\log(\text{exports in year t}) - \log(1983 \text{ exports})] = G \text{ (year t)}, \tag{3}$$

where the range of "t" was 1984 through 1988. For two-thirds of the countries listed in Table 9.3, the regression coefficient, G, serves as a measure of EXPGRO. For the remaining eight countries, however, EXPGRO was held to zero because the null hypothesis regarding G was accepted with a confidence interval of 90 per cent.

Calculation of the third independent variable in the regression equation involved two steps. First, FAO's (1989a) index of crop production in 1982 was divided by crop land in the same year (FAO 1989a) to obtain yields for 1982. Yields for 1987 were obtained in the same fashion. Second, a procedure like the one described in equation (2) was applied to identify annual yield growth during the intervening five years:

$$\text{YLDGRO} = \frac{100\left[\log(1987 \text{ yields}) - \log(1982 \text{ yields})\right]}{5} \tag{4}$$

Based on a comparison of current agricultural land-use and land-use capabilities (see preceding section), the value of the land constraint dummy variable, NOLAND, was set equal to one for the following eleven countries: Bolivia, Costa Rica, the Dominican Republic, El Salvador, Guatemala, Haiti, Honduras, Jamaica, Nicaragua, Peru, and Uruguay.

Other than dummy variable values, all data used in the regression analysis are presented in Table 9.3. As can be seen, AGLNDGRO varies considerably from country to country. Between 1982 and 1987, agriculture's extensive margin actually receded in Argentina, Jamaica, and Uruguay. In several other countries, frontier expansion was negligible. Compared to EXPGRO and YLDGRO, POPGRO does not exhibit much variation. Only one country, Guyana, lost population, due to heavy emigration. Between 1980 and 1988, population growth exceeded 2.5 per cent a year in seven countries. Annual rates of increase were between 1.5 per cent and 2.5 per cent in nearly half the sample.

Values of EXPGRO and YLDGRO are widely scattered. Agricultural exports declined in countries that suffered civil conflict, maintained policies that discouraged crop and livestock production, or both. In light of increased domestic consumption of agricultural commodities (and, in many countries, increased exports), yield trends have been disappointing. Only in Brazil, Chile, Jamaica, and Venezuela did annual percentage yield increases exceed rates of population growth. The ratio of crop and livestock output to agricultural land actually declined in nine countries.

Other than a weak correlation between EXPGRO and YLDGRO, multicollinearity is not a major problem in the data set. It is particularly interesting to note that there is no strong linkage between YLDGRO and the dummy variable indicating the presence of a serious land constraint (NOLAND). The governments of countries where that constraint holds have apparently been slow to encourage formation of substitute assets in the agricultural sector.

9.4 Regression results

Indices of crop production being unavailable for Belize, that country had to be deleted from the sample used in the regression analysis. With data for the remaining twenty-three countries (Table 9.3), ordinary least squares estimation yielded the following results:

$$
\begin{aligned}
\text{AGLNDGRO} = 0.463 \quad &+ \quad 0.249 \text{ POPGRO} \quad + \quad 0.031 \text{ EXPGRO} \\
(0.161) \quad & \quad (0.066) \quad \quad (0.014) \\
(2.876) \quad & \quad (3.773) \quad \quad (2.214) \quad\quad\quad (5) \\
&- \quad 0.198 \text{ YLDGRO} \quad - \quad 0.641 \text{ NOLAND} \\
& \quad (0.033) \quad \quad (0.205) \\
& \quad (-6.000) \quad \quad (-3.127)
\end{aligned}
$$

ADJ R^2 = 0.669,   DW = 2.065,   SSR = 3.489,   F = 12.098

For a cross-sectional study, an adjusted R^2 of 67 per cent is very good, particularly since aggregate national-level data for a heterogeneous group of countries have been used. Dummy variables for war, inclement weather, inflation risks, and the like could have been introduced. But to maintain a sharp focus on linkages between frontier expansion and agricultural development, this was not done. That the F-statistic exceeds 8.290 – which is the minimum value for rejecting the hypothesis that there is no linear relationship between AGLNDGRO and the four independent variables (99 confidence interval) – re-enforces the conclusion that this chapter's simple model is a satisfactory framework for analyzing encroachment on tropical forests and other natural environments in Latin America.

The signs of all parameter estimates are consistent with what one expects. The two rows of figures under the regression coefficients are standard errors and t-statistics, respectively. Using a two-tail test and a 99 per cent confidence interval, one rejects the null hypothesis for the coefficients of POPGRO, YLDGRO, and NOLAND. At a 95 per cent confidence interval, the null hypothesis is rejected for EXPGRO's coefficient as well.

Interpretation of the coefficients is straightforward. In a country where natural conditions do not favour frontier expansion (i.e., where the value of the dummy variable, NOLAND, is 1 instead of 0), the annual increase in cropland and pasture is expected to be 0.641 percentage points lower than would be the case if soils that lend themselves to crop or livestock production were "unoccupied". Of more direct relevance to the focus of this paper is that a Z per cent increase in yields offsets nearly four-fifths of the frontier expansion otherwise induced by Z per cent population growth. Alternatively, the same Z per cent yield increase can compensate for 6Z per cent growth in agricultural exports.

9.5 How to contain agricultural colonization

If shifts in agriculture's extensive margin were driven exclusively by increasing or decreasing demands for agricultural commodities, the prospects for containing frontier expansion in Latin America would be very bleak indeed. Throughout the region, populations are overwhelmingly young. With numbers of women capable of bearing children expected to rise for many more years, continued population growth is inevitable, even with the decline in fertility rates currently taking place in nearly every part of the Western Hemisphere (World Bank 1990c). As the number of people needing to be fed increases, pressure on natural resource inputs to agricultural production will mount.

Unfortunately, this scenario is being played out in many parts of the Western Hemisphere. Consider, for example, the case of Ecuador, which is the most crowded country in South America. As reported in Table 9.3, Ecuador's population has been growing by nearly three per cent a year and, between 1983 and 1988, annual increases in agricultural exports amounted to 11.4 per cent. The latter rate was exceeded only in Chile and Mexico. Demand growth has not been matched by productivity increases in Ecuadorian agriculture. Indeed, yields actually declined between 1982 and 1987. As a result, dedicating more land to crop and livestock production has proved to be essential. At 2.0 per cent per annum, Ecuador had the second highest rate of frontier expansion in Latin America between 1982 and 1987. Surinam's rate (3.2 per cent a year) was higher only because its initial base of cropland and pasture was tiny.

Unlike El Salvador, Nicaragua, and a few other nations, Ecuador cannot pin the blame for stagnating productivity on civil conflict. Disappointing yield trends were instead a consequence of limited investment in non-land assets. In particular, the scientific base underpinning crop and livestock production is weak in the country. As Whitaker (1990) points out, research and extension networks are highly fractured in Ecuador. Separate entities created for agriculture, forestry, and other sectors of the rural economy do not co-operate on basic scientific research. Similarly, co-ordination among narrowly focused divisions of the extension service is limited. In addition, funding is meagre. Real spending on agricultural research, for example, declined 7.3 per cent a year from 1975 through 1988. Having fallen to 0.17 per cent of agricultural GDP, research expenditures compare poorly with spending by neighbouring countries (Whitaker 1990). Given the state of Ecuadorian agriculture's scientific base, yields are low in the country. This means that growing demands for crops and livestock have to be met by bringing more land, which is usually of marginal quality, into production. Two-thirds of the increased crop production occurring in Ecuador between the middle 1960s and the middle 1980s, for example, were accounted for by frontier expansion. Improved productivity explained only the remaining third (Whitaker & Alzamora 1990).

Contrasting with what has taken place in Ecuador is the recent performance of Chile's agricultural economy. If yields had not risen in the latter country during the 1980s, 17.5 per cent annual growth in agricultural exports combined with 1.7 per cent annual population growth would have induced frontier expansion exceeding 1.0 per cent a year. But because yield increases were also impressive, agriculture's extensive margin remained stable.

Chilean agriculture has become more productive because application of a full range of non-land inputs has grown more intense. Farming has become more mechanized and investments in irrigation have been made. In addition, use of chemical inputs has increased. Between 1985 and 1988, for example, nitrogen, phosphate, and potash imports rose 154 per cent, 120 per cent, and 355 per cent, respectively, in real terms. Since agricultural research and extension are strong in Chile, yield increases have also been achieved through wider use of improved varieties and cultivars (Arensberg et al. 1989).

In light of the possible impacts of mechanization, irrigation, and chemical use on soil and water quality, the last source of productivity growth deserves special attention. Indeed, it is possible that investment in the scientific base underpinning crop and livestock production can allow Latin America to satisfy growing demands for agricultural commodities without frontier expansion or deterioration of renewable natural resources within agriculture's extensive margin.

This possibility is illustrated by recent initiatives of the Brazilian Corporation for Agricultural Research (EMBRAPA). At EMBRAPA's National Soil Biology Research Center, scientists have isolated nitrogen-fixing bacteria for a number of crops grown in hot and acidic soils. Most Brazilian soybean farmers now plant seeds inoculated with those bacteria. As a result, high soybean yields are maintained and annual expenditures on fertilizer have been reduced by about $1 billion (Mangurian 1990). Another initiative has to do with biological pest control. Scientists at EMBRAPA's National Soybean Research Center have isolated a virus that kills the velvet bean caterpillar. Applying that virus to a soybean field costs 75 per cent less than spraying pesticides. In addition, toxic chemicals are not released into the environment (Mangurian 1990).

9.6 Implications for conservation strategies

Some economists attempting to explain the loss of natural habitats in the developing world fall into a habit of analysis that is nearly as old as the discipline itself. Like those who advocate acreage controls to reduce agricultural commodity surpluses in the United States, they underestimate the degree to which land and other inputs to the production of crops and livestock are interchangeable. If the option of substitution is ignored, then the predictions of a simple Ricardian model of the agricultural economy hold. That is, frontier expansion is the only possible response to market or demographic "shocks". To be sure, formation of non-land assets should reflect an agricultural economy's factor endowments (Hayami & Ruttan 1985). For example, investment in yield-enhancing technology is not particularly urgent where land and other natural resources are abundant. Unfortunately, investment of that type continues to be marginal in many Latin American countries where the prospects for frontier expansion are limited. Put another way, agricultural underdevelopment and encroachment by farmers and ranchers on fragile environments go hand in hand in the region.

In part, enhancing agricultural productivity requires governmental involvement. For example, many of the products of research, technology transfer, and education are "public goods". Accordingly, public spending on those activities is essential. However, policies that discourage farmers and ranchers from adopting productivity-enhancing measures also need to be reformed. Obviously, yields are bound to be low in any country where food price controls and currency distortions discourage production for domestic and international markets.

Increased productivity is not a panacea for all environmental problems arising in the agricultural sector. As indicated by the Chilean experience (see preceding section), one way to reduce farmers' encroachment on natural habitats is to increase fertilizer and pesticide applications on existing cropland. However, this can amount to solving one environmental problem by exacerbating another. Fortunately, there is an alternative. Investing in research, extension, and education allows for crop and livestock yields to increase without adverse impacts on renewable natural resources.

Finally, development of the technology needed for an environmentally sound response to mounting demands for agricultural commodities will not be enough to save Latin America's natural habitats. Schneider et al. (1990) point out that extending roads into previously inaccessible tropical forests almost always leads to timber extraction, depletive agriculture, and related "nutrient mining". It is also important to emphasize that the future of natural habitats depends on a re-ordering of property rights. Vast stretches of the region's tropical forests continue to be open access resources in which individuals can secure property rights by removing natural vegetation (Mahar 1989a, Southgate et al. 1991). As long as this tenurial regime remains in place, continued deforestation is inevitable.

III
Country case studies

10

The causes of tropical deforestation: a quantitative analysis and case study from the Philippines

David Kummer and Chi Ho Sham

10.1 Introduction

Tropical deforestation has become a topic of importance in many disciplines. In fields as diverse as climatology, biology, economics, and geography, tropical deforestation is now the object of concerted research efforts. One topic of importance to academics and policy makers alike is the cause of tropical forest decline. The purpose of this chapter is to review the limited quantitative work that has been done and to assess its relevance to understanding recent tropical deforestation. In addition, an original study of deforestation in the Philippines will be presented and contrasted with existing quantitative research. Lastly, we would like to raise some questions about the way deforestation is measured; in particular, we have doubts as to the use of forest cover as the dependent variable in statistical tests, since it cannot capture the dynamic nature of tropical forest removal.

By quantitative research, we mean those studies which are characterized by statistical tests of hypotheses involving a group of selected dependent and independent variables. These can be contrasted with descriptive analyses of the causes of deforestation. The latter, while informed by study and years of observation, do not include any explicit statistical tests. The majority of writings on deforestation belong to this category.

The first section will review the findings of recent quantitative research on the causes of tropical deforestation. The following sections analyze studies of tropical deforestation with forest cover and deforestation as the respective dependent variables. A quantitative study of the causes of deforestation in the Philippines is presented and then the relationship between population and deforestation is discussed.

10.2 A review of quantitative work on deforestation

The seven works reviewed in this section (an eighth is presented in the next section) have all appeared in the past ten years (see Table 10.1). All are examples of cross-sectional analysis; instead of using data obtained for the same unit of analysis over different points of time, as in time series analysis, these studies use data from different units of analysis at one point in time. Cross-sectional analysis is often used when there is a lack of time series data. The unit of analysis is the nation for five studies and province and municipality for the other two. Only the major findings are reported in Table 10.1. The dependent variable is either percentage forest cover (%FC) or absolute deforestation in hectares (DEF). One study, instead of using percentage forest cover, used percentage deforestation (%DEF); this is just the obverse of percentage forest cover and is a measure of percentage forest cover lost.

In all seven cases, a population variable was found to be important. Population increases were positively correlated with deforestation and population density was negatively correlated with forest cover. In short, there seems to be general agreement about the correlation of tropical deforestation with population. However, we shall argue below that the results are fundamentally flawed and the rôle of popula-

Table 10.1 Variables found to be important in quantitative work on tropical deforestation

Unit/area	Sample size	R^2	Dependent variable	Independent variables
Municipality (Brazil)[a]	165	0.84	%DEF	+ population density, + road density, + crop area
Province[b] (Thailand)	64	0.77	%FC	− population density, − wood price, + provincial income, + distance from Bangkok
Nation[c] (Caribbean)	30	0.65	%FC	− population density, − energy use density
Nation[d] (global)	60	0.60	%FC	− population density
Nation[e] (global)	36	0.77	DEF	+ population growth, + availability of capital
Nation[f] (global)	43	0.52	DEF	+ population increase, + logging
Nation[g] (global)	39	0.11–0.50	DEF	+ population increase, increase in farmland, + wood use

Sources: a) Reis & Margulis (1990); b) Panayotou & Sungsuwan (1989). Four other variables were included in the final equation but none of these were significant at the 0.05 level. As such, the final r-square is overstated. They used pooled cross-section and time series data; c) Lugo et al. (1981); d) Palo et al. (1987); e) Rudel (Ch. 7); f) Grainger (1986); g) Allen & Barnes (1985).

tion overstated. The two major problems with the studies enumerated above is that four use an inappropriate dependent variable and the other three are based on datasets which are of doubtful validity.

Studies of tropical deforestation with forest cover as the dependent variable

Even when the dependent variable in cross-section analysis is not deforestation itself, most users have then gone on to make statements about deforestation on the basis of the cross-sectional analysis (CSA). We believe that the use of forest cover as a dependent variable is inappropriate for an examination of recent tropical deforestation and the remarks that follow are directly applicable to the first four studies in Table 10.1.

The forest cover (percentage or absolute) that exists in an area is a function of two factors: the amount of original forest cover and the forest cover lost since the deforestation process began. The original forest cover can be augmented by re/afforestation but for the present discussion this factor will be ignored. In other words, the forest cover that exists today is the result of a history which may be as long as three thousand years (parts of China) or three months (parts of Amazonia). The use of percentage or absolute forest cover as a dependent variable cannot capture these different histories. Additionally, the use of forest cover as a dependent variable does not account for the fact that the initial forest covers were not the same and, in any case, are almost certainly unknown. Thus, for example, a CSA across countries could include Country A which is 70 per cent forested and whose deforestation started a hundred years ago and Country B which is 70 per cent forested and whose deforestation started 10 years ago. When it is considered that it is unlikely both countries had an equal percentage of forest cover when the deforestation process was initiated, it should be apparent that to make statements about present day deforestation on the basis of present day forest cover is misleading. In other words, percentage forest cover represents cumulative deforestation and, as such, is not relevant for a study of recent deforestation.

For CSA to work with forest cover as the dependent variable, two major assumptions have to be made: first, all geographical units included in the analysis had the same amount of forest cover to begin with and, second, the deforestation process started at the same time in all of the units. Since both of these assumptions are unrealistic, we are forced to conclude that CSA with forest cover as the dependent variable is not capable of making meaningful statements about present day deforestation. An additional consideration has to do with the lack of correspondence between the time spans of the dependent and independent variables. The independent variables contained in Table 10.1, and the statistical results for the Philippines to be examined in this chapter, all represent contemporary periods, for example, population in 1957 or agricultural area in 1970. However, since forest cover in 1957 is the result of historical processes which began hundreds of years before 1957, it is impossible for a contemporary independent variable to explain the past behaviour of the dependent variable. For example, forest cover in the Philippines in 1957 was partly the result of the expansion of commercial agriculture in the mid-nineteenth century as the Philippines was incorporated into the global economic system (Lopez-Gonzaga 1987, McClennan 1980) and it is impossible for a variable in 1957 to have been the "cause" of that past change. In other words, even if we assume that different regions began with equal forest cover and that deforestation started in both areas at the same time, a cross-sectional analysis with forest cover as the dependent variable still would not be meaningful unless data on the independent variables covered the same time frame as the deforestation process. Given that tropical deforestation has a long history in most parts of the world, this necessarily means that the use of CSA with forest cover as a dependent variable will be relevant only to those few areas where data exist on the independent variables since the process of forest removal began.

The first four studies referred to in Table 10.1 use percentage forest cover (or percentage forest cover lost) as the dependent variable. As such, one would expect that statements about deforestation were made on the basis of an examination of the variation in forest cover and this, in fact, is what happened. As an example, Panayotou & Sungsuwan (1989: 21) state that their concern is ". . . to assess the relative importance of the different determinants of the forest cover (*or inversely, causes of deforestation*). . ." (italics added). As the above discussion has made clear, however, present day deforestation is not the inverse of present day forest cover. As an example of the type of mistake that can be made by confusing forest cover and deforestation, the authors observed that forest cover was positively correlated to distance from Bangkok and, therefore, concluded that deforestation was greater the closer one was to Bangkok. Unfortunately, this misrepresents the reality of deforestation in present day Thailand. On going deforestation in Thailand is occurring in those areas which still have forest cover and, in general, these are the areas furthest from Bangkok in the Northeast and Northwest sections of the country (Bhumibhamon 1986, Feeny 1984, Hirsch 1987).

Palo et al. (1987: 63) make a similar mistake in their study because ". . . forest coverage as a 'negative' proxy for deforestation was chosen as the dependent variable. . .". Given the reservations that we have expressed above about using forest coverage as the dependent variable, their conclusion that "increasing population pressure" is the driving force of the deforestation process was not supported by their statistical tests. Similar comments can be made about the results of Reis & Margulis (1990), and Lugo et al. (1981). In short, the results of the first four studies in Table 10.1 cannot be used to make statements about tropical deforestation; they can be used to make statements about the relationship between present day forest cover and the chosen independent variables or, alternatively, they can be used to make statements about forest cover lost since the deforestation process began but these are of doubtful relevance to contemporary deforestation.

The importance of the critique above is highlighted by ongoing work at the Food and Agricultural Organization (FAO). As part of their "Forest Resources Assessment 1990", the FAO is developing a deforestation model which is based on the theoretical framework of M. Palo (Scotti 1990). Palo's work, as presented in Table 10.1, is based on the fact that a negative relationship exists between population density and percentage forest cover. Scotti (1990) uses a sample size of 47 subnational geographic units drawn from eight different countries in conducting his cross-sectional statistical tests. He argues that this is appropriate because "representing forest area by a density value and substituting time by corresponding population density values enables direct comparisons of countries of different size and at different growth stages" (p. 3). However, this line of reasoning ignores the fact that, as demonstrated above, percentage forest cover is not an appropriate dependent variable to use when analyzing deforestation. As a result, it would be difficult to place much confidence in FAO predictions of forest area cover based on population growth or changes in population densities.

There are two additional features of a CSA with forest cover as the dependent variable that are worth mentioning. First, the results are entirely expected. By definition, very few human settlements are found in forests and very few forests are found in human settlements; as such, we would expect that there will always be a negative relationship between forest cover and, for example, roads and population. This is simply logical but it does not say anything about the causes of ongoing deforestation. Secondly, if forest cover is the dependent variable then variables which represent commercial logging such as size of concession and timber volume harvested cannot be included as independent variables (to be discussed in a later section in more detail). Since large scale commercial logging has been so prominent in Southeast Asia and parts of Western Africa, this is a serious restriction.

Studies of tropical deforestation with deforestation as the dependent variable

The three remaining studies presented in Table 10.1 (Allen & Barnes 1985; Grainger 1986; Rudel, Ch. 7) have actual area deforested as the dependent variable. As such, one would expect these works to be much more relevant than the four studies reviewed above. Rudel argues that there are two major causes of tropical deforestation: population growth and large capital-intensive development projects in countries with large tracts of forest, for example, road building in Amazonia. Allen and Barnes conclude that tropical deforestation is caused by increases in population and land under agriculture, and wood production.

Grainger determined that population increases and logging were the major causes of forest decline.

Unfortunately, all of these studies are based on datasets which are of doubtful reliability. Rudel and Grainger have used the data provided by the FAO (1981a). However, recent studies (World Resources Institute 1990: 101) have provided evidence which ". . . points to much higher rates of depletion than were measured in the late 1970s. . .". As an example, the FAO/UNEP estimate of deforestation in Brazil from 1981 to 1985 was 14,800 km² a year but it is now considered to be in the order of 80,000 km². Since it is difficult to believe that deforestation rates in Brazil increased by more than five times in less than ten years, there is a strong possibility that Brazil's rate of deforestation was under-estimated by the FAO/UNEP. Numerous other examples could be given. In the case of the Philippines, not only were deforestation rates too low but actual forest cover was over-estimated (Kummer 1992). If the new studies (World Resources Institute 1990: 102) are accurate, they are ". . .79 per cent over FAO's 1980 estimate." Under these circumstances, it may be the case that the earlier estimates were not accurate (as in the case of the Philippines) and it would be difficult to place a great deal of confidence in the statistical results derived from these data. Allen and Barnes, on the other hand, used the FAO production yearbooks. This dataset is considered by most researchers to be inappropriate for purposes of examining tropical deforestation due to the broad definition of forest cover used and the fact that the data are of questionable reliability (Palo – personal communication 1990, Rudel Ch. 7).

In effect, the availability of data has determined, to a large extent, the type of analysis that could be conducted. The FAO production yearbooks and the FAO/UNEP study provided numbers, but the numbers themselves are so unreliable that any statistical results derived from them can be dismissed as not being based on a strong enough database. This highlights what to us is one of the more glaring aspects of the study of tropical deforestation: much of the discussion has not been informed by appropriate data. The failure of remote sensing, until recently, to provide accurate information regarding tropical forest decline and the lack of national studies are two aspects of the poor data problem.

In short, of the seven quantitative examinations of the causes of tropical deforestation, four use an incorrect dependent variable and three are based on weak datasets. Even if there were no serious problems with the data used by the authors just discussed, what do their results tell us? Can regression analysis across nations provide the detail that we need to understand and control the process of deforestation? For instance, Allen and Barnes found that export of wood was related to deforestation but large-scale logging and exports of wood products has only occurred in a handful of countries. Most of the tropical world has had high rates of deforestation and, yet, wood exports have been minimal. Under these circumstances, a CSA of 39 nations which finds wood exports to be significant would be wrong if applied to the majority of the individual countries in the sample. In other words, the statistical results across the sample of nations cannot be applied to individual nations within the sample; as such, they provide virtually no information that can be useful to a particular country. In terms of policy recommendations, cross-national analysis would appear to be at a decided disadvantage. Lastly, even if a cross-section analysis does determine a strong positive relationship between deforestation and increases in population, this is not proof that the relationship in fact exists. Without geographic information proving that deforestation in the individual countries is occurring where there are increases in population, the relationship determined by the CSA must be considered to be a weak association at best.

The previous two sections have presented a critical assessment of previous quantitative work as to the causes of tropical deforestation. We do not feel that the studies referred to tell us much about deforestation and, in fact, the data and conceptual problems may be insurmountable. The material to be presented below is an attempt to highlight the difference between using forest cover and deforestation as the dependent variable in regression analysis of the causes of deforestation. In addition, it is an attempt to articulate the immediate causes of deforestation more clearly than is possible in cross-national studies.

10.3 A Philippine case study of the causes of deforestation

The following material draws primarily from Kummer (1992). Deforestation in the post-war Philippines was analyzed in two ways: first, a cross-sectional analysis for the years 1957, 1970 and 1980 was conducted and, secondly, a panel analysis from 1970 to 1980 (unfortunately, due to changes in provincial boundaries, it was not possible to conduct a panel analysis from 1957 to 1970). The basic data unit was a province with sample size varying from 52 to 72.

In the cross-sectional analysis, the dependent variable is the absolute amount of forest cover per province in hectares. This was used because the authors feel that the two major proximate causes of deforestation in the Philippines have been logging which converted the primary forest to a degraded secondary forest and subsistence agriculture which converted the secondary forest to farmland. For both loggers and agriculturalists, the main concern has been hectares of forest. It is not obvious to us that percentage forest cover is an important motivating factor to loggers or subsistence farmers.

In the panel analysis, the dependent variable is the absolute loss of forest cover in hectares between 1970 and 1980. This is a direct measure of deforestation. At this point we simply note that the cross-sectional and panel analyses are measuring two different relationships: the cross-sectional analysis is concerned with the relationship between absolute levels of forest cover and the independent variables, and the panel analysis with the relationship between deforestation and the independent variables.

An additional consideration is that the choice of dependent variable will have an effect on the independent variables that can be included in the statistical analysis. In the cross-sectional analysis, variables which represent commercial logging such as volume of timber harvested, size and number of concessions, and annual allowable cut, cannot be considered as independent variables because they are a function of the dependent variable; that is, they are a function of forest size. As an example, the area of logging concessions is a function of the size of the forest but, in a static analysis, concession area cannot influence forest area. In the panel analysis this is not the case and data on commercial logging can be included as an independent variable. In short, not only are the dependent variables in the two types of analyses different, but the choice of dependent variable determines which independent variables are appropriate.

Since the main interest of this section is with the correct dependent variable to be used in analyzing deforestation, we will not discuss the background to the cross-sectional results presented in Table 10.2; rather, the interested reader is urged to consult Kummer (1992). The level of significance has been set at five per cent and all variables included in the equations meet this criterion. All coefficients have been standardized and t-statistics are in parentheses below the corresponding variable. Adjusted r-square, F, chi-square (test for heteroscedasticity, where p greater than 0.05 indicates that the results are significant at the 0.05 level) and Durbin–Watson (test for serial correlation) statistics are to the right of each equation. Equations 1 and 2 are in double logarithm form (the L represents logarithmic) but Equation 3 is not. There would appear to be no major statistical problems with any of the equations.

Table 10.2 Multiple regression results for 1957, 1970, and 1980 showing the importance of road density (RDK), population density (POPK) and change in roads from 1970 to 1980 (RD7D8).

$L(FA57) = -0.87 \, L \, (RDK)$	$r^2 = 0.76, F = 159$	(1)
$N = 52 \qquad (-12.6)$	Chi-square = 4.69	
	DW = 1.70	
$L(FA70) = -0.54 \, L \, (POPK) - 0.28 \, L \, (RDK)$	$r^2 = 0.58, F = 48$	(2)
$N=68 \qquad (-4.7) \qquad\quad (-2.4)$	Chi-square = 11.0	
	DW = 1.66	
$FA80 = 0.32 \, RD7D8 - 0.38RDK - 0.29 \, POPK$	$r^2 = 0.41, F = 17$	(3)
$N=72 \qquad (3.5) \qquad\quad (-3.5) \qquad (-2.6)$	Chi-square = 10.9	
	DW = 1.92	

According to Equation 1 in Table 10.2, forest area in 1957 (FA57) is negatively related to road density (RDK). The sign of the variable is as expected and road density accounts for approximately 75 per cent of the variation in provincial forest cover. Forest area in 1970 (FA70) is negatively related to road (RDK) and population density (POPK). Once again the signs are as expected and these two variables account for 58 per cent of the observed variation in provincial forest cover. Forest area in 1980 (FA80) is negatively related to road (RDK) and population density (POPK) and positively related to the increase in roads from 1970 to 1980 (RD7D8). The signs of the density variables are as expected and the positive association with increases in road networks may indicate continued movement to frontier areas.

The results of the cross-sectional analysis appear to be unambiguous: absolute provincial forest cover in 1957, 1970 and 1980 is negatively related to the presence of human settlements (roads and people). Before the results of the panel analysis are presented, two observations should be noted: first, agricultural area does not appear in any of the equations in Table 10.2 and, secondly, the fact that all three equations contain different variables may mean that each separate cross-section is capturing a different aspect of the relationship between forest cover and the independent variables.

Equation 4, below, presents the results of the panel analysis. Since accurate data on logging are not available, we decided to use the provincial annual allowable cut (AAC) as a proxy for commercial logging. In the post-war period, annual allowable cut has consistently been greater that recorded logging volumes and, as such, should be a more accurate measure of actual timber removal. The initial equation estimated was deforestation in the 1970 to 1980 period as a function of distance from Manila (a measure of the lack of control of the forest removal process or corruption), change in agricultural land from 1970 to 1980, change in population from 1970 to 1980, change in length of roads from 1970 to 1980 and annual allowable cut in 1970. However, in the initial equation and subsequent variations, changes in roads and population and distance from Manila were not significant and the final equation contained only two independent variables.

$$\text{Deforestation} = 0.41 \text{ AAC70} + 0.41 \text{ Change in agricultural area} \qquad (4)$$
$$(4.79) \qquad (4.40)$$

$$N = 67, F = 34, \text{adjusted r-square} = 0.50, DW = 1.99, \alpha = 0.06.$$

Tests for multicollinearity were conducted which indicated that it was not present, as such; there would appear to be no major statistical problems with Equation 4. Equation 4 demonstrates that absolute deforestation in hectares from 1970 to 1980 is positively related to annual allowable cut in 1970 and to the spread of agriculture from 1970 to 1980. The signs of the variables are as expected.

A comparison of the cross-sectional and panel analyses is revealing: they are completely dissimilar. None of the independent variables in the cross-sectional equations are found in the panel analysis and vice versa. Based on the discussion in the second section, we conclude that the different results flow directly from the fact that the dependent variables are themselves different. Furthermore, it would be incorrect to make statements about current deforestation on the basis of the cross-sectional analyses for 1957, 1970 and 1980. On the other hand, the panel analysis, with deforestation as the dependent variable, is capable of making statements about the process of deforestation itself. In short, this empirical case study from the Philippines confirms the earlier discussion regarding the inappropriateness of using forest cover as a dependent variable when analyzing ongoing tropical deforestation. Given the shortcomings of the seven studies in Table 10.1 and the FAO work discussed above, it may be reasonable to question their common conclusion that population growth is the major cause of tropical deforestation.

10.4 Population growth and tropical deforestation

Equation 4 does not include a population variable which suggests that factors other than population were more important in explaining deforestation from 1970 to 1980. Even more surprising is that when the absolute change in population from 1970 to 1980 is regressed by itself against deforestation, the r-square is only 0.05 and when the change in population density is regressed against deforestation the r-square is 0.02 and the result is not statistically significant. In short, the results of the panel analysis cannot support the contention of most scholars of tropical deforestation, that population growth is one of the leading causes of deforestation (however, for an analysis that argues that migration to the uplands is a major cause of deforestation in the Philippines, see Cruz & Cruz 1990). This result is even more striking when one realizes that the Philippines has one of the highest population densities of any tropical country and a current population growth rate of 2.6 per cent (Population Reference Bureau 1990).

The panel analysis indicates that deforestation in the Philippines is a two-stepped process: logging first converts the primary to a secondary forest and the secondary forest is then converted to farmland by agriculturists. Neither of these two separate but linked processes necessarily require high population densities. This is not to say that increasing populations have no effect whatsoever on deforestation (someone is cutting down the trees); rather, the picture that emerges, at least for the Philippines, is of a deforestation process which is more complex than commonly imagined. Population is important but it cannot be assumed that there is a one-to-one correspondence between increases in population and deforestation. At a more basic level, much ambiguity surrounds the use of the term "population pressure".

Several standards are available to measure population pressure: total population, population density, physiological density (number of people per unit area of arable land), percentage increase in population, and absolute increase in population. An additional measure could be in-migration, since it may be the case that migrants, as opposed to local inhabitants, are moving to frontier areas. Alternatively, it could be the case that socio-economic conditions in the sending areas are causing poor farmers to leave and, as such, measures of landlessness or unemployment could be considered to be measures of population pressure. Thus, there are potentially seven methods available to measure population pressure. It is not clear which measure is the most appropriate for studies of deforestation. In the studies listed in Table 10.1, Grainger, Allen & Barnes, and Rudel used absolute population growth, Lugo et al., Panayotou & Sungsuwan, and Reis & Margulis used population density, and Palo et al. used both.

It is important to keep in mind that the two measures of population pressure used in the studies in Table 10.1 are not necessarily consistent. For instance, an area experiencing rapid population growth could have either a low population density (forested area) or a high population density (urban area). Deforestation, in general, occurs where population densities are low since, logically speaking, large-scale forests preclude large-scale human settlements. As such, an area of low population density and high population growth, which is experiencing deforestation, does not demonstrate that population pressure is a problem within the area; rather, it may demonstrate that forces outside the area are causing people to migrate. An example of this is what is occurring in the Brazilian Amazon Basin (Millikan 1988). In short, invoking the population pressure argument, without providing the details of how it actually works, does not further our understanding of the deforestation process.

Another consideration is that there is such a wide range of population densities in the tropical Third World that we feel it is impossible to draw any firm conclusions regarding the effect of population on deforestation. For example, population per square kilometre in 1990 for five selected countries was as follows: Brazil (18), Indonesia (100), Malaysia (54), Thailand (108), Philippines (220) (Population Reference Bureau 1990). Absolute rates of deforestation are higher in Indonesia, Malaysia and Thailand than in the Philippines; yet, Thailand has only 50 per cent and Malaysia 25 per cent of the population density of the Philippines. Brazil, which has the most extensive ongoing deforestation in the world, has a population density which is only eight per cent of that of the Philippines. These examples do not exhaust the possibilities; many countries in South America and Africa with population densities less than Malaysia are experiencing rapid deforestation. Needless to say, these large variations in population density and rates of deforestation get swept under the rug in cross-national statistical tests.

An additional reason for doubting that population pressure is the main cause of tropical deforestation has to do with the rapidity with which deforestation has occurred in the post-war period and why it occurs at certain times and places. The clearest example of this phenomenon today is the province of Rondônia in Brazil. The rates of deforestation that have been observed are impossible to reconcile with demographic factors (Fearnside 1986b). While in-migration has been rapid, the resulting deforestation is not a function of the population density or population growth rate of the Brazilian nation. Rather, it involves millions of extremely poor people in other parts of Brazil who are desperate to provide their families with at least a subsistence living. Road building and migration meet the needs of a Brazilian elite anxious to lessen tensions in the impoverished Northeast and cities of Southern Brazil. The colonization of the Amazon and the resulting deforestation is a substitute for reform within the larger Brazilian society (Millikan 1988).

The rapid increase in deforestation in Indonesia after 1966 was not the result of population pressure. Rather, after President Sukarno's ouster, it reflected President Suharto's views on development in general and utilization of forest resources in particular. Rapid deforestation in Malaysia started around 1960 after the Communist Insurgency was put down; not because of exploding population pressure. Even more telling is the recent history of forest destruction in Malaysian Borneo. Deforestation is now rapid in Sarawak and Sabah, provinces with extremely low population densities. In short, population pressure cannot explain the rapid initiation of deforestation at various places around the globe in the post-war period. If the population pressure argument was correct, one would expect to see steady declines in forest cover correlated with steady increases in population; since this does not appear to be the case, the population pressure argument by itself is incapable of explaining the very rapid rates of deforestation that occur in different places and times. Finally, we note that increasing populations as a cause of deforestation does not apply to a large number of areas, particularly North America, Western Europe and Japan. In other words, it is not inevitable that increases in population must lead to deforestation.

While population obviously plays a rôle in tropical deforestation, a discussion which ignores the context in which population growth is occurring is incomplete. The major timber producing countries of Southeast Asia have had rapid population growth but, in addition, they have experienced elite control of government, and deliberate government policies which have encouraged large-scale commercial forest exploitation. Blanket statements about population pressure which disregard the skewed distribution of income, poverty and vested interests of the elites are of very little help in our attempts to understand the causes of tropical deforestation.

10.5 Conclusion

It is clear to us that what is needed is more in-depth studies of tropical deforestation in individual countries. In fact, one of the most striking aspects of the literature on tropical deforestation is the almost complete lack of detailed studies at the national level (Grainger 1986, Rudel Ch. 7). General statements about the causes of tropical deforestation which are based on cross-national statistical tests are of little value since that they virtually nothing to say about deforestation within individual countries. In addition, cross-sectional regression analyses with forest cover as the dependent variable have not stated the problem correctly.

National case studies of tropical deforestation are needed to capture the unique aspects of each country. Commercial logging is very important in all of Southeast Asia; it is negligible in Brazil (although increasing). Indonesia and the Philippines are archipelagos, Brazil is a large, compact country. The forests of Southeast Asia are "owned" by their respective governments, in the Brazilian Amazon Basin many of the forest lands have been titled and given or sold to settlers. Lastly, population densities in countries undergoing deforestation range from low to high. We do not accept that population pressure is the most important driving force of deforestation and we also do not feel that its proponents have proven their case. As the case study from the Philippines has revealed, the two most important factors in deforestation from 1970 to 1980 were logging in 1970, and the spread of agriculture. Both the granting of concessions and the spread of primarily subsistence agriculture are reflections of a development process which has concentrated resources in the hands of a small elite and left the majority of Filipinos in poverty. While we are aware of the severe data problems faced by researchers in the field of tropical deforestation, we feel that many of these could be overcome if more strenuous efforts were directed toward detailed studies at the national level. General statements to the effect that increasing populations are the sole or main cause of tropical deforestation are simply inadequate in terms of their ability to articulate the process of deforestation clearly and/or to suggest remedies.

11

Incentives for tropical deforestation: some examples from Latin America

Dennis Mahar and Robert Schneider

This chapter aims to analyze the nature and impact on tropical deforestation of various policy and non-policy incentives in the Latin American context. The chapter also examines areas of possible policy reform and new approaches to environmental management at the frontier. The analysis is based largely on two independent studies conducted by the authors on the effects of certain government policies on deforestation in the Brazilian Amazon (Mahar 1989a, and Schneider 1992b). The majority of discussions and examples come from these studies, although similar work for other countries is drawn upon whenever possible.

11.1 Migration to the frontier

High overall population growth rates, and extensive migration, have been distinguishing demographic characteristics of post-war Latin America. Between 1950 and 1990, the region's population nearly tripled, rising from 166 to 448 million inhabitants. Brazil alone accounted for about a third of this increase, its population rising from 53 million in 1950 to 150 million in 1990. Even though total fertility rates have generally declined in Latin America, the momentum of population growth will continue due to the young age structure and the large numbers of women still of child-bearing age.

Mounting population pressures on the land in areas of older settlement have contributed to migration to, and the deforestation of, remaining frontier areas. However, most research on this issue indicates that population pressure, by itself, has probably not been a major factor. (Although the situation in some of the more densely-populated countries of Central America is perhaps the clearest exception to this generalization). In most of Latin America, "surplus" rural populations have, in fact, been much more likely to emigrate to the cities than to the agricultural frontier. In Brazil, for example, 16 million rural inhabitants migrated to the cities in the 1970s versus only 770,000 to the Amazon frontier; given further declines in fertility and high rates of urbanization which occurred during the eighties, it is likely that demographic pressures to migrate to frontier areas in Brazil continue to ease. As a result of a combination of high underlying population growth rates and intense rural-to-urban migration, the urban population of Latin America has increased from 26 per cent of the total in 1950 to around 70 per cent today, making it by far the most urbanized of the world's developing regions.

However population growth has played a rôle in explaining migration to the frontier and the deforestation that has ensued. In many countries, underlying demographic pressures in rural areas have been superimposed on skewed land distributions, poverty, inadequate health and education services, poor infrastructure and lack of access to agricultural technology. Under such conditions, Latin American governments have frequently encouraged migration to frontier areas in an effort to deflect mounting pressures on the region's burgeoning cities and to diffuse potential social tensions. In countries with extensive unprotected borders, like Brazil, a desire to achieve effective national sovereignty over remote frontier areas has been an additional driving force.

One other factor that has tended to "push" migrants to frontier areas has been reductions in employment opportunities in older settlement areas brought on by technological change. For example, the rapid spread in the mid-1970s of a mechanized system of soybean production in southern Brazil was an important factor leading to the sharp increase in migration to Rondônia observed in subsequent years. A survey carried out in one of the larger colonization projects in that state found that 70 per cent of the settlers from the southern state of Parana (site of a major shift from labour-intensive coffee production to mechanized soybean cultivation during the 1970s) had migrated to the frontier because of agricultural mechanization or the inability to support their family on a small piece of farmland.

While technological improvements in southern Brazil provided a strong impetus for farmers to migrate to the Amazon frontier, in other countries of Latin America improving agricultural productivity has actually resulted in a sharp reduction in the incentive to migrate to frontier areas. This is a major conclusion of a recent paper by Southgate (1991). In this paper the author showed, through regression analysis, that countries (such as Chile) experiencing major productivity gains in agriculture were, through this process, able to contain potential pressures on frontier areas resulting from rising demands for agricultural exports coupled with ongoing population growth. Ecuador, with inadequate investment in agricultural research, was unable to do so: between the mid-1960s and mid-1980s, two-thirds of the increased crop production in that country were accounted for by frontier expansion.

Technological change and other factors can intensify land price differentials within a country or region. When these prices vary significantly, farmers with small parcels in high-price areas have a strong incentive to sell their small plot (particularly when there are other pressures to sell) in order to buy a larger plot in a low-priced area. The larger the differential, the larger the economic incentive to make the transfer. For example, when road building in Brazil created an abundance of cheap land in the Amazon region, the pronounced skewing of relative north-south land prices created extraordinary incentives to migrate. A small farmer in the southern part of Brazil contemplating selling out to his large neighbour in 1970 could double the size of his farm by moving north. Five years later, however, he could expand his farm size tenfold through such a transaction. This differential remained about 1 to 10 for the next eleven or so years, even reaching as high as 1 to 15 in 1982.

11.2 The paramount importance of roads

Little migration to frontier areas, and hence little deforestation, takes place in the absence of roads. Before the first interregional highways were built in the mid-1960s, for example, the population of the Brazilian Amazon was overwhelmingly concentrated in the two major regional cities, Belem and Manaus. The hinterland was a virtual demographic void. Although the Amazon region's centuries-old isolation from the more dynamic parts of the country had arguably retarded its economic development, it had also protected the rainforest (as late as 1975, only about 0.6 per cent of the Brazilian rainforest had been cleared). Similar situations prevailed in other frontier areas of Latin America before the advent of roads. The reasons for this are not difficult to fathom. As Bromley aptly notes in reference to the Oriente of Ecuador: ". . . the colonist in roadless areas not only suffers great social isolation, but also a marked economic disadvantage relative to producers in other parts of the country" (Bromley 1981: 22).

By and large, the settlement and subsequent clearance of frontier lands in Latin America have closely followed the expansion of the road network. Indeed, we would argue that road-building is the single most powerful element in the deforestation of frontier areas in Latin America, and it should be noted that road-building is not always carried out exclusively by governments. In Ecuador, for example, the early penetration roads into the environmentally fragile Oriente region were largely built by multinational oil companies (Bromley 1981). Based on observations in Panama, one author has estimated that 400 to 2000 hectares of deforestation occurs for each kilometre of roads built in forested areas. Moreover, the effects of providing all-weather, overland access to frontier areas are often cumulative and irreversible. That is, the increase in population associated with the completion of primary roads usually generates demand for secondary and feeder roads, which in turn attract more population. This process has been documented throughout the Brazilian Amazon, in the eastern lowlands of Ecuador, Peru and Bolivia, as well as in Central America.

Several examples of the effects of road building in the Brazilian Amazon may be cited. The population increase in the zone of influence of the region's first inter-regional highway (Belem-Brasilia), completed in the mid-1960s, is estimated to have been between 300,000 and 2,000,000 during the 1970s. The number of migrants entering Rondônia annually more than doubled after the state's main highway was paved in the early 1980s. In both cases, the surge in migration following road construction was followed by massive deforestation; according to official estimates, the deforested area of Rondônia increased from 4000 km^2 in 1978 to 30,000 km^2 a decade later. In the Brazilian state of Para a highway opened in the late 1960s resulted in the cleared area increasing from 300 km^2 in 1972 (0.6 per cent of the area), to 1700 in 1977 (3.6 per cent), to 8200 (17.3 per cent) in 1985 all within an area of 47,000 km^2 (small by Amazonian standards, but about the size of Switzerland).

11.3 Who owns the land?

Land tenure rivals road construction in terms of its impact on decisions to deforest. Secure titles to the land are relatively rare in frontier areas of Latin America. Estimates for the Brazilian Amazon, the largest frontier area in the region, indicate that only about 11 per cent of the land was titled as of the early eighties (Ozario de Almeida 1992). This physical abundance of land with essentially "open access" makes the land practically "valueless" from a private economic standpoint. In other words, land in remote frontier areas can usually be obtained at little or no financial cost to the settler or buyer. In a market economy, when land is cheap and abundant, and access is not controlled, it is in the economic best interest of the individual to "mine" the land and move to new land when the first parcel has become depleted. In private financial terms, this approach makes more sense than managing the land sustainably (at a lower rate of production) or importing artificial nutrients to increase production at a sustainable rate.

The open access nature of land-abundant frontier areas is due to lack of clear rights to property. When land is abundant and essentially a "free good", it generally does not make sense from a private economic perspective to invest much in demarcating, protecting and defending any particular parcel of land. Under such conditions, it is usually cheaper in terms of time, energy, or dollars simply to occupy another parcel. Few conflicts would exist in any case, as persons would find it easier to occupy newly-opened lands than to dispute existing claims.

When government land tenure services and mechanisms of settling disputes are weak or non-existent, one relatively cheap method of signalling to others one's intent to put the land to economic use is the clearing of boundary zones or some other permanent marking of the land. As population pressure on the land increases, it is no longer sufficient to mark the land. It then becomes necessary to monitor whether or not the land is being invaded. In forest areas, this monitoring is greatly facilitated when the land is cleared of trees, which both improves visibility and increases the prospective owner's presumptive claim on the future use of the land.

A policy that has surely encouraged inappropriate land-use is the acceptance by land reform and titling agencies of deforestation as evidence of land improvement. This typically occurs when a migrant in an official settlement project or an invaded area can obtain rights of possession simply by clearing the forest. Settlers who wish to engage in extractive activities that do not disturb the forest, such as rubber tapping or the gathering of nuts, are particularly disadvantaged by this policy. Moreover, it encourages the indiscriminate deforesting of both good and poor lands, particularly if the purpose is simply to acquire land for speculative purposes.

Land-titling regulations which recognize deforestation as evidence of effective occupation are (or were) in effect in Costa Rica, Honduras, Panama, Brazil, Ecuador, and other Latin American countries. Prior to 1986, for example, settlers in Costa Rica did not have to prove that their land was acquired legally if less than 50 per cent of the parcel was covered by forest. In cases where documentation proving legal ownership could not be obtained, claimants would simply cut 50 per cent or more of the forest cover (Peuker 1992: 9–10, and Southgate et al. 1991, for examples from Ecuador). In much of Brazil, the national land settlement agency still determines the geographical extent of possession rights by multiplying the cleared area by three. Once obtained, these rights of possession can be sold either formally or informally, depending on whether the migrant has occupied the land long enough to qualify for a definitive title.

Land speculation, which is frequently a major reason for obtaining effective possession of land in frontier areas, is yet another factor which can encourage excessive deforestation. When land is initially abundant and cheap, demand and migration is high, and the government continues to invest in improvements and infrastructure, the potential gains from land speculation can be very high. For example, the real prices of land in the Brazilian state of Rondônia soared in the mid-1980s, largely in response to continued migration and improvements in roads and other infrastructure financed through a major regional development programme known as POLONOROESTE. Calculations made by the FAO/World Bank Cooperative Program (FAO/CP) show that it was possible during the 1980s for speculators to net the equivalent of US$9000 if they cleared 14 hectares of forest, planted pasture and subsistence crops for two years, and then sold the rights of possession acquired by doing so. This constituted a large sum of money in Rondônia where the average daily farm wage was equivalent to less than US$6.

11.4 Directed settlement

As already mentioned, government policies and programmes to colonize frontier areas, particularly those along international borders, have often stemmed from a desire to secure remote regions in the face of real or perceived threats to national security through foreign intrusion. For example, a key motive for "Operation Amazonia", a scheme in the mid-1960s by the new Brazilian military government to develop and occupy the region, was largely geopolitical. Several neighbouring countries (particularly Peru and Venezuela) had already initiated programs to occupy and develop their respective Amazon regions, and Brazil's military leaders were anxious to ensure national sovereignty. A 1970 decision by Brazilian President Emílio Garrastazú Médici to construct a 5400 km east–west transamazon highway was partly due to fear of foreign domination in the region.

Colonization schemes, which generally accompany or follow road construction, can be an important factor in deforestation. When the settlement area has at least some high quality soils, and demographic and other pressures are strong, colonization schemes can act as a powerful attraction to migration. The likelihood that socially-marginal deforestation will occur increases substantially when provision of infrastructure in such schemes is not preceded by good land-use capability studies, and when government institutions are not strong or well-funded enough to control where and how settlement occurs.

There are relatively few examples of successful directed colonization projects in frontier areas of Latin America (see Nelson 1973, Jones 1990). A case in point is the massive transamazon colonization program in Brazil which was started in the early 1970s. Initial plans called for settling 70,000 families, mainly drawn from the country's chronically-poor Northeast, between 1972 and 1974. However, only about 5700 families were effectively settled during this period, and only about 8000 by the end of the decade. Many factors contributed to this shortfall. Perhaps most importantly, planning for the project was carried out hastily and was not based on good land-use capability studies. Migrants were thus frequently settled on plots with poor quality soils and topography unsuitable for agriculture. As a result, cleared land eroded rapidly, which necessitated the clearing and burning of additional forest to restore lost soil fertility. Ironically, the failure of this project to attract anywhere near the target number of settlers ultimately served to limit its impact on the forest.

11.5 Policy-induced price distortions

Policy-induced price distortions have also encouraged excessive deforestation in Latin America, although their importance has often been exaggerated in the literature. The nature of these policies varies from country to country, but generally falls into the categories of fiscal incentives, directed (frequently subsidized) credit, minimum purchase prices for crops, and uniform fuel pricing. Each of these is discussed briefly below.

Special *fiscal incentives* have been employed by several Latin American governments to promote the establishment of various types of private undertakings in frontier areas, including industrial, agricultural and livestock projects. In Brazil, legislation passed in 1963 allowed registered corporations to take up to a 50 per cent credit against their federal income tax liability if the resulting savings were invested in projects located in Amazonia. The tax credit mechanism proved very attractive to investors, and by late 1985 over 600 livestock projects were benefitting from this legislation (Gasques & Yokomizo 1986). These projects have been an important source of deforestation in certain parts of the Brazilian Amazon. However, their relative contribution to deforestation in Amazonia as a whole has clearly been much smaller; probably less than 10 per cent of the total. In recent years, the use of fiscal incentives for livestock development in Amazonia has been sharply curtailed, although not totally eliminated, by the Brazilian government.

Other features of the income tax system can also encourage deforestation. For example, when agricultural income is practically exempt from taxation, demand for land among individuals with high incomes (as a means of sheltering income) is increased. These provisions thus contribute to a more rapid conversion of forest to agricultural uses as well as to land price appreciation and concentration of land ownership (Binswanger 1991). In Brazil, the differential between the taxation of agriculture and the taxation of profits from other sources created a powerful incentive for corporations and high-income individuals to engage in agriculture as a tax shelter. First, favourable income tax treatment made private and corporate investors willing to accept a smaller pre-tax rate of return in agriculture than in other enterprises. Second, it created a tendency to capitalize into the value of land the difference between the post-tax profit in agriculture and other enterprises.

Special provisions of income tax laws are not the only fiscal incentives to deforestation. In many countries of Latin America, land taxes are constructed in such a manner as to ostensibly promote a more "productive" use of land. In practical terms, this is translated into differential tax rates, with "unused" forested areas being taxed at a higher, or even the highest, rates. This tends to encourage deforestation, if only for conversion to "low-maintenance" or "low-management" pasture. For example, the rural land tax in Brazil, administered by the national government, is currently assessed at a maximum rate of 3.5 per cent of the market value of the land; a required 50 per cent forest reserve is exempt from taxation. Reductions of up to 90 per cent in the basic rate are given according to the degree of utilization of land (that is, the proportion cleared) and certain "efficiency" indicators – for example, crop yields, cattle stocking rates, rubber extraction per hectare – established by the government. Fortunately, the land tax has little influence on the rate of deforestation in the Brazilian Amazon, mainly because the landowners themselves declare the value of their land as well as the efficiency of its use. For example, only about half of all registered landowners in Rondônia paid any rural land tax in 1986, and the average payment of those who did was equivalent to only US$5.

Subsidized credit lines, like fiscal incentives, increase private rates of return to investment and thereby encourage activities – and by extension, deforestation – which would not be undertaken if credit were priced at market rates. In Brazil, subsidized credit was employed on a large scale during the 1970s in an effort to stimulate the economy after the first oil crisis. The volume of such credit committed in Amazonia increased tenfold in real terms between 1974 and 1980. Although much of this credit went to support crop production, the livestock sector also received large increases. The availability of subsidized credit undoubtedly facilitated the acquisition and deforestation of large tracts of land in Amazonia, particularly in the latter half of the 1970s when such credit was frequently used in tandem with fiscal incentives. Subsidized rural credit was effectively phased out in the 1980s as a result of efforts to restore internal balance in the Brazilian economy.

Credit has also been a favoured instrument for encouraging the formation of cattle pasture in Panama, a major cause of recent deforestation in that country (Ledec 1992b). Both the absolute and relative amounts of credit going to livestock have been very high. During the 1970s alone, government and commercial banks provided $542 million in credit to cattle ranchers. It is estimated that 7–10 per cent of Panama's annual deforestation is due directly to government bank cattle credit; an unspecified additional amount of deforestation is financed through commercial credit to the cattle sector. In contrast to Brazil, livestock credit in Panama has been a more important factor in the loss of remaining forest fragments than forest conversion in frontier areas. However, given the high environmental value of these remnants, this credit-financed deforestation constitutes a serious policy issue.

In Brazil, and elsewhere in Latin America, governments have often supported marginal or uneconomic farming through *guaranteed minimum purchase prices* for certain types of crops. In frontier areas, high transport costs would normally make agricultural production for distant markets in the more developed parts of the country uncompetitive with "locally" produced food. But price supports can make farming (and the prior necessary forest conversion) in remote areas privately profitable, even if they are economically inviable. In the Brazilian Amazon, a tremendous locational disadvantage has been neutralized historically through crop purchases under a minimum price programme. However, the fragility of such a system became apparent in 1989 when resource constraints resulted in major declines in government purchases of rice and maize in the frontier states of Rondônia and Mato Grosso. Without these purchases, farmers found that they were left with much smaller markets.

The *uniform price system for fuels* is an additional wedge between economic viability and private profitability, and is yet another mechanism used by the Brazilian government to offset the higher transport costs from frontier areas to market. Under this programme, the price of fuel is regulated so as to be the same in all regions of the country, regardless of the cost of transportation. This serves as a fiscal subsidy to those firms or individuals which choose to locate in remote areas, where the cost of transporting fuel would be a significant part of its purchase price. As the price of fuel is probably higher for everyone (to incorporate the costs of transporting a proportion of it to frontier areas), those in more settled areas are also in effect subsidizing those in areas far from markets.

11.6 Lack of support for parks and reserves

In most of Latin America, the designation of a given area as a park or reserve has provided little real protection to the forest. Indeed, Ledec (1992a) concludes that, among Latin American nations, only Costa Rica has achieved some semblance of success in limiting deforestation by setting aside legally protected areas. Many parks and reserves in Latin America have not even been demarcated, and agencies entrusted with their management tend to be chronically under-funded.

In Brazil, for example, the government has officially designated over one million square kilometres, approximately one-fifth of the Amazon region, as either reserves or parks. However, at present there is only one park guard per 6161 km^2 of parks and reserves. Even in Costa Rica, many of the legally protected areas are said to be threatened due to weak enforcement capabilities and under-funding. Only a few of these protected areas earn revenues and there are no programmes to share these revenues with local populations. In addition, cleared land within protected areas is acquired by government at a price about twice that paid for forested land. This latter feature quite clearly encourages prospective sellers to deforest in anticipation of receiving a higher price for their land (Peuker 1992, Lutz & Daly 1990).

11.7 Recommendations for policy reform

As the previous discussion shows, the most important government activity affecting deforestation in frontier areas is the provision of effective overland access to new land. Although some private road-building takes place, primarily in conjunction with forestry, road construction and maintenance would be extremely limited without government assistance. In future, any decision to construct a road into a new area should be carefully evaluated against a set of guiding principles. Although an exhaustive review of these principles is beyond the scope of this paper, the basic guidelines would include:

(a) the road should be economically justified on the basis of benefit–cost analysis that allows for environmentally sustainable activities and includes domestic environmental costs;

(b) mechanisms should be sought to include an analysis of global environmental costs and benefits (i.e., global warming, biodiversity loss), and to allow for the financing of global benefits through international routes (e.g., the Global Environment Facility);

(c) alternatives should be identified that may give the same or similar benefits, but at lower environmental (domestic or global) costs;

(d) any analysis should be preceded by thorough land-capability studies and biological inventories, and should explore the environmental implications of various alternatives;

(e) new roads should be consistent with any zoning or land-use restrictions, including biological and indigenous reserves.

Without secure tenure, or the ability to exclude others, a settler has little incentive to conserve the land or to enhance its productivity through various fixed investments. This, together with a policy of recognizing deforestation as a form of land improvement, and therefore grounds for granting possession, has tended to support deforestation trends. The strengthening of government land titling services, including enforcement of claims, should therefore be of highest priority. As part of this strengthening process, all policies which recognize simple deforestation – that is, deforestation for the sake of deforestation – as a form of "land improvement", and therefore grounds for granting possession or title, should be abolished. In addition, the land-use potential or conservation value of lands should always be considered before rights of possession or use are granted. Areas with high conservation value should be declared reserves, with no granting of use; intermediate areas, with poor soils but with potential for extractive activities, should be available for use under long-term concessional arrangements, or ownership with restrictions on use. This approach could be a part of a land-use zoning exercise or simply a policy of land-titling agencies.

Land-use zoning as a policy tool has begun to take hold in many countries of

Latin America and is now being carried out by several Brazilian states. The basic idea is that decisions to provide roads and other infrastructure, or to promote certain activities through special financial incentives, should be based on reliable data on the physical environment. Detailed land surveys are commonly used to determine the productive potential of the land and its conservation value. Under an environmentally oriented zoning plan, lands found to have limited agricultural potential, high conservation value, or both, would normally not be provided with roads or other infrastructure. This would in effect put such lands off limits to most types of development. The intensity of development permitted in other areas would be inversely related to the environmental fragility of these areas.

One variant of the land-use zoning concept currently in vogue in Latin America is the *extractive reserve*. The inspiration for such reserves came from Brazilian rubber-tappers who viewed the accelerated frontier expansion and associated deforestation of the seventies and eighties as a major threat to their livelihoods (Allegretti 1988). Under such schemes, the government usually first assumes ownership of a given area, which may be a large stand of rubber or Brazil nut trees. It then cedes long-term exclusive use rights to practitioners of traditional extractive activities living within the area. At the same time, steps are also taken to improve processing and marketing facilities. Extractive reserves have enjoyed great favour among some environmentalists who see them as a socially responsible way of preserving the rainforest. More recently, however, they have been criticized by others as being economically unsustainable (see for example, Browder 1992).

Although land-use zoning is a useful tool, it has yet to prove its usefulness for resolving land conflicts and for halting inappropriate deforestation in fragile areas. To be successful, land-use zoning requires the existence of strong agencies to implement policies affecting land-use – for example, agencies administering fiscal incentives (to monitor their use), agencies entrusted with the protection of reserves, agencies entrusted with forest concessions, agencies entrusted with tax collection, and so forth. This, in turn, implies an adequate and timely flow of funding from public coffers. Ultimately, however, the success or failure of the current attempt to apply agro-ecological zoning in Amazonia will depend on the strength and depth of political support at both the national and local levels.

Fiscal incentives and differential income tax rates have probably contributed to increasing demand for agricultural land in countries like Brazil, particularly among corporate investors looking for a method of hiding income. Moreover, certain features of land tax codes may have increased the tendency to clear rural land investments to prove "productive use". These policies should either be eliminated or modified with a view toward strengthening environmental objectives. For example, land taxes should be revised to account for the environmental services provided by forests and other natural ecosystems as "productive use". Moreover, taxes should be structured so as to penalize those who engage in environmentally damaging or unsustainable activities. Of course, tax reform would have only limited effectiveness in the absence of parallel improvements in tax administration.

As frontier areas are usually far from markets, uniform fuel pricing policies can serve to subsidize activities (including agricultural and other activities requiring deforestation) in these areas. Although the link between uniform fuel pricing and deforestation may be indirect, it does contribute to making some frontier activities personally profitable that may not be economically viable, much less environmentally sustainable. In the long term, uniform fuel pricing policies distort transport costs and promote the uneconomic location of enterprises. They should as a rule be abolished.

Many development activities can result in an unequitable distribution of the environmental costs and benefits between individual countries and the global community, with countries reaping most of the benefits and the global community bearing the brunt of the cost of global environmental externalities. Examples would include global warming from the accumulation of greenhouse gases, and the loss of biodiversity – both products of deforestation. Conversely, the costs of conservation – in both direct costs of management and indirect costs of opportunities foregone – are normally borne by the country while certain benefits accrue globally.

One way to rectify the situation is through an examination of global incremental benefits and costs, as well as domestic ones, for any development project, particularly large projects in frontier or relatively virgin tropical forest areas. This analysis should include:

(a) an accounting of global environmental externalities, as well as domestic ones, as a result of development activities;
(b) an examination of the global benefits of conservation versus the benefits to the country of the development activity;
(c) an identification of alternatives which would maximize both global and domestic benefits, relative to global and domestic costs. When this requires that an individual country must sacrifice benefits, or incur "extra" costs, the global community should pay these incremental costs.

Mechanisms are becoming available to finance global benefits. The Global Environment Facility – administered by the World Bank in co-operation with UNEP and UNDP – is one such mechanism. There are also precedents, such as in the Netherlands where an energy company was permitted to avoid costly pollution control costs at home by investing in pollution control in Poland and in tree plantations in Latin America. Likewise, industries could invest in forest conservation or park protection. There is no mechanism, and no global consensus, for a country to pay the global community for damage to, or loss of, global environmental services and benefits.

11.8 Conclusions

While forest conversion continues unabated in many parts of Latin America, recent evidence suggests a dramatic reduction in the pace of deforestation in the Brazilian Amazon. Estimates based on satellite images show that deforestation in that region, which peaked at around 40,000 km^2 in 1987, has fallen to an annual rate of around 10,000 km^2 in the 1990s. Some government officials have attributed this decline in annual deforestation to changes in the policy framework such as the curtailment of fiscal incentives for livestock in environmentally fragile areas, and to improved enforcement of environmental legislation. These policy related measures have certainly had some positive impact. However, it seems more likely that the recent slowdown in deforestation is related to demographic trends – especially the fall in absolute terms of the country's rural population which has drastically reduced the pool of potential migrants to the Amazon region – coupled with economic recession. This second factor has, among other things, sharply reduced the availability of public resources for building roads and other physical infrastructure in forested areas and elsewhere.

While the declining pace of deforestation in Brazil is good news indeed, it should not lull decision makers into a false sense of security. As long as "fundamental" incentives to deforest continue to exist in a country – particularly "open access" situations – most changes in government policy aimed at reducing deforestation (with the exception of eliminating new road construction in environmentally sensitive areas) are likely to be of limited effectiveness. Therefore, governments seriously interested in reducing deforestation would be better advised to concentrate on implementing policies aimed at maximizing quality employment opportunities in regions far removed from the rainforest. Such policies might include land reform in areas of older settlement; greater investment in education, health and family planning; and the elimination of tax and credit systems which favour capital.

12

An econometric model of
Amazon deforestation

Eustáquio Reis and Rolando Guzmán

12.1 Introduction

Based upon cross-section data at municipal level, this chapter specifies, estimates, and simulates an econometric model of Brazilian Amazon deforestation. Since one of the main concerns relating to deforestation is the release of carbon dioxide (CO_2), the chapter also simulates the contribution of Amazon deforestation to carbon releases. The model consists of three blocks of equations: in the first, deforestation – distinguished by vegetation types – is determined by major economic activities; in the second, the relationship between vegetation type and biomass content determines carbon dioxide emissions caused by deforestation; finally, in the third, the growth rates of population and of major economic activities are assumed to depend only on their respective spatial densities, thus allowing projections and simulations of the geographic distribution of economic activities, and their effects on deforestation and CO_2 emissions.

The chapter improves Reis & Margulis (1991) in three major aspects: the theoretical specification of the model; the database, which was enriched by better information on agricultural output, vegetation cover, transportation conditions, as well as on the spatial characteristics of data; and the regression analysis which takes into account spatial autocorrelation phenomena, thus allowing a better diagnosis and treatment of problems resulting from the omission of variables, measurement errors, and improper specification.

The chapter is organized in seven sections. After the introductory remarks on the contribution of Amazon deforestation to CO_2 emissions, the first section surveys early econometric results on tropical deforestation. The second section derives the basic equations of an economic model of Amazon deforestation. The third section discusses estimation issues, with particular attention to spatial autocorrelation. The fourth section describes the database used and is followed by the presentation of

Table 12.1 Amazon deforestation and carbon dioxide emissions.

Year	Area (km^2)	(%)	Annual increase (km^2)	(%)	Annual CO$_2$ emissions (10^9t)	(% of world)
1978	152,910	3.1	–	–	–	–
1988	377,633	7.7	22,472	9.5	0.31–0.45	4.4–6.2
1989	401,433	8.2	23,800	6.3	0.33–0.48	4.6–6.6
1990	415,251	8.5	13,818	3.4	0.19–0.27	2.7–3.8
1991	426,351	8.7	11,100	2.7	0.15–0.22	2.2–3.1

Source: INPE 1992; carbon dioxide emissions estimated by the author.

the estimation and simulation results. The concluding section suggests research extensions and further developments of the model.

The climatic and ecological consequences of Brazilian Amazon deforestation are among today's leading global environmental concerns. The main reasons for concern are the contributions of tropical deforestation to CO$_2$ emissions and to the loss of biodiversity. In what follows we address only the first of these issues.

Evidence of the importance of Brazilian Amazon deforestation to CO$_2$ emissions is presented in Table 12.1. The contribution of Amazon deforestation to global carbon emissions is specially significant if one considers that agricultural activities in the region represent less than one per cent of Brazilian Gross Domestic Product (GDP). This makes the slowdown of Brazilian Amazon deforestation one of the most cost-effective ways to reduce carbon dioxide emissions (Nordhaus 1991, Hoeller 1991, Mors 1991), although this kind of statement tends to underestimate the costs of implementing incentives schemes to compensate local population for the losses in economic opportunities (for example, Reis & Margulis 1991).

12.2 Early econometric results on tropical deforestation

Despite the importance of tropical deforestation to the greenhouse effect, econometric analysis of demographic and economic factors is lacking. Projections are usually based upon naive extrapolations of past trends, often leading to significant overestimates (INPE 1990, Schneider 1992a). As a consequence, considerable uncertainty exists with regards to both future rates of deforestation and the costs of halting it.

To take an authoritative example, in the IPCC report (IPCC 1991) the driving mechanism of the projection model is the simple assumption of a unit elasticity of deforestation in relation to population (lagged by 20 years). Moreover, the distribution of deforestation between closed broadleaved and other kinds of forests is assumed to be in proportion to the area covered by each kind of forest. In both cases, it is only the lack of knowledge concerning relevant parameters that justifies the simplistic assumptions adopted.

Econometric results on elasticities of deforestation are scanty. Table 12.2 provides an incomplete survey of them, which shows major differences of specification, sampling, variables, geographic aggregation, and measurement of data. Furthermore, in most cases, parameters were not explicitly derived from theoretical models, thus making comparisons even more difficult. Note also that equations are under-identified. Thus, population elasticities reflect both the effects of the supply of labour and of the demand for output.

Despite these problems, the results seem reasonable at first sight. Thus, comparing Latin America and Southeast Asia, population growth and logging exert greater pressure on deforestation in Southeast Asia, while elasticities of the road variable are greater in Latin America; the impact of agriculture is practically the same in both regions, if cattle raising elasticity is not taken into account. These differences can be justified by the lower population densities and the more recent settlement of tropical forests of Latin America. As a caveat, however, note that the larger geographical

Table 12.2 Survey of econometric results for elasticities of deforestation in relation to major economic activities.

Author	Panayotou	Southgate	Kummer	Kummer	Reis et al.
Region	Thailand	Latin America	Philippines	Philippines	Brazil Amazon
Dependent variable	Deforestation	Agricultural area	Deforestation	Deforestation	Deforestation
Geographical unit	Municipality	Country	Province	Province	Municipality
Period	1973–82	1970–80	1970–80	1970	1985
Data	Panel	Panel	Cross-section	Panel	Cross-section
Method	OLS	OLS	OLS	OLS	OLS
Specification	Log-Log	Growth rates	First differential	Log-Log	Logistic
Variable	*Elasticity estimates (t values in parentheses)*				
Population	1.51(9.7)	0.25(3.8)		0.54(na)	0.30(2.7)
Roads	0.11(1.4)		0.23(2.4)	0.28(na)	0.28(4.7)
Agriculture	0.32*(1.7)		0.41(4.2)		0.40(3.2)
Logging	0.41*(4.1)		0.32(3.2)		0.04(1.0)
Productivity	−0.38(1.9)	−0.20(6.0)			
Cattle herd					0.11(1.83)
R^2	0.80	0.67	0.49	0.58	0.84
D.F.	55	18	64	66	165

Note.: * assuming a supply price elasticity equal to 1.
Sources: Panayotou & Sungsuwan 1989, Southgate et al. 1992, Kummer 1991, Reis & Margulis 1991.

units of the Latin American samples tend to weaken the relationship between population and deforestation, introducing a downward bias in the value of elasticities.

The results surveyed suggest that the IPCC (1991) assumption of a unit elasticity of deforestation in relation to population is probably exaggerated. Indeed, most of the remaining tropical forests are in Latin America where population elasticities seems to be significantly lower than unity.

In any case, the lesson to be derived from this brief survey points to the precarious state of the art of the economics of tropical deforestation, and to the contribution which could be brought by econometrics. In this way, it indicates the urgent need for further research efforts on data gathering, model specification and estimation techniques.

12.3 An economic model of Amazon deforestation

Three blocks of equations comprise the model. The first one relates deforestation to economic activity; the second block links deforestation to vegetation cover and to CO_2 emissions; and the third one specifies generating functions for the spatial growth of population and economic activities.

The first block is based upon an aggregate production function for major agricultural activities - supposedly the main source of deforestation in Brazilian Amazon. The derived demand for cleared land in agriculture is determined by profit maximization. Output prices are considered exogenous to the model; wages are determined by demand and supply of labour, and land prices by clearing costs. A logistic function relates deforestation to land cleared in agriculture and to the land requirements of other economic activities. A Cobb–Douglas production function relates agricultural output (Q) to the inputs of labour (L) and cleared land (C). Profit maximization – given output prices, wages (w), and clearing costs (k) – leads to the following derived demand for cleared land (C_d) and labour (L_d), respectively:

$$C_d = ((1-a)/a)^a . w^a . k^{-a} . Q \qquad (1)$$

$$L_d = (k. a/(1-a). w)^{1-a} . Q, \qquad (2)$$

where $0 < a < 1$.

Output (Q) is a long run concept defined as a composite index of cattle herd (H) and trend output for temporary and permanent crops (Y) as follows:

$$Q = H^h . Y^{1-h}, \qquad (3)$$

where $0 < h < 1$.

The supply of labour (L_s) is assumed to increase with population (N) and wages, and to decrease with transport costs proxied by a vector of variables specified as spatial discount factors which include the distances to local and national (M) markets, and the networks of roads and rivers (R):

$$L_s = w^b.\ N^g.\ \exp(e_1.\ R - m_1.\ M), \tag{4}$$

with $b, g, e_1, m_1 > 0$.

Equilibrium in the labour markets leads to:

$$w = (k.\ a/(1-a))^{j.\ (1-a)}.\ Q^j.\ N^{-j.\ g}.\ \exp\{-j.\ (e_1.\ R - m_1.\ M)\}, \tag{5}$$

where $j = 1/(1+b-a)$.

Furthermore, the free availability of land in the region makes it legitimate to assume that deforestation decisions are short sighted (Panayotou & Sungsuwan, 1989). Accordingly, land prices are assumed to depend solely on clearing cost which, in turn, is assumed to depend only on the vegetation cover as proxied by the density of forest (F/A), and transport cost which is specified in the same way as in (4):

$$k = \exp(f.\ (F/A)+m_2.\ M - e_2.\ R), \tag{6}$$

with $f, e_2, m_2 > 0$,

where A is the geographic area of municipalities.

Substituting (5), (6) and (3) in (1), and taking logarithms, the reduced form of the derived demand for cleared land in agriculture is:

$$C = a/(1-a)^{a.\ (j.\ (1-a)-1)Q j.\ (a+1)}.\ N^{-j.\ a.\ g}. \tag{7}$$

$$\exp\{j.\ (e_2.\ (1-a)-e_1).\ R + j.\ (m_1+ m_2.\ (1-a)).\ M\}.$$

$$\exp\{a.\ (j.\ (1-a)-1).\ f(F/A)\}$$

or

$$C = B_0.\ Q^{B_1}.\ N^{B_2}.\ \exp(B_3.\ (F/A)+B_4.\ R+B_5.\ M), \tag{7'}$$

where:

$B_0 = ((1-a)/a)^{ab/(1+b-a)} > 0$ $B_3 = -a.\ b.\ f/(b-a+1) < 0$

$B_1 = (b+1)/(b-a+1) > 0$ $B_4 = a(be_2-e_1)/(b-a+1) > 0$ if $e_2.\ b > e$

$B_2 = -a.\ g/(b-a+1) < 0$ $B_5 = -a(bm_2-m_1)/(b-a+1) > 0$ if $m_2.\ b > m$.

Finally, the extent of deforestation *(D)* is determined by land clearing in agriculture, logging activities *(W)* for timber and firewood production, and by all kinds of urban activities as proxied by urban population *(U)*. The specification adopted is a logistic function relating density of deforestation *(d)*, defined as the relation between deforested area and total geographic area, to the economic activities described above:

$$\log(d/(1-d)) = B_0 + B_1.\, c + B_2.\, u + B_3. + B_5.\, a \qquad (8)$$

where small letters refers to logarithms.

The logistic form is used to describe deforestation as a process which tends toward saturation within a given geographic area. In other words, in the early stages of settlement and deforestation, the effect of economic activities on the density of deforestation is high. As the remaining forest area dwindles, the impact of economic activities on deforestation diminishes, eventually dying out in totally deforested areas. Mathematically, this is shown by the fact that the elasticity of density of deforestation is equal to $A.\, (1-d)$, and therefore, in absolute value it decreases monotonically from A, in areas with no deforestation, to zero, in areas where the density of deforestation is one.

According to the above model, Amazon deforestation is the result of profit maximizing behaviour in a static framework. Dynamic considerations related to the rôle of land as an asset, to land price speculation and to wealth maximization are ruled out by the assumptions embodied in equation (6). Institutional considerations related to the open access to land and to the weakness of government institutions in Amazonia are also ruled out from the model. These motivations make deforestation a means of securing property rights in land, and as a consequence, cleared land tends to exceed land requirements for agricultural purposes, especially in areas where land conflicts are pervasive (Southgate 1989). An *ad hoc* test of the institutional hypothesis would be to include proxies for tenure conditions like population of squatter farmers *(S)*, and land area in the public domain *(V)*, as additional variables in equation (8).

A reduced form specification of the model is obtained by substituting (7) in (8):

$$\log(d/(1-d)) = A_0' + A_1'.\, u + A_2'.\, n + A_3'.\, q + A_4'.\, w \qquad (9)$$
$$+ A_5'.\, a + A_6'.\, (F/A) + A_7'.\, M,$$

where, again, small letters refers to logarithms.

The second block of equations uses an identical logistic specification to estimate the distribution of deforested areas by major types of vegetation. Thus:

$$\log (d_j/1-d_j) = D_0 + D_1 \cdot q + D_2 \cdot n + D_3 \cdot R + D_4 \cdot M \qquad (10)$$
$$+ \Sigma_j (D_{5j} \cdot F_j) + D_6 \cdot D$$

where j denotes the six main types of Amazon vegetation – dense forest, seasonal forest, savanna, campinarana, wetland and ecological transition – and d_j = deforested share in vegetation type j.

Based upon the biomass content of each major type of vegetation, CO_2 emissions are determined, as follows:

$$CO_2 = \Sigma_j \, q_j \, c_j \, (b_j - b_0) \, D_j \qquad (11)$$

where

CO_2 = CO_2 emissions (in tons)

q_j = per cent of biomass which is burnt in vegetation j

b_j = biomass content (t/ha) of vegetation j

b_0 = biomass (t/ha) content in deforested areas (converted or abandoned)

c_j = per cent of CO_2 in vegetation j.

Estimates of the biomass content in major types of vegetation cover of Brazilian Amazon are presented in Table 12.3. For estimation and simulation purposes, they were aggregated into two types: forests (which includes dense, seasonal and ecological transition) and savannas (including campinarana and wetlands). Within each of these two types, deforestation was distributed in proportion to areas of each type of vegetation in the municipality. Finally, for all kinds of vegetation it is assumed that the biomass is completely burnt, that is:

$$q_j = 1, \text{ for all } j \qquad (12)$$

Table 12.3 Estimates of above ground and total biomass for major types of vegetation covers of Amazonia (t/ha).

Vegetation type	Area %	Above ground min.	Above ground max.	Roots[a] min.	Roots[a] max.	Totals min.	Totals max.
Dense rainforests	69.53	188	300	54	100	242	400
Open forests	3.03	112	186	37	62	160	247
Ecological transition[b]	5.11	75	112	25	37	100	148
Savanna	13.97	6	75	6	32	12	107
Campinarana	6.34	6	120	6	45	12	165
Wetlands	2.01	6	115	6	38	12	153
Average[c]	100.00	139	240	–	–	180	322

Notes: a) assumed to be 1/3 of above ground biomass; b) biomass content assumed to be between the maxima for savannas and seasonal forests; c) weighted by area.

Sources: author estimates for areas and various sources for biomass.

178

The third block of equations consists of the generating functions for the spatial distribution of major economic activities. For population, crop output, cattle stock, logging, and roads, the assumption is that rates of growth in municipality i and time t depends only on the spatial density of the respective activity at time $t - 1$. Thus:

$$\hat{x}_{ki,t} = C_{0k} + C_{1k} \cdot \log(x_{ki,t-1}), \tag{13}$$

where k is population, agriculture, cattle, and logging, $\hat{x}$ is growth rate, and x is the spatial density, that is, the relation between activity level and geographic area.

Equation (13) describes the patterns of spatial concentration of economic activities over time. An activity k will show increasing spatial concentration if C_{1i} is greater than zero, whereas it will show increasing spatial dispersion if C_{1i} is less than zero.

12.4 Estimation issues: spatial autocorrelation (SAC) and seemingly unrelated regression (SURE)

The model is designed to make secular projections and simulations of the ecological and climatic consequences of tropical deforestation. Reliable estimates of long-run elasticities of deforestation in relation to major economic activities are crucial for this purpose. A major obstacle, however, is the lack of a time series sufficiently long to characterize long run equilibrium solutions. This is particularly true for deforestation data. Fortunately, cross-section data for Brazilian Amazonia are especially suited to estimate long-run elasticities. The sample includes more than 300 municipalities in very diverse stages of demographic and economic settlement, thus encompassing a wide range of configurations concerning the geographic densities of deforestation, population, and economic activities. Metaphorically speaking, the data mimic long run equilibria situations where differences between countries represent decades or centuries (Pindick 1979). Additionally, the availability of panel data for major economic and demographic variables allows more rigorous dynamic analysis.

Deforestation, population settlement and economic activities are simultaneous processes taking place in the same geographic space. This brings the possibility of two major econometric problems, namely residual covariance across different equations and spatial autocorrelation of residuals in each equation – both of them deserve careful consideration since, otherwise, estimates of long-run elasticities are likely to be biased and inconsistent. The simultaneity and interdependence of economic decisions give rise to Seemingly Unrelated Regression (SURE) problems.

Thus, equations describing population settlement, forest clearing, cropping, cattle raising, and logging are likely to show stochastic dependence, and therefore, residual covariance across them. The stochastic dependence can result from common generating mechanisms, latent variables and/or adding up restrictions not explicitly recognized in the model. Techniques to deal with these problems are well known (Zellner 1962).

In turn, spatial or geographic contiguity gives rise to phenomena like contagion and/or spatial inertia across observations (neighbouring municipalities in this case) which can lead to the presence of spatial autocorrelation of residuals in each equation. Spatial autocorrelation is usually a signal of missing variables, improper structural form, or measurement error. Therefore, its diagnosis can be a strong tool for improving model specification. Its identification requires a contiguity matrix, and usually, its correction is made by the use of Generalized Least Square (GLS) or Maximum Likelihood (ML) methods (Miron 1984, Cliff & Ord 1973, 1981). Moran (1950) and Geary (1954) coefficients are the usual statistics to test the presence of spatial autocorrelation. For a variable X, with normal deviates z, the Moran coefficient (M) is:

$$M = (n/1. \, W. \, 1). \, (Z'. \, W. Z/Z'Z), \tag{14}$$

and the Geary coefficient (G) is:

$$G = ((n-1)/2(1'. \, W. \, 1)). \, (Ew_i. \, p_i/Z'Z), \tag{15}$$

where:

n = number of observations (municipalities in this case)

W = contiguity matrix $(n \times n)$ with elements w_{ij} equal to 1 if i and j are spatially contiguous observations, and equal to zero otherwise

1 = column vector with all elements equal to 1

Z = matrix $(n \times n)$ with elements $z_{ij} = (x_i - x_j)$

p_i = ith line of matrix P $(n \times n)$ where $p_{ij} = (x_i - x_j)^2$.

It is possible to demonstrate that both M and G are asymptotically normal under weak assumptions (Cliff & Ord 1981). Both coefficients can also be used to test residual autocorrelation in regression equations. For more than two independent variables, however, test statistics are not straightforward.

Table 12.4 presents both the Moran and Geary coefficients for the main variables of the model. Standard errors were calculated under the normality assumption. Spatial autocorrelation for the logarithms of densities of deforestation, population and major economic activities are about the same and significantly higher than the coefficients obtained for growth rates.

In the model, problems of spatial autocorrelation as well as of residual covariance

Table 12.4 Spatial autocorrelation for logarithms of geographic densities of main variables.

	Moran		Geary		
	Value	Standard error[a]	Value	Standard error[a]	Sample size
$\log(dt{-}1)$	0.717	0.049	0.266	0.070	153
$\log(ct{-}1)$	0.740	0.032	0.226	0.048	336
$\log(nt{-}1)$	0.700	0.033	0.267	0.048	336
$\log(ht{-}1)$	0.706	0.032	0.289	0.048	336
$\log(qt{-}1)$	0.724	0.032	0.259	0.048	335
$\log(wt{-}1)$	0.487	0.032	0.432	0.048	335
$\log(nt/nt{-}1)$	0.400	0.033	0.658	0.048	336
$\log(ht/ht{-}1)$	0.079	0.033	0.886	0.048	336
$\log(qt/qt{-}1)$	0.350	0.033	0.718	0.048	335
$\log(wt/wt{-}1)$	0.282	0.035	0.705	0.052	295

Notes: a) standard error assuming a normal distribution
1. small letters refer to geographic densities: d = deforestation; c = agricultural land; n = population; h = herd; q = crop output; w = logging.
2. period t is 1985 and $t{-}1$, 1980, except for logging, where they refer to 1982 and 1987, respectively.

across equations are likely to be specially severe for equation (13). The reason for this is the parsimonious and naive specifications used for the generating functions of spatial distribution of population, agriculture, cattle raising, and logging. For equations (7–9), specifications are supposed to be theoretically more rigorous and to include a good number of the relevant spatial factors. To that extent, the damage caused by improper specification or omitted variables is smaller. Moreover, since the specification of deforestation and deforestation by vegetation type are practically the same, SURE techniques are not likely to make significant improvements compared to OLS results. Therefore, in these cases, it is fair to neglect the problems posed by spatial autocorrelation and residual covariance across equations.

A general specification for the presence of spatial autocorrelation in a model is (Case, 1991):

$$Y = p.W.Y + Z.B + u \tag{16}$$

$$u = r.W.e + e,$$

where Y = vector $(n{\times}1)$ of dependent variable, Z = a matrix $(n{\times}k)$ of explanatory variables, B = a vector $(k{\times}1)$ of coefficients, u = vectors $(n{\times}1)$ of residuals, e = vectors $(n{\times}1)$ of residuals, p = intensity of spatial autocorrelation in the dependent variable, r = intensity of spatial autocorrelation in the residuals.

The three possible cases of spatial autocorrelation and their respective implications are: (a) if $p \neq 0$ and $r = 0$, spatial autocorrelation occurs in the dependent variable but not in the residuals and least square estimators will be biased and inconsistent; (b) if $p = 0$ and $r \neq 0$, spatial autocorrelation in the residuals but not in the dependent variable and OLS estimator of B will be unbiased but inefficient; and (c) if $p \neq 0$ and $r \neq 0$, spatial autocorrelation occurs both the in the dependent and in the residuals, and in this case maximum likelihood methods are suggested for estimation.

When spatial autocorrelation in residuals combines with seemingly unrelated regression, the specification of equation (16) becomes:

$$Y_{(mn \times 1)} = Z_{(mn \times mk)} \cdot B_{(mk \times 1)} + u_{(mn \times 1)}, \qquad (17)$$

$$u_{(mn \times 1)} = (P_{(m \times m)} \cdot ox. \ I_n) W_{(mn \times mn)}$$

$$= W \cdot_{(mn \times mn)} \cdot u_{(mn \times 1)} + E_{(mn \times 1)},$$

$$E(E) = 0, \ E(EE') = H_{(mn \times mn)}$$

where subscripts in parentheses indicate the number and rows and columns of respective matrices, ox is the Kronecker product, and m is the number of dependent variables or equations in the model, n is the number of observations, k is the number of independent variables or coefficients, and p is the diagonal matrix with elements p_i ($i = 1, 2, \ldots, n$).

12.5 The data

Deforestation (D) data were derived from the Landsat satellite images plotted at municipal level. They are available for a reduced sample of municipalities, and in a single point in time. The images are from 1983 for some observations, 1985 for others, and 1987 for the remaining. Dummy variables for 1983 (DU83) and 1987 (DU87) were included in regressions to reduce bias introduced by these measurement errors.

Geographical areas of major types of vegetation cover (V_j) – closed forest, seasonal forest, savanna, wetlands, campinarana, and ecological transition – as well the extent of deforestation in each of them (D_j) come from estimates of IBDF–IBGE also based upon Landsat images.

Cleared land and output in agriculture come, respectively from the 1985 Agricultural Census, and the Annual Agriculture Production Municipal Surveys available from 1977 to 1987. Cleared land (C) is a stock variable, the area allocated to various economic uses (excluding natural pastures) at census data. For cattle raising, output *(H)* is also measured by a stock variable: the size of herds at census date. Since temporary and permanent crop outputs are annual flows, to make time dimensions less disparate, the variable used in the estimations was trend output *(Q)*, defined as the average quantities produced in the five-year periods centred on census year. This concept smooths the yearly fluctuations in crop output, and accounts for leads of deforestation in relation to output, particularly relevant for permanent crops.

Rural *(R)* and urban *(U)* population come from the 1980 and the 1991 (preliminary data) Demographic Census. Figures for 1985 are geometric interpolations assuming the same rural/urban composition as in 1980. Urban population is used as a proxy variable for all kinds of urban activities.

Logging *(W)* for timber, charcoal and firewood comes from the Extractive Production Country Surveys available from 1982 to 1987. In equations (8) and (9), the variable used is aggregate cumulative production flows from 1982 to 1987 for timber and charcoal. In equation (13) rates of growth of logging refer to the 1982–87 period.

Access and transportation conditions are described by the distance to state capitals and to Brasilia – as proxies to local and national markets – and by the extension of major roads and rivers in each municipality. Roads (federal and state, paved and non-paved) and rivers (deeper than 2.1 m 90 per cent of the time) come from maps and are available only for 1985/6. For the purposes of the model, it is reasonable to neglect feeder roads since they are simultaneously determined with deforestation and population settlement.

12.6 Estimation results

Table 12.5 reports the estimation results for the reduced form in equation (9). Alternative estimation procedures were used: ordinary least squares (OLS), seemingly unrelated regression (SURE), and maximum likelihood (ML) which was applied with the additional assumptions of spatial correlation in the dependent variable (SACD) or in the residuals (SACR). The statistics of the top of the table are, in that order, the number of observations (N. Obs.), the degrees of freedom (D. F.), the adjusted R^2 (R^2adj.) and the root mean square residual (Rmse), in the OLS equation, replaced by the maximum value of the likelihood function (LM) and the asymptotic standard errors in the remaining equations; and finally, the Moran coefficient for the residuals, and the spatial autocorrelation coefficient computed by OLS for residuals (Rho) or for the dependent variable (Thau).

Figures for adjusted R^2 and LM show a good fit for equation (9). With OLS,

Table 12.5 Estimates for agricultural clearing and deforestation.

Eq. Number	(8)	(9)	(9)	(9′)	(9′)
Dependent	Ag.Cle	Def.Den	Def.Den	Def.Den	Def.Den
Specif.	Log-Log	Logistic	Logistic	Logistic	Logistic
Method	OLS	OLS	OLS	OLS	ML
N.Obs	305	151	151	151	151
D.F.	290	145	144	132	132
R^2 adj.	0.99	0.72		0.81	
RMSE	0.55	0.36	0.34	1.77	0.62
Moran residual	0.23	0.30	0.14	0.21	0.11
Rho residual	0.61	0.80	0.38	0.72	0.30

Variable Coefficient (standard errors in parentheses)

Variable	(8)	(9)	(9)	(9′)	(9′)
Intercept	–	9.05	6.48	–	–
		(1.60)	(0.98)		
Agricultural clearing	–	1.13	0.97	–	–
		(0.13)	(0.09)		
Rural population	0.23		0.52	0.60	
	(0.06)		(0.29)	(0.33)	
Urban population		0.11	0.04	–0.10	–0.14
		(0.12)	(0.07)	(0.16)	(0.19)
Crop output	0.32		0.64	0.34	
	(0.03)		(0.14)	(0.16)	
Cattle herd	0.41		0.20	0.22	
	(0.03)	(0.09)	(0.10)	(0.14)	(0.16)
Logging		–0.44	–0.12	–0.49	–0.20
		(0.12)	(0.08)	(0.13)	(0.17)
Geographical area	0.06	–1.65	–1.61	1.76	–1.97
	(0.03)	(0.09)	(0.10)	(0.14)	(1.52)
Paved road	–0.28			1.76	2.77
	(0.39)			(1.57)	(2.46)
Non-paved road	0.33			1.27	0.63
	(0.12)			(0.50)	(0.62)
Rivers	–0.53			0.37	–3.89
	(0.37)			(1.31)	(3.36)
Dist. state	0.19			–0.26	0.13
	(0.12)			(0.65)	(1.06)
Dist. federal	–0.23			–0.92	–0.92
	(0.06)			(0.24)	(0.63)
Rainforest	–3.20			6.13	2.84
	(0.48)			(2.76)	(3.68)
Open forest	–3.09			6.66	3.81
	(0.48)			(2.99)	(3.20)
Savanna	–3.20			5.52	2.23
	(0.44)			(2.37)	(2.93)
Ecological transition	–3.12			6.12	2.98
	(0.49)			(2.78)	(2.97)
Wetlands	–4.80			2.97	0.90
	(0.49)			(2.68)	(3.69)
Campinarana	–3.06			5.19	2.06
	(0.62)			(3.05)	(14.8)

approximately 80 per cent of variance in the geographic density of deforestation of Amazonian municipalities is explained by the model. OLS estimation shows that most of the coefficients have the expected sign, and are significantly different from zero at the 95 per cent confidence level. With ML procedures, however, the coefficients are not significantly different from zero.

Spatial autocorrelation of residuals is weak, specially if compared with the strong correlation observed for deforestation density which is the dependent variable of the model. Moran coefficients for OLS and ML residuals are 0.21 and 0.13, respectively, and 0.77 for the geographic density of deforestation. Thus, OLS estimates are likely to be biased and inefficient. Maximum likelihood estimates are not significantly affected by the assumptions of SACD or SACR, except for the coefficients of river, the distance, and road variables.

Though not significant, the coefficients of population and logging came out with theoretically unexpected signs. The elasticity of population is expected to be negative if land and labour are substitutes, as is the case in a Cobb–Douglas production function, and if the supply of labour in different municipalities displays a "normal" response to real wages levels. The coefficients, however, are likely to show a simultaneous equation bias to the extent that deforestation, population, and major economic activities like cropping and cattle raising are mutually interdependent.

Furthermore, ". . . one should not be surprised to find large wage differences from one district or village to another and apparent disequilibrium in frontier labour markets – especially if one must reside on one's land to retain ownership" (Kazmer 1977: 432). In a context of land abundance, high real wages in a given municipality could simply mean relatively enlarged possibilities for establishing an independent farm and, therefore, reduced supply of labour. On the other hand, low real wages ". . . may indicate only that new settlers arrive to claim land faster than they can be absorbed into employment" (Kazmer 1977: 432).

The above arguments show that the model is under-identified and point to the need for more careful specification of the dynamics of labour supply and land settlement. This, however, can only be made with a combination of time series and cross-section for which no data are presently available. For the time being, therefore, results will be accepted as they are based upon the assumption that labour will continue to be the binding factor for the expansion of the Amazon agricultural frontier.

Against theory and intuition, the coefficient of logging came out negative and, in OLS estimations, significantly different from zero. The problem seems to be rooted in the use of annual flows of logging to measure the cumulative impact of logging on deforestation. This leads to problems of dynamic specification, since logging flows are at the same time cause and consequence of forest clearing. Thus, they tend to be large in relatively unsettled regions where the agricultural frontier is expanding fast. The suggestion for eliminating this negative bias is to specify panel data and to specify a system of equations where logging flows are simultaneously determined with changes in land clearing.

Since equation (9) is a logistic, the elasticity of deforestation density in relation to logarithmic variables, like the density of population, crop output, cattle herding and logging, is defined by:

$$E_{d,x} = (1-d). A'_x \tag{18}$$

where A'_x is the value of the coefficient of the variable x in case. The absolute value of the elasticity decreases monotonically with the deforestation density of municipalities being equal to A' when there is no deforestation, and one when the density of deforestation equals one.

For variables like distance, roads and river, the elasticity of deforestation is defined by:

$$E_{d,x} = (1-d). A'_x. x \tag{19}$$

The value of the elasticity decreases with deforestation density and increases with the value of the variable in the municipality in question. Thus, if deforestation density shows a positive relationship with the variable in question, like roads, the value of the elasticity will first increase and later decrease. When the variable has a negative relation, the absolute value of the elasticity will decrease monotonically.

The values of the coefficients, though not statistically different, imply that cropping has a stronger effect on deforestation than cattle raising activities. This corroborates the results of Reis & Margulis (1991), with the difference that now cropping is measured by the output of crops and not by crop area as was done earlier.

Table 12.6 reports results for equation (10) which determines the share of deforestation which takes place in forest areas (including dense forests, seasonal forests, and ecological transition) as opposed to areas covered by other kind of vegetation (savannas, wetlands, and campinaranas). The same specification as in equation (9) is used in this case, except for the substitution of geographic area for deforested area as a normalizing variable.

Table 12.7 reports estimation results for the generating functions of spatial distribution described by equation (12). The specification used is too simple to explain the variances in growth rates of population and economic activities, thus resulting in extremely low correlation coefficients. However, all the activities show small standard errors for the slope coefficients which quantify the relation between its geographic density and subsequent growth rate. The results show a spatial dispersion of economic activities typical of frontier areas, with growth rates proving lower in areas of less dense economic activity. According to the estimates of C_{k1} obtained by combining spatial autocorrelation of residual and seemingly unrelated regressions (SACR+SURE), this pattern is stronger for logging, followed by cropping and cattle

Table 12.6 Estimates for the share of deforested areas in forests.

Eq. number	(10)	(10)
Dependent	Clearing in forest	Clearing in forest
Specifi.	Logistic	Logistic
Method	OLS	ML
N. Obs	70	151
D.F.	51	132
R^2 adj.	0.89	
RMSE	0.52	4.55
Rho	−0.09	0.79
Moran	−0.00	0.31
Dependent variable	*Coefficient (standard errors in parentheses)*	
Rural population	−3.01	−0.05
	(1.74)	(2.51)
Urban population	1.59	0.16
	(0.96)	(1.23)
Crop output	0.54	0.37
	(0.88)	(1.15)
Cattle herd	1.84	−0.18
	(0.61)	(0.80)
Logging volume	0.89	−0.42
	(0.81)	(0.99)
Deforested area	−4.08	0.30
	(0.03)	(2.26)
Paved road	0.01	0.001
	(0.02)	(0.02)
Non-paved road	−0.003	−0.0002
	(0.005)	(0.006)
Rivers	−0.04	−0.01
	(0.02)	(0.02)
Dist. state	−0.009	−0.005
	(0.005)	(0.006)
Dist. federal	0.016	−0.001
	(0.003)	(0.005)
Rain forest	−9.72	−0.45
	(16.9)	(26.5)
Open forest	6.87	4.06
	(15.2)	(22.3)
Savanna	−15.7	−1.67
	(13.2)	(19.2)
Ecological transition	−7.98	2.76
	(13.9)	(20.3)
Wetlands	−23.9	−1.01
	(22.8)	(26.6)
Campinarana	−1.43	−11.3
	(0.63)	(123)

Table 12.7 Estimates for the generating functions of the spatial distribution of population, crop output, herd and logging.

Model	OLS	SURE	SACD	SACR	SURE + SACR
Dependent: population growth, 1980/85					
C_{10}	3.412	3.397	1.585	3.218	1.48
	(0.264)	(0.232)	(0.292)	(0.207)	(3.52)
C_{11}	−0.722	−0.752	−0.420	−0.699	−0.625
	(0.111)	(0.100)	(0.114)	(0.101)	(0.16)
Rho/Tau			0.55	0.55	0.95
			(0.059)	(0.059)	
R^2/Lm	0.123	0.15	−254.2	−252.4	
RMSE	0.030		0.001	0.001	
N.Obs	293	326	336	336	326
Dependent: cattle herd growth, 1980/85					
C_{20}	9.245	8.498	7.583	8.427	11.3
	(0.798)	(0.714)	(0.965)	(0.691)	(2.29)
C21	−2.516	−2.341	−1.973	−2.444	−4.47
	(0.332)	(0.282)	(0.356)	(0.306)	(0.45)
Rho/Tau			0.05	0.25	0.74
			(0.083)	(0.076)	
R^2/Lm	0.161	0.15	−716.2	−712.5	
RMSE	0.100			0.013	0.012
N.Obs	294	326	335	335	326
Dependent: crop output growth, 1980/85					
C_{30}	1.396	−4.467	0.902	1.073	1.58
	(0.655)	(2.939)	(0.627)	(0.606)	(1.70)
C_{31}	−3.462	−0.557	−2.532	−3.600	−4.46
	(0.315)	(0.976)	(0.440)	(0.330)	(0.47)
Rho/Tau			0.35	0.45	0.70
			(0.077)	(0.066)	
R^2/Lm	0.268	0.15	−681.9	−677.4	
RMSE	0.106		0.099	0.097	
N.Obs	326	326	335	335	326
Dependent: logging growth, 1982/87					
C_{40}	−4.325	−4.467	4.277	9.306	15.8
	(2.963)	(2.939)	(2.900)	(2.727)	(19.2)
C_{41}	−0.614	0.557	−2.334	−5.721	−6.94
	(0.988)	(0.976)	(1.072)	(0.982)	(0.96)
Rho/Tau			0.70	0.75	0.94
			(0.046)	(0.041)	
R^2/Lm	−0.002	0.15	−942.8	−932.7	
RMSE	0.295		0.049	0.045	
N.Obs	326	326	328	328	326

Note: Rho and Tau refers to spatial autocorrelation of residuals and dependent variables, obtained in ML estimation.

raising, and much lower for population. This is a reasonable finding if one considers that, in the case of population, centripetal forces related to frontier expansion are offset by agglomeration phenomena such as urbanization and industrialization.

Comparing the different estimation procedures, OLS values for the slope coefficient are significantly lower than values obtained assuming spatial autocorrelation in residuals (SACR), especially when it is combined with seemingly unrelated regressions (SACR+SURE). Differences are specially significant in the case of logging growth but for population growth they are not significant. Finally, note that SACR+SURE make the slope coefficients for crop and cattle almost identical, suggesting that they are subject to the same determinants.

12.7 Simulation results

Table 12.8 presents assumptions, projections and simulations of Brazilian Amazon deforestation and its contribution to CO_2 emissions for 1990–2090. Needless to

Table 12.8 Simulations for Brazilian Amazon deforestation, 1990/2090.

	A Basic Scenario	B Growth 1% high	C Herds 1% high	D Crops 1% high	E Roads 1% high	F OLS Param.
Assumptions on average annual growth rates for 1990/2090 (%)						
Population	1.4	2.5	1.4	1.4	1.4	1.4
Crops	2.9	4.0	2.9	2.9	4.0	2.9
Cattle	6.7	7.8	7.8	7.8	6.7	6.7
Logging	6.8	7.8	6.8	6.8	6.8	6.8
Paved road	3.3	4.4	3.3	3.3	4.4	3.3
Non-paved road	2.0	3.1	2.0	2.0	3.1	2.0
Annual growth rate for deforested area (%)						
1990/2025	3.7	4.6	3.8	3.9	4.4	3.9
2025/2090	0.2	0.6	0.2	0.2	0.6	0.4
1990/2090	1.4	1.9	1.5	1.5	1.9	1.6
Percentage of geographic area deforested						
1990	7.1	7.1	7.1	7.1	7.1	7.1
2025	25.2	34.1	26.0	26.7	31.7	26.6
2090	28.3	50.2	29.2	30.0	46.8	34.5
Cumulative carbon dioxide emissions, in 10^9 tons						
1990/2025	10.8	15.8	11.3	11.6	14.4	11.7
2025/2090	1.6	7.6	1.8	1.6	7.3	5.8
1990/2090	12.4	23.4	13.1	13.2	21.7	17.4

Note: For lack of information, estimates for deforestation and carbon dioxide emissions exclude the geographic area of the state of Maranhão which represents approximately 6% of Legal Amazonia. *Source: author's estimates.*

say, secular projections should be taken *cum grano salis*. As Theil said, ". . . models are to be used, not to be believed." In any case, they are surely better than the naive extrapolations usually made.

Projections were made using the maximum likelihood estimations of equations (9) and the SACR+SURE estimations for equation (13). The benchmark year for projections was 1990. The basic scenario, presented in column A, assumes a secular slowdown in population growth which declines from an average 3.1 per cent p.a., in 1980–90, to 2.1 per cent, in 1990–2025, and 1.1 per cent, in the 2025–90. In per capita terms, agricultural GDP (including crops, cattle raising and logging) is assumed to grow at an average rate of 3.0 per cent p.a. in all subperiods. These high rates for secular growth characterize an optimistic economic scenario which is broadly comparable to Scenario E in the IPCC (1991).

It also assumes that, in per capita terms, paved and non-paved roads, grow at constant rates of 1.9 per cent p.a. and 0.5 per cent p.a., respectively, which implies a gradual substitution between them. For the sake of comparison, paved and non-paved roads in the first half of the eighties increased 4.7 per cent p.a. and 3.5 per cent p.a., respectively, while population grew 3.5 per cent p.a. during the 1980s.

Finally, it assumes that the patterns of spatial dispersion of economic activities in the next century is the same as the one observed for the estimation period, 1980–85. In other words, in the projections, the slope coefficients (C_{k1}) in equations (13) were kept constant. Note that constant coefficients (C_{k0}) were adjusted to make growth rates of municipalities compatible with the aggregate rates assumed above.

From an environmental perspective, the results of the projections are alarming, though still far from the catastrophic scenarios usually depicted for the Amazon. According to the projections, approximately 45 per cent of the geographic area of Brazilian Amazon will be deforested by the end of next century. For the greenhouse effect, the consequences will be to push cumulative carbon dioxide emissions to something around 21.7 billion tons, which by itself means an increase of 6.8 per cent in the current level of concentration in the atmosphere. Assuming that, in the absence of drastic policies, carbon dioxide concentration in the atmosphere will grow at around 0.5 per cent to 1.0 per cent p.a., the Amazon's cumulative contribution would stay somewhere between 3.7 per cent and 2.6 per cent of global emissions.

Most of the deforestation takes place in the 1990–2025 period, reflecting, on the one hand, the higher rates of growth assumed for this sub-period and, on the other, the saturation effect in deforestation implicit in the logistic functional form. The trade-offs between growth and deforestation are roughly estimated by comparing the basic scenario with the alternative scenario presented in column B, where secular rates of growth of population, economic activities, and roads are all increased by one per cent.

One per cent more growth leads to a 0.4 per cent increase in the average rate of growth of deforestation for 1990–2090, an additional 20 per cent more in the share of deforested areas in 1990, and an additional 11 billion tons in CO_2 emissions. Growth has a significant impact on deforestation and high rates of growth, without changes in land-uses and/or technologies, are unsustainable.

Columns C, D, and E present comparative dynamic exercises were secular rates of growth of herd, crop output and roads in the are increased by one per cent, *ceteris paribus*. They show that road expansion is, by far, the most important single factor for deforestation with an average elasticity close to 0.3. Both cattle raising and cropping have much smaller effects, with elasticities which are close to 0.1.

Finally, column F shows that projections made with OLS estimates would give much more conservative results.

12.8 Concluding remarks

The model and the simulations presented in this paper are useful tools for the appraisal of cost and benefits of sustainable development of Brazilian Amazon. Reliable projections, however, require some improvements in the methodology and data. In concluding, we point to the most critical aspects for research extensions and further developments.

The most disturbing aspect is perhaps the simultaneous equation biases introduced by the fact that population growth and economic activities are, at the same time, causes and consequences of deforestation. Thus, they should be treated as endogenous variables in the reduced form of the model. This is probably the reason for both the wrong signs obtained for the coefficients of population, and logging, and the large instability of parameter values in estimations. The proposed solution is to estimate fixed effect and random effect models based upon panel data. Since census data at municipal level are available every five years since 1970, and since deforestation can be proxied by land-uses defined in the census, road network at municipal level is the crucial missing variable for panel analysis.

Another aspect deserving careful specification is the long run determinants of technical progress in Amazon agriculture, and their relationship to changes in the geographic densities of population and of other economic activities. Panel data analysis is again a crucial requirement to estimate the parameters related to efficiency and technical progress. Furthermore, the analysis should be made at more relevant levels of aggregation like cattle raising, temporary and permanent crops, as well as reforestation and fallow lands.

Finally, two other dynamic aspects need to be modelled. The first is the determinants of population growth, migration and urbanization. The suggestion here is to complement econometrics with demographic techniques. The second aspect concerns the specification of dynamic relationships between the fate of carbon stocks and land-use changes.

13

An econometric analysis of the causes of tropical deforestation: the case of Northeast Thailand

Theodore Panayotou and Somthawin Sungsuwan

13.1 Introduction

Tropical deforestation is one of the major issues of global concern for the 1990s. It is widely believed to be: the cause of destruction of vast areas through erosion and flooding; the cause of species extinction and loss of biological diversity; and a major source of greenhouse gases and associated climatic change. There is also a consensus that tropical deforestation is excessive and must be reversed. Yet, our understanding of the problem of tropical deforestation is limited and more the product of casual observation than rigorous analysis.

Deforestation cannot be excessive except by comparison to some optimum. What is the optimum rate of deforestation or, inversely, the optimum forest cover? We do not know, and apparently do not care. Otherwise we would have seen more research than advocacy. The presumption is that all deforestation is bad and excessive, which is equivalent to saying that opportunity costs do not matter. As it is shown in a related paper (Panayotou and Sungsuwan 1989), an optimal rate of deforestation can be determined and used as a yardstick for determining whether current rates of deforestation are excessive.

A second presumption is that we understand the causes of tropical deforestation. Therefore, it is a simple matter of stemming those causes. Those who hold the view that logging is the cause of deforestation call for logging bans. For example, recently Thailand declared a ten-year comprehensive nationwide ban on all logging in response to devastating floods in parts of southern Thailand that were blamed on deforestation. Others hold the view that log exports are the culprit and declare log export bans as Thailand, the Philippines, and Indonesia have done and others are encouraged to follow. Yet others believe that shifting cultivation is the cause of destruction and thus advocate resettlement or policing by the army.

We contend here that activities such as logging, fuelwood collection, and land clearing for sedentary or shifting agriculture are *sources* but not the *causes* of deforestation. Unless we understand the root causes of deforestation we will be treating the symptoms with little hope for a sustainable improvement. The objective of this study is to use economic theory to determine the theoretically valid causes of deforestation, and to test the model for the case of Northeast Thailand.

13.2 A theoretical model of tropical deforestation

Deforestation is defined as an abrupt change in forest land-use from forestry to something else. Land previously under forest and now under shifting or permanent agriculture, plantation, or crops is considered deforested even if the new land-use is more beneficial than forestry. In contrast, forest degradation is a gradual deterioration in the quality of the forest cover and its ecosystem, including soil. Forest degradation results from selective and destructive logging (for example in Indonesia and Malaysia), excessive fuelwood collection, fodder harvesting, and overgrazing of animals (for example in Nepal and India), and uncontrolled fires (for example the 1982 Kalimantan fire). Of course, degradation may eventually lead to deforestation by either conversion to agriculture and other non-forestry uses or degeneration into wasteland such as the millions of hectares of *Imperata cylintrica (alang alang)* lands in Southeast Asia.

Tropical deforestation is the result of three major, distinct, but related demands: (a) the demand for logging (both legal and illegal) by logging firms and illegal loggers that have export licenses or log processing plants; (b) the demand for fuelwood and fodder and other forest products by the local population for "direct" consumption; and (c) the demand for land for both shifting cultivation and permanent agriculture including forest plantations, tree crops, annual crops, and livestock raising (ranching). A fourth demand is that for forest land by the central government and the local administration for construction of public infrastructure such as roads and irrigation systems. We will discuss these demands in the above order.

13.3 Demand for logging

The demand for logging is a derived demand. It is derived from logging firms' profit-maximizing behaviour subject to constraints. The demand for logging can be understood at two levels: the demand for logging concessions by logging firms (an asset demand), and the demand for standing logs or trees by concessionaires and illegal loggers (a variable input demand). Here we will be using logging demand in the latter sense, but the demand for concessions is implicit in the dynamic optimization from which the demand for standing logs is derived.

The objectives and the constraints of the logging firms, and therefore the arguments of the demand function of trees for logging, depend on the institutional framework in which they operate. If concessionaires have secure long-term concessions extending beyond one felling cycle, it is reasonable to postulate that their objective is to maximize the net present value of the forest over the life of the concession, subject to a harvesting production function, a stock constraint, and a regeneration function and the corresponding prices of outputs and inputs including transportation costs. The demand function for logging derived from such constrained present value maximization incorporates a user cost argument that ensures that current logging is not at the expense of future harvests beyond the point justified by the opportunity cost of capital (see Panayotou and Sungsuwan (1989) for the derivation):

$$X_G = G(P_G, W_G, \lambda; Z) \tag{1}$$

$$\frac{\partial G}{\partial P}, \frac{\partial G}{\partial Z} > 0; \quad \frac{\partial G}{\partial w}, \frac{\partial G}{\partial \lambda} < 0$$

where:

X_G is the volume of standing logs (trees) demanded for logging per unit of time;

P_G is the price index of harvested logs at the market;

W_G is the price index of a composite logging input;

λ is the shadow price, or user cost, of standing trees;

Z is a vector of fixed factors or spatial variables representing accessibility.

Equation (1) states that an increase in the price of logs or wood or a reduction in harvesting costs would stimulate the demand for trees to be logged. Increased accessibility of the forest through road expansion would have a similar effect. The rôle of λ here is critical. The scarcity value or user costs of the standing trees, λ, is expected to be rising over time as logging (and other sources of deforestation) deplete the stock of standing trees. Higher λ would be reflected in higher future timber prices. Thus, every tree cut and sold at today's lower prices involves an opportunity cost, λ, in terms of foregone future revenues. whether this intertemporal opportunity cost is taken into account in the present logging demand depends on the length and security of the concessions. For long-term concessions that extend over several logging cycles the user cost is internal to the logging firm and is, therefore, part of its economic calculus. In the absence of regeneration and growth, optimality requires that the rate of logging is such that the user cost grows at the rate of interest. A higher rate of interest would lower the user cost in the future. Hence one could expect both future price expectations and interest rate changes to affect the current demand for logging via the user cost term (λ).

Under short and insecure concessions, as is the case with most tropical timber producers, the user cost is external to the logging firms and hence not part of their profit-maximizing calculations. Firms with 29-year concessions have no interest in values generated or opportunity incurred 35 or 50 years hence. Yet these are the minimum felling and rotation cycles possible for tropical forests. In fact, the insecurity of tenure and general uncertainty surrounding logging concessions – under constant threat of fire, encroachment, and cancellation – precludes any interest in revenues beyond the first few years of the concession. Thus, under the prevailing concession system, the only reasonable objective we can postulate for concessionaires and illegal loggers is that of static profit maximization, in which case the user cost (λ) drops out from the demand for logging:

$$X_G = G(P_G, W_G, Z) \tag{1$'$}$$

Because $G/\lambda < 0$, $\lambda > 0$, and $\lambda/t > 0$ the absence of λ from the logging demand function results in excessive demand of trees for logging and consequent excessive deforestation. Inclusion of λ through longer-term concessions would ensure that this source of excessive deforestation is eliminated but, alone, it would not ensure the social optimum rate of logging if there were unaccounted externalities (such as downstream erosion or flooding) resulting from logging or divergence between the private and social rate of discount. Additional policy instruments such as a logging tax and a replanting subsidy would be necessary for achieving the socially optimal rate of logging.

In conclusion, insecurity of logging concessions is identified theoretically as a potential cause of excessive and destructive logging leading to deforestation. This hypothesis can be empirically tested by comparing logging rates between short- and long-term concessions while controlling for other factors. Unfortunately, long-term concessions are extremely rare in the tropics.

13.4 The demand for fuelwood

The demand for fuelwood is used as a representative demand for forest products by the local population that might lead to forest degradation and deforestation. Other demands in this category are the demand for fodder and the demand for construction poles for local use. The demand and harvest of other forest products such as fruits, vegetables, flowers, medicinal plants and resin, are not considered as sources of deforestation.

Because fuelwood is largely demanded for direct consumption, the demand function is derived from a model of consumer utility maximization subject to a budget constraint, given the price of fuelwood and the prices of close substitutes and other goods competing for the consumer's budget:

$$X_F = F(P_F, P_K; Y) \qquad (2)$$

$$\frac{\partial F}{\partial P_F} < 0; \quad \frac{\partial F}{\partial P_K} > 0; \quad \frac{\partial F}{\partial Y} > 0$$

where:

X_F is the quantity of fuelwood demanded (collected);
P_F is the price of fuelwood;
P_K is the price of substitute;
Y is income.

A number of differences, however, exist between the demand for fuelwood and the demand for a "normal" good. First, the price of fuelwood, as such, often does not exist because fuelwood is often not traded in formal markets. It is collected by family members (usually women and children) from open and free access forests. The only relevant cost that can serve as a price for fuelwood is travel and collection costs, which consist simply of the opportunity cost of labour engaged in fuelwood collection. In poor regions with surplus labour and stagnating real rural wages (such as Northeast Thailand) the opportunity cost of family labour is extremely low and constant, and unlikely to have significant explanatory power.

One variation that does occur across space and over time, however, is accessibility. Logged-over forests are generally more accessible to fuelwood collectors, and dry matter littering the floor of logged-over forests reduce the cost of collection and transport. We may thus postulate $P_F = P_F (X_G)$ with $\partial P_F/\partial X_G > 0$. On the other hand, as nearby forests become depleted fuelwood collectors have to travel longer distances and spend increasingly more time searching and collecting fuelwood. We can, therefore, construct a shadow price for fuelwood that varies in space and time. Alternatively, we could capture this spatial effect through shifts of the demand function caused by changes in the accessibility, variable Z.

Second, fuelwood is likely to be an inferior good but unlikely to be a Giffen good because the proportion of the household's budget spent on fuelwood is minimal if not zero. Note, however, that autonomous increases in income (not income effects from a price decline, which are not relevant in this case), are certain to reduce the demand for fuelwood, the fuel of the "poor", and increase the demand for its closest substitute, kerosene, the fuel of the "rich". Of course, this is an empirical question but our presumption is for a negative income coefficient and therefore a positive relationship between income level and forest cover.

Third, in an aggregate demand function for fuelwood, population growth or population density will be an important determinant of demand for two reasons: (a) the larger the population, the larger the number of fuelwood consumers for a given level and distribution of income; and (b) the higher the population growth/ density, the more surplus family labour is likely to be available for fuelwood collection and therefore the lower the implicit "price" of fuelwood is likely to be.

Thus, the modified fuelwood demand function becomes:

$$X_F = F[P_F(P_O; X_G), P_K; Y, N, Z] \qquad (2')$$

$$\frac{\partial F}{\partial N}, \frac{\partial F}{\partial Z} > 0; \quad \frac{\partial X_F}{\partial X_G} > 0$$

where:

P_F is fuelwood collection cost;
N is population growth/density;
Z is accessibility of forests;
P_O is opportunity cost of labour.

13.5 Derived demand for agricultural land

In developed countries there is an active market for agricultural land. Forest land is no longer available for conversion to agricultural land. In the tropics, rural land markets are more limited than in developed countries and forest land is generally available for agriculture by virtue of its semi-open access status. In countries such as Thailand, the Philippines, Indonesia, and Nepal there is a strong demand for forest land to convert into shifting and permanent agriculture. In Thailand agricultural land area almost doubled over the past 30 years through forest conversion.

The demand for agricultural land is derived from the farmer's profit-maximizing behaviour subject to production function constraints and parametric input and output prices. Because forest land clearing for agriculture is considered illegal encroachment, squatters have no legal title and hence no security of ownership. Therefore, the most reasonable objective to postulate is that of static profit maximization. Profit maximization is justified further by the fact that most forest land clearing has been for cash crops such as maize, rubber, and cassava rather than for subsistence crops such as rice, although part of shifting cultivation is for upland rice. Consider the following variable profit function for farming:

$$\pi_A = \pi(P_A, P_L, P_O; Z)$$

$$\frac{\partial \pi}{\partial P_A}, \frac{\partial \pi}{\partial X} > 0; \quad \frac{\partial \pi}{\partial P_L}, \frac{\partial \pi}{\partial P_O} > 0$$

where:

π is variable profit;

P_A is price index of agricultural products (price of upland crops relative to rice);

P_L is rental price of forest land (relative to rental price of existing agricultural land);

P_O is opportunity cost of other inputs (mainly labour);

Z is a vector of fixed factors including spatial variables and fixed farm assets (if such exist).

Assuming π has all the necessary properties to be a well-behaved profit function dual to a transformation function we can derive input demands and output supplies using Shepard's lemma by partially differentiating π with respect to the corresponding prices. Then the demand for land, which is here treated as a variable input (because of the insecurity of tenure) is:

$$L_A = \frac{\partial \pi}{\partial P_L} = A(P_A, P_L, P_O; Z) \qquad (3)$$

with

$$\frac{\partial A}{\partial P_A} > 0, \quad \frac{\partial A}{\partial P_L} < 0, \quad \frac{\partial A}{\partial Z} > 0$$

where:

L_A is the area of forest land demanded for conversion to agriculture.

Equation (3) states that the demand for forest land for conversion to agriculture rises with an increase in the relative price of agricultural cash crops and accessibility of the forest and declines with a rise in the "price" of land and the price of other inputs. Because there is no market for forest land to convert into agriculture, there is no market price for forest land (P_L). As in the case of fuelwood, however, there is an implicit price determined by the cost of clearing of forest land and its accessibility. It is considerably more costly for farmers (and shifting cultivators) to enter and clear a virgin rainforest than a logged-over forest from which the large trees have been removed and dry matter covers the forest floor. Furthermore, logging firms open up access roads that make forests more accessible to farmers and shifting cultivators, who literally follow the loggers, slashing and burning the logged-over forests. Generally, the more intensive and destructive the logging, the lower the cost of clearing of forest land for agriculture, i.e.:

$$P_L = P_L(P_O; X_G), \quad \frac{\partial P_L}{\partial X_G} < 0$$

where:

P_O is the opportunity cost of labour, the main input used in land clearing; and X_G is the rate of logging.

At the aggregate level, demand for forest land for conversion to agriculture (encroachment) would also depend on the population growth/density, N, because in more densely populated areas we would expect higher pressure on land. It is not, however, population *per se* but population relative to the alternative opportunities available for employment and income generation that determines the pressure on land. For a given population, the higher the income level the lower the encroachment of forest land for agriculture. We may use aggregate income level as an indicator of the level of poverty and the availability of non-agricultural employment opportunities.

With the above modifications the demand function for forest land to convert to agriculture becomes:

$$L_A[P_A, P_L(P_O; X_G), P_O; Y, N, Z] \tag{3'}$$

with $\quad \dfrac{\partial A}{\partial X_G} = \dfrac{\partial A}{\partial P_L} \cdot \dfrac{\partial P_L}{\partial X_G} > 0; \quad \dfrac{\partial A}{\partial N} > 0; \quad \dfrac{\partial A}{\partial Y} < 0$

Equation (3') postulates that logging and population growth stimulate the demand for forest land to convert to permanent and shifting agriculture while economic growth discourages it.

13.6 A deforestation function

To obtain a deforestation function for all sources we must combine the three distinct but related demand functions (1'), (2'), and (3'). In doing this we are faced with two difficulties. First, the demand for logging is in terms of cubic metres of standing logs; the demand for fuelwood is in terms of tons of biomass; and the demand for land is in terms of hectares. To combine them we must first convert them all to the same unit, preferably hectares of land area. We do this through the following conversion functions:

$$L_G = \theta(X_G) \qquad\qquad d\theta/dX_G > 0$$

$$L_F = \phi(X_F) \qquad\qquad d\phi/dX_F > 0$$

where:

L_G is degraded forest area due to logging; and

L_F is degraded forest area due to excessive fuelwood collection.

Second, the three sources of deforestation are not additive. While all deforestation due to conversion is final under tropical conditions, not all forest degradation due to logging and excessive fuelwood collection leads to irreversible deforestation. There is also a considerable overlap between the three sources, especially between degraded logged-over areas and areas converted to permanent and shifting cultivation. To avoid double counting we use the area converted to agriculture as a base to which we add: (a) the fraction, α, of logged-over area that neither regenerates nor is converted to agriculture but degenerates into wasteland such as *alang alang* or compacted soils; (b) the fraction, β, of degraded forest area due to excessive fuelwood collection that becomes similarly deforested without having been converted to agriculture; and (c) forest area displaced by public infrastructure, L_I, such as roads and irrigation systems:

$$DF = L_A + \alpha L_G + \beta L_F + L_I(I)$$

where DF is the area deforested annually.

The two major types of public infrastructure that affect the rate of deforestation are roads and irrigation systems. Both have direct (forest displacement) and indirect (deforestation inducement) effects. Both also have positive and negative effects on deforestation. Roads make forests more accessible to logging, encroachment, and conversion but they may also promote migration, rural industry, and regional growth, thus reducing the pressure on forests. Irrigation systems displace both forests and people who, having lost their lands, clear new forest land for cultivation. On the other hand, irrigation raises productivity on existing land, thus making forest encroachment by lowland farmers less attractive. On balance we expect both rural roads and irrigation systems, net of their productive raising effect, to exacerbate deforestation.

Given the interdependence between different sources of deforestation we could define an aggregate deforestation function in terms of the seven prices and four shifters or fixed factors identified as theoretically important in the individual demand functions:

$$DF_t = DF(P_G, W_G, P_P \quad P_K, P_A, P_L, P_O; Y, N, Z, I).$$
$$(+) \quad (-) \quad (-) \quad (+) \quad (+) \quad (-) \quad (-) \quad (-) \quad (+) \quad (+) \quad (+)$$

The sign under each variable indicates our expectations as to the sign of the corresponding coefficients.

Alternatively, we may define the forest cover FC as:

$$FC_t = FC_0 - \sum_{i=0}^{t-1} DF_i$$

and rewrite the equation as:

$$DF_t = DF(P_G, W_G, P_F, P_K, P_A, P_L, P_O; Y, N, Z, I) \qquad (4)$$

$$(-) \ (+) \ (+) \ (-) \ (-) \ (+) \ (+) \ (+) \ (-) \ (-) \ (-)$$

with reversed sign expectations.

Because $P_F = P_F(P_O; X_G)$ and $P_L = P_L(P_O; X_G)$, substitution into (4) yields:

$$FC_t = FC(P_G, W_G, P_F, P_O, P_L, P_A; Y, N, Z, I) \qquad (4')$$

Given an appropriate functional specification, a stochastic framework and data on the rate of deforestation, the seven prices, and the four fixed factors equation (4') can be econometrically estimated through ordinary least-square regression techniques. The estimated parameters can be used to test various hypotheses as to the causes of deforestation, to rank these causes in terms of magnitude and significance, and to make projections about future rates of deforestation.

13.7 The causes of deforestation in Northeast Thailand

The theoretical model of tropical deforestation summarized in equation (4') was tested for the case of Northeast Thailand. The Northeast covers an area of 169,000 square kilometres or one-third of Thailand's total land area. It is Thailand's most populous region with almost 40 per cent of the population and a population density of 110 persons per square kilometre. The quality of the soil is poor and the rainfall unreliable. With 90 per cent of the region's population engaged in farming, much of it rain fed, it is no wonder that the Northeast is Thailand's poorest region. It is also the source of a large flow of seasonal migration to other regions. The average income per capita in 1982 was only 6390 baht compared to 51,000 baht in the Central region. While the share of agriculture in the Gross Domestic Product for the country as a whole has dropped to below 20 per cent, its share in the Northeast continues to be about 40 per cent.

The Northeast is also known to be the region with the most rapid forest loss. Its forest cover, as a percentage of the region's area, has decreased from 60 per cent in 1952 to 15 per cent in 1982 (Klankamsorn & Adisornprasert 1983). The most severe forest destruction has taken place in Kalasin (at 7.5 per cent per annum),

Mahasarakrn, Udon Thani, and Knon Kaen. The most seriously affected forest is the *Dipterocarpus spp.* which grows in relatively fertile soils suitable for agriculture. The most important forest species in the Northeast are *Dipterocarpus tuberculatus, Dipterocarpus obtusifolius, Shorea obtusa, Shorea taluna, Pentacure suavis,* and *Pterocarpus pavuifolius.* Other types of forests covering smaller areas are dry evergreen, hill evergreen, and pine forests.

The government has attempted to stem deforestation by creating forest reserves and national parks but has had little success. By 1980 42 per cent of the declared forest reserves and 16 per cent of the national parks have been cleared and occupied by farmers. Rice, the staple food crop, is planted on 80 per cent of the region's cultivated area. The main upland (cash) crops are cassava, maize, and kenaf. Of the Northeast's 80,000 square kilometres of agricultural area, only 5800 is irrigable and of this only 4400 is already irrigated.

The Northeast, as the region with the largest population and the lowest income, is the largest consumer of fuelwood, accounting for almost 50 per cent of total fuelwood consumption in the country estimated at 24 million cubic metres in 1970. Fuelwood is collected from the forests and used mainly for home consumption.

13.8 Data and method of estimation

Data availability and reliability are a problem. Moreover, a large number of observations is necessary if we are to test our model with sufficient degrees of freedom and without undue functional restrictions. Particularly limiting is information on forest cover and rates of deforestation. There have been only four remote sensing (LANDSAT) surveys – conducted in 1973, 1976, 1978, and 1982 – obviously an insufficient sample for estimating a time-series equation. Similarly, the cross-section observations from 16 Northeast provinces are inadequate for estimating a cross-section equation.

Because the forest cover equation (4') to be estimated includes several time-related explanatory variables, it is appropriate to pool cross-section and time-series data to secure sufficient degrees of freedom and efficient parameter estimates (Pindyck & Rubinfield 1981). This is done at a cost, however. Combining cross-section and time-series creates difficulty in model specification because the disturbance term might consist of time-series-related disturbances, cross-section disturbances, or a combination of the two. There are several alternative methods for pooling data. The selected method is the co-variance model using the ordinary least squares with dummy variables to allow the intercept term to vary over time and across cross-section units.

The model is specified as follows:

$$\ln Q_{kt} - \ln a_0 + \sum_{i=1}^{m} a_i \ln X_{ikt} + \frac{1}{2}\sum_{i=1}^{m}\sum_{j=1}^{m} a_{ij} \ln X_{ikt} \ln X_{jkt}$$
(5)

$$+ r_2 V_{2t} + r_3 V_{3t} + \cdots + r_{16} V_{16t}$$

$$+ \delta_2 T_{2k} + \delta_3 T_{3k} + \delta T_4 V_{4k} + \varepsilon_{kt}$$

$$k = 1,2,3, \ldots, 16$$
$$i = 1,2,3, \ldots, m$$
$$j = 1,2,3, \ldots, m$$
$$t = 1,2,3,4$$

where Q_k = forest as a percentage of provincial area;

X_{ik}, X_{jk} = individual independent variable;

$\ln X_{ik}. \ln X_{jk}$ = when ij is designed to capture the interaction effects between independent variables; and when $i = j$ the term $\ln X_{ik}. \ln X_{jk}$ allows for a quadratic relationship in the model;

V_{kt} = 1 for individual t; $k = 2, 3,\ldots 16$; 0 otherwise;

T_{kt} = 1 for time period t; $t = 2, 3, 4$;

a = coefficients;

ε = error term.

It is noted that the terms V_{1t} and T_{kl} are omitted in order to avoid the problem of perfect collinearity among explanatory variables. The total degrees of freedom are

$$kt - 2 - (k - 1) - (t - 1)$$

To gain further additional degrees of freedom we aggregated crop prices (P_A) into a price index using consistent aggregation (Divisia Index) and we dropped variables that exhibited no significant variability (P_O, G_W) or for which there were no market data (P_L, P_F). Since the variability of these two variables is driven by accessibility Z and the logging rate X_{G3}, however, they are implicitly included in the estimated equation.

13.9 Some preliminary results

In order not to restrict the functional form unduly and to allow for interactions between the explanatory variables a translog function was specified as described by equation (5). The interaction terms as well as the dummy variables for time, however, turned out to be statistically insignificant and were dropped for the estimated equation, which collapsed into a log linear function.

A second estimation problem was a high multicollinearity between kerosene price, crop price, and the price of wood. To avoid the problem, two separate models were estimated: Model I with the prices of wood and crops, and Model II with the price of kerosene. As expected, there was only marginal difference in the explanatory power of the two models, and the coefficients of the non-price variables changed little between the two models:

Model I

$$\ln FC = -0.41 \ln P_G - 0.32 \ln P_A + 0.42 \ln Y - 1.51 \ln N + 0.70 \, Z_1$$
$$ (-4.1) \qquad (-1.7) \qquad (4.0) \qquad (-9.7) \qquad (4.8)$$

$$-0.11 \ln Z_2 - 0.22 \ln I + 0.38 \, R$$
$$(1.4) \qquad\quad (-1.02) \qquad (1.9)$$

$$R^2 = 0.80; \; R^{-2} = 0.77, \; F = 32, \; df = 55$$

Model II

$$\ln FC = -0.37 \ln P_k + 0.41 \ln Y - 1.49 \ln N + 0.67 \ln P_k + 0.41 \ln Y$$
$$ (-3.06) \qquad (3.8) \qquad (-10.2) \qquad (4.9) \qquad (-1.6)$$
$$-1.49 \ln N + 0.44 \ln R$$
$$(-1.52) \qquad (2.5)$$

Model I is used as the basic model for policy analysis. The signs of all coefficients are consistent with our *a priori* expectations. Over 75 per cent of the spatial and temporal variation of the forest cover was explained by the included variables. All variables except the irrigation infrastructure are statistically significant at a reasonable level (0.10). Four critical variables – wood price, income level, population density, and remoteness (distance from Bangkok) – are statistically significant at the 0.01 level of significance.

In order to assess the relative importance of the different determinants of the forest cover (or inversely, causes of deforestation), we normalize their coefficients by calculating the 13-coefficient and rank them accordingly, as shown in Table 13.1.

Table 13.1 Ranking of causes of deforestation according to ß-coefficients (Model I).

Variable	ß-coefficients	Rank
Population, N	−0.65	1
Wood price	−0.35	2
Income, Y	0.29	3
Distance	0.27	4
Rural roads	−0.12	
Rice yields, R	0.12	
Irrigation infrastructure	−0.08	
Crop price	−0.08	

Population density emerges as the single most important cause of deforestation in Northeast Thailand. This is to be expected for Thailand's poorest and most populous region; population growth is high, non-farm activities are limited, and soil fertility low. While there is a large flow of seasonal and permanent migration out of the Northeast, for most new entrants into the labour force clearing forest to acquire agricultural land continues to be the most attractive alternative as it requires no cash or special skills. As to shifting cultivation, population growth has forced shifting cultivators to return to their plots more often, thus reducing the fallow cycle. The shorter the fallow cycle, the less time is available for forest regeneration and inevitably part of the swidden lands degenerate into *alang alang* lands unsuitable for either agriculture and forestry. In response, shifting cultivators move on to clear new plots of forest land, leading to increasing deforestation. Furthermore, farmers, accounting for 90 per cent of the Northeast's population, find it more economical to use fuelwood than other sources of energy such as kerosene, as evidenced by the negative relationship between kerosene price and forest cover (Model II). The forest cover was, therefore, found to be very elastic with respect to population density: a 10 per cent increase in population density results in a 15 per cent decrease in the forest cover.

The second most important determinant of the forest cover was found to be the price of wood (logs). A 10 per cent increase in the price of wood leads to a 4.1 per cent decrease in the forest cover. This is the result of the increase in the profitability of logging. As expected from our theoretical model, the higher the price of logs the higher the demand for logging. With secure ownership of forest land or long-term concessions the higher wood price would induce additional tree planting and even forest conservation if the price is expected to continue rising. But under the conditions of insecure land ownership, semi-open access forests, and short-term concession such incentives do not exist.

The third most significant variable was aggregate provincial income, a variable that does not only capture any income effects but also serves as a surrogate for the availability of alternative employment opportunities and the general level of development. Income is positively related with the forest cover. The higher the aggregate income (GDP) of a province, the less deforestation takes place. A 10 per cent increase in income results in a 4.2 per cent increase in the forest cover. This results from both a decrease in the demand for forest land for conversion to agriculture and from a decrease in the demand for fuelwood.

Distance from Bangkok is the next most important determinant of the forest cover. The more remote a province is, the higher the forest cover because of a decreased accessibility and lower demand for land to grow cash crops. As one moves away from Bangkok a 10 per cent increase in distance (from Bangkok) results in a seven per cent increase in the forest cover.

The remaining four variables are less significant determinants of the forest cover.

A 10 per cent expansion of village road density would lead to only a 1.1 per cent decrease in the forest cover. While this is a relatively low elasticity of response it is quantitatively important if we consider that the village road system during the 1970s and 1980s expanded at the annual rate of 12 per cent. Rice yield appeared to be as important as rural roads except that they have the opposite impact. The higher the rice yields the larger the forest cover, because farmers have higher income from existing cropland and less incentive to encroach on the forest to obtain land for planting upland crops. A 10 per cent increase in the rice yield results in a 3.8 per cent increase in the forest cover. Recall that "rice yields" was introduced as a surrogate of the beneficial effects of irrigation infrastructure on the forest cover via the increased productivity of existing agricultural land and higher incomes.

It was found that a negative effect of irrigation infrastructure on the forest cover did exist but it was a relatively small one: a 10 per cent increase in irrigated area results in a two per cent decrease in the forest cover, presumably because of the displacement of forest by the irrigation systems. The price of upland crops relative to rice was negative and of about the same importance as irrigation infrastructure. Farmers found the annual one per cent increase in the relative price of upland crops to rice during 1973–82 a sufficient attraction to seek new agricultural land for upland crops in the forest. A 10 per cent increase in the relative crop price leads to a 3.2 decrease of the area under forest.

13.10 Projections

Projections were attempted based on the estimated elasticities and growth rates of the independent variables estimated through trend equations (see Table 13.2). The

Table 13.2 Elasticities, growth rates, and base values (Model I).

Independent variable	Elasticity	Growth rate (% per annum, 1982)	Base value
Gross provincial income	0.42	6.0	2,850 (million baht)
Irrigated agricultural area	−0.2	6.0	4.4
			(% of agricultural area)
Price of wood	−0.41	16	1,559 (baht/m^3)
Village road density	−0.11	12	0.0913 (km/km^2)
Yield of rice	0.38	0.3	188 (kg/rai)
Relative price of upland crops to rice	−0.32	1.0	0.0339 (%)
Population density	−1.51	2.5	99 (persons/km^2)
Distance from Bangkok	0.70	–	521 (km)

Table 13.3 Projections (based on Model I).

Percentage	Year of forest cover							
	1973	1976	1978	1982	1985	1990	1995	2000
Actual	30.01	24.57	18.49	15.33	–	–	–	–
Simple trend	–	–	–	–	12.06	8.08	5.42	3.63
Projection 1[a]	29.25	18.47	14.86	12.23	9.25	6.32	4.41	3.35
Projection 2[b]	29.25	18.47	14.86	12.23	9.57	6.49	5.11	4.10
Projection 3[c]	29.25	18.47	14.86	12.23	10.04	7.86	6.29	5.33

Notes: a) projecting past growth rates of independent variables; b) assuming a 3 per cent rice yield increase; c) assuming a 5 per cent decrease in the price of upland crops relative to rice in addition to a 3 per cent increase in rice yields (such a scenario requires a rice price support or subsidy).

results are reported under different scenarios in Table 13.3 and compared with the actual figures and a projection based on a sample trend of the dependent variable.

The first projection is based on the estimated trends of the independent variables without allowance for possible policy changes. The projected values differ considerably from actual and simple projections in earlier years but they converge in the later years. In the second projection we assume a three per cent rather than a one per cent increase in yield as originally assumed. The area under forest in the year 2000 is 24 per cent higher than in the base projection. In projection (3) we introduce – in addition to a three per cent increase in yield – a rice price support that would lower the ratio of the price of upland crops to rice by five per cent. The forest area in Northeast Thailand is almost 30 per cent higher than in the basic projection. Yet, as a percentage of the region's land area, the forest cover in the year 2000 is only slightly over five per cent, which is only one-third of the 1982 forest cover. Alternative scenarios were also projected for population, income, and the price of wood. The results indicated that no single policy instrument can reverse deforestation. A combination of policies is needed as discussed below.

13.11 Summary and policy implications

The purpose of this study was to develop a theoretically valid and empirically estimable model of tropical deforestation. This was done by introducing three demand functions: the demand for logging, the demand for fuelwood, and the demand for conversion to agricultural land. These individual demand functions derived from private optimizing behaviour provided the clues as to which variables to include in a reduced deforestation equation. The theoretically critical variables are the price of wood, the price of upland crops relative to rice, the price of kerosene (a substitute for fuelwood), and several demand shifting factors such as population, income, road density, irrigation infrastructure, rice yields, and distance from markets.

The model was then empirically estimated for the case of Northeast Thailand. All our *a priori* expectations as to the signs and significance of the variables included in the model were met. Not surprisingly, population followed by the price of wood and income level emerged as the most important causes of deforestation in Northeast Thailand. Accessibility factors such as population, income, road density, irrigation infrastructure, rice yields, and distance from markets were also important. The price of kerosene was also significant but because of collinearity with other prices it could only be tested in a separate model.

As these results are only preliminary and tentative it would be premature to derive policy implications. Certain policy instruments seem to present themselves, however, even at this early stage. A first critical step is the establishment of secure and long-term property rights over land and forests to ensure full consideration of the user cost in harvesting decisions. Second is the promotion of migration to other regions of the country and improved education, especially at the secondary level, to induce mobility to other regions and other occupations. Increased school attendance and higher labour participation of women would also help reduce family size and population growth.

Increased migration by itself would help increase incomes in the region by tightening the labour market and by generating increased remittances from other regions. Income growth is not only important in containing deforestation directly (by reducing both the demand for land from shifting cultivation and the demand for fuelwood), but also in reducing the desired family size, and hence population growth. Therefore, more steps need to be taken to increase income levels in the Northeast by encouraging off-farm employment and rural industry which is now discouraged by sectoral and macroeconomic policies that favour Bangkok. This should be done ideally by reforming those policies rather than through public projects or subsidies.

Northeast Thailand is the location of many wood-using industries. Wood is also in high demand for the construction of houses, shops, bridges, and other infrastructure. The rapid deforestation, the remoteness of the Northeast, and the high transportation cost of logs because of bulkiness has resulted in rapidly rising wood prices (16 per cent per annum during 1973–82). With open-access forests this led to increased deforestation, as our results show. Reducing wood prices, however, will not necessarily reduce deforestation because what is gained in reduced logging might be lost in increased conversion of forest land to other uses. The relationship between wood prices and tropical deforestation is not a monotonic one because it depends fundamentally on the institutional framework (property rights) and dynamic factors not captured by our model which is essentially a static one (for a dynamic model see Panayotou and Sungsuwan 1989).

Yet, the direction of policy reform with regard to wood prices should be clear.

To stem deforestation, current consumer prices must be "high" to induce substitution away from wood, while current producer prices must be "low" to reduce the incentive to liquidate the forest. The wedge between the two prices can be driven by a tax on logging that creams off resource rents. The lower the producer prices, however, the higher the incentive to convert the forest to some other use or – in the absence of private property rights over the forest land – to abandon the forest to encroachment. A combination of three factors can prevent this from happening beyond the optimal level: (a) long-term secure concessions extending beyond the next felling cycle; (b) rising price expectations (forest product prices are the only resource prices that have risen in real terms over the long haul); and (c) reforestation subsidies or tax incentives commensurate with the social benefits from forests not captured by private concessionaires. Such benefits include, among others, water, soil, and genetic resource conservation.

With regard to infrastructure, our findings suggest that whatever the social benefits from public infrastructure such as roads and irrigation systems, there are also social opportunity costs beyond the resources used in construction. These additional opportunity costs consist of both forest displacement and inducement of deforestation through displacement of settlements by irrigation reservoirs and increased accessibility of forests by road expansion. These costs should be taken into account in project evaluation at the planning stage of public infrastructure. We expect this to result in a number of adjustments: (a) more careful planning of scale, location, and timing of projects; (b) prior resolution of institutional problems (e.g., lack of property rights over forest lands to be opened up by new roads); and (c) to the extent that it is technically and economically feasible, justifiable attenuation of certain projects to gain more project-bound information on their environmental impact.

The case of Tung Jula Rong Hal in Northeast Thailand is a case in point. Only fifteen years ago the lower Northeast region of Thailand was covered by undisturbed forest. Then the area was made accessible by the construction of a major highway. According to Thailand's National Economic and Social Development Board (NESDB, 1982: 233): "landless farmers. . . from around the area and elsewhere have moved in and cleared the land for cultivation, resulting in the destruction of forest land (and watersheds) of 5.28 million rai (one million hectares) between 1973 and 1977. The sporadic immigration to clear new land for cultivation has given birth to 318 villages in the past nine years." Today the area is totally devastated by salinization and soil erosion, making both forestry and agriculture unsustainable. Had private and/or communal property rights been issued before the opening up of the area, both agriculture and forestry could have been sustainable.

Another important finding of the study is the significantly positive effect of rice yield on the forest cover. Because rice is the most important crop (80 per cent of cultivated area) in the Northeast, high rice yields reduce the need for extensifi-

209

cation of agriculture into the forest for the cultivation of upland crops. Again, here the Northeast is behind the rest of the country, with only 60 per cent of the rice yield attained in other regions of the country. This is because rice in the Northeast is mostly grown under rain-fed conditions and no significant advancements in rain-fed rice technology have yet been achieved. Given the limited scope for irrigation expansion in the region, more emphasis should be placed on agricultural research to develop drought- and flood-resistant high-yielding varieties for rain-fed rice.

A last variable that was found to be a significant determinant of forest cover is the relative price of upland crops to rice. The more profitable upland crops are relative to rice, the more deforestation takes place. Yet, historically Thai governments have taxed rice very heavily through export taxes, export premia, and rice reserve requirements. In contrast, upland crops (cassava, maize, kenap) have been left largely untaxed and, in fact, have received an implicit labour subsidy through rice taxation and an implicit land and "fertilizer" subsidy through open and free access to public forests. While the rice export premium has been suspended in recent years, the implicit "fertilizer" and land subsidy remains as long as the forest land is effectively open to encroachment.

In conclusion, the present study has provided some useful, albeit tentative, insights into the causes of tropical deforestation, particularly in Northeast Thailand. Further research is needed to corroborate the findings with additional empirical evidence and to determine the cost-effectiveness of alternative policy interventions for reversing tropical deforestation where it is determined to be excessive under the prevailing ecological, social, and economic conditions.

14

Deforestation in Thailand

Chiara Lombardini

14.1 Forests in Thailand

The Thai economy has been growing very rapidly since the end of the Second
World War, averaging a 4.4 per cent increase in real GDP between 1950 and 1990,
with annual growth rates of 10 per cent throughout the 1980s. Economic growth
has been accompanied by the gradual depletion of Thailand's natural resources.

Thailand is the country in Southeast Asia that recorded one of the highest rates
of deforestation in recent years (Hirsch 1990a). In 1961, 53 per cent of Thailand
was covered with forest. In 1989 only 28 per cent of the country's forest cover
remains. This means that Thailand has lost some 47 per cent of its forests over a 26-
year period.

The shortage of timber resources and fuelwood, the loss of biological diversity
and the destruction of watersheds are among the consequences of deforestation. In
1988 devastating floods related to the destruction of watersheds hit parts of southern Thailand, killing 350 people. In response to these floods the Thai government
declared a ten-year comprehensive nationwide ban on all logging starting from
January 1989, the effects of which are now being investigated (Sadoff 1992).

The actual target of the Thai National Forest Policy adopted in 1985 is to reach
and maintain a 40 per cent forest cover, of which 15 per cent will be devoted to
conservation forests and 25 per cent to productive forests. This target is the result of
a revision of the national policy target of 50 per cent forest cover set in the 1950s.
Given the actual 74,000 hectare afforestation annual target, and provided that deforestation can be totally stopped, it is estimated that it will take 100 years to attain
a 40 per cent forest cover.

In Thailand 45 per cent of land area in 1989 was *formally* registered as the
national forest reserve belonging to the crown. A significant portion of the national
forest reserve, however, has no trees at all, having suffered heavy encroachment. At

least 1.2 million families, approximately 20 per cent of Thai farmers, are estimated to rely on national forest land for their livelihood (Hirsch 1990b).

In the absence of adequate controls over occupation, state forest land becomes an open access resource. However, to view Thai forest as an open access resource, although correct in strictly legal terms, does not appropriately reflect reality. A number of policy practices by the Thai state have, in fact, led to a very ambiguous situation, so that most of the small farmers occupying state forest reserve land regard it as their own. First, agricultural land taxes are collected by the Lands Department of the Ministry of the Interior on its land. The receipts of these taxes are frequently used as title deeds, both in land transactions between farmers and in negotiation with state authorities in order to establish prior occupancy. Moreover, since many of those who in the past had illegally occupied forest reserve land were later granted legal tenure, most farmers expect that "sitting tight" will be sufficient in order to obtain legal tenure on the encroached forest reserve land (Poffemberg 1990).

14.2 The causes of deforestation in Thailand

The purpose of the model presented here is to assess quantitatively the causes of deforestation for Thailand as a whole. The model differs from the existing studies both in the choice of proxy explanatory variables (for example, see Katila 1992a) and in the regional unit of observation – Thailand as a whole as opposed to North-east Thailand (Panayotou & Sungsuwan, Chapter 13, this volume).

From the existing qualitative and quantitative analyses of deforestation in Thailand, it appears that the main driving force behind deforestation in Thailand is the demand for agricultural land fuelled by the need to feed and employ the growing population, as well as by the desire to produce cash crops to earn foreign exchange. The government's attitude favouring extensive agriculture and cash crop development has, in the past, further encouraged such demand. An example is given by the government subsidies to cassava farmers to increase export earnings which led to widespread loss of Thailand's Eastern forests (McNeely & Dobias 1991).

As a measure of cash crop production, the model uses the aggregate value of the exports of cassava, maize, rubber and sugar cane. Most of the deforested land in Thailand has been converted to production of these crops, although part is devoted to the production of rice and *kenaf*.

As measures of the demand for agricultural land, the model uses the absolute level of population, and the share of the labour force employed in agriculture (a proxy for off-farm employment opportunities). Increased opportunities for off-farm employment are expected to reduce population pressure on the forests. The share of the labour force employed in agriculture seems a more satisfactory proxy than the share of services and manufacturing of total GDP proposed by Katila (1992a). In fact, whereas the agricultural share of Thai GDP has steadily decreased (one-third in 1961 compared to one-sixth in 1988), the share of the population employed in the agricultural sector did not show a similar decrease (from 77 per cent in 1961 to well over 60 per cent in 1989). A third proxy could be found in the rate of urbanization, on the assumption that people in the cities will not support themselves by working in the agricultural sector. The reliability of this variable could be checked by examining migratory trends to and from the cities.

Agricultural productivity is expected to affect the demand for agricultural land – in which direction, however, it is not clear. On the one hand, increased agricultural productivity may mean that less land is needed to meet the basic food requirements of subsistence farmers, but on the other hand increased productivity may well result in a greater demand for agricultural land for cultivating export oriented cash crops, as Katila's (1992a) findings imply. As a proxy for agricultural productivity, the average cassava yield per cultivated hectare is used.

The demand for forest land for commercial timber harvesting and for fuelwood seems to have also played an important rôle in the deforestation process (Hafner 1990, Panayotou & Sungsuwan 1989, Katila 1992a). Log prices, government rent capture, population, income growth, and accessibility are all believed to affect this demand. Unfortunately, no measures of government rent capture are available. Accessibility both encourages logging by decreasing its production costs, and is increased by logging through the building of roads to penetrate the forests for harvesting timber. In Thailand road construction has been formulated in response to national security objectives, such as the eradication of insurgency groups located in inaccessible forest areas. Increased accessibility allows encroachers to penetrate the forest more easily, encouraging more forest to be converted into agricultural land. In fact, the migration of farmers to forests reclaimed from rebel forces following the opening of roads has been actively encouraged by the Thai government through the Accelerated Rural Development Programme, the ARD. Population directly affects the demand for fuelwood and wood for construction, the latter being also stimulated by economic growth.

Poverty, as measured by GDP per capita, appears to have a negative impact on forest cover, i.e. the higher the per capita income, the lower the deforestation (Panayotou & Sungsuwan 1989, Katila 1992a). The results of Tongpan et al. (1990) however, are in contradiction with this thesis. This may be due to the fact that per capita GDP is not an appropriate proxy for poverty, but for economic growth only, which without appropriate redistribution may be compatible with increasing or constant levels of poverty. The lack of sufficient data on income and land distribution did not permit the introduction of an adequate proxy for poverty in the model.

Economic growth is expected to affect forest cover. The direction of its impact is, however, ambiguous. On the one hand, per capita income growth normally implies an increase in the share of GDP of the manufacturing and service sectors and therefore the growth of employment opportunities in sectors other than agriculture. More resources are available to enhance agricultural productivity and part of the domestic demand can be satisfied through imports of agricultural products. However, per capita income growth can initially increase the demand for traded food, thus stimulating an increase in the area under permanent agriculture. As income grows, the demand for fuelwood is likely to decrease since kerosene can be substituted for fuelwood (Panayotou & Sungsuwan 1989), whereas the demand for timber for house and furniture construction is likely to increase, especially among urban populations and rural elites (Hirsch 1987). Finally, higher incomes may also increase the demand for environmental quality.

As a measure of economic growth per capita GDP is used in the model.

We specify the demand for forest land as follows:

LOGFC = INPT + b1 LOGPCGDP + b2LOGPOP + b3 LOGLABAGR

+ b4 LOGAGEXP + b5 LOGAGPR + b5LOGPRLOGS + b6LOGROADS

where:

LOGFC is the logarithm of forest cover expressed in millions of rai (1 rai = 0.16 hectares);

INPT is the intercept;

LOGPCGDP is the logarithm of real per capita GDP in millions of 1980 baht;

LOGPOP is the logarithm of population (in thousands);

LOGLABAGR is the logarithm of the share of the labour force employed in agriculture, a proxy for off-farm employment opportunities;

LOGAGEXP is the real f.o.b. value of the exports of rubber, sugar, maize, and tapioca products (in millions of 1980 bahts);

LOGAGPR is the logarithm of the average yield of cassava in metric tonnes per cultivated hectare, used as a proxy for agricultural productivity;

LOGPRLOGS is the logarithm of the price of logs (in constant 1980 US$ per cubic metre);

LOGROADS is the logarithm of paved and unpaved kilometres of roads.

Particular attention needs to be paid to the choice of the dependent variable. Most econometric studies on deforestation use forest cover, often a percentage rather than absolute forest cover, as a dependent variable, the reason being lack of accurate data on deforestation and too few degrees of freedom. Some authors, however (for example, Kummer 1991), have strongly argued against the use of this variable. Two regressions are thus estimated using both dependent variables, absolute forest cover and percentage forest cover.

14.3 Data and methods of estimation

The source of data on forest cover is the Thai Office for Agricultural Economics as reported by the Royal Forest Department (Katila 1992a). Data on GDP are obtained by the World Bank *World Tables*. The *Statistical Yearbook for Asia and the Pacific* published by the UN provides the data on population, roads, agricultural exports and agricultural productivity. The World Bank *World Tables* and the FAO *Production Yearbook* are the sources for data on the labour force active in agriculture. Finally, data on log prices are obtained from the World Bank *Commodity Trade and Price Trends*, 1986 edition.

The model was estimated by OLS. Several functional form specifications were tested.

14.4 Regression results and indications for further research

The results of the regression are presented in Table 14.1, and the data used in the analysis in Table 14.2. Only the coefficients of per capita GDP and of the labour force employed in agriculture are significant. The negative relationship between forest cover and per capita GDP, although explainable, contrasts with the findings of the previous studies. Moreover, the coefficient of the labour force employed in agriculture is, against our expectations, positive and significant. The equation was therefore re-estimated by using another proxy for off-farm labour opportunities, namely urban population. Once again, against our expectations, the coefficient was negative (although non-significant). The regression results thus seem to refute the thesis that the greater the employment opportunities outside the agricultural sector, the higher the forest cover.

The equation was re-estimated by using deforestation as a dependent variable and, surprisingly, the result was that the null hypothesis, that all the equation coefficients were simultaneously equal to zero (F-statistic), could not be rejected.

Table 14.1 Deforestation in Thailand: regression results.

LOGFC = −21.62   − 0.621 LOGPCGDP★ + 0.9757 LOGPOP
  (−2.04)   (−3.433)   (1.56)

  + 0.0030 LOGAGREXP   + 0.015 LOGAGRPR   + 0.0053LOGPRLOGS
  (0.1164)   (0.363)   (0.199)

  − 0.0012 LOGROADS + 3.06 LOGLABAGR★
  (−0.137)   (3.253)

Notes: Adj. R^2 = .99; N = 18; DW = 1.0867; F(7,10) = 376.65
★ significant at the 0.01 level

Table 14.2 Data for regression analysis.

	FC	POP	LABBAGR	PCGDP	AGREXP	AGPR	PRLOGS	ROADS
1967	156.48	33,000	80.9	8.1505	36,392.0	16.6290	113.4	13,410
1968	152.80	34,040	80.5	8.6024	4,235.0	20.0923	117.0	14,953
1969	149.10	34,678	80.1	9.1087	5,353.9	20.7692	107.3	15,697
1970	145.42	35,745	79.8	9.6573	5,517.7	15.3170	106.6	16,293
1971	141.88	36,884	78.8	9.8321	5,812.6	14.1545	103.3	17,105
1972	138.32	38,017	77.8	9.8989	6,758.0	12.1159	94.0	17,686
1973	134.71	39,142	76.9	10.567	11,240.0	13.1204	141.4	18,672
1974	134.56	40,257	76.1	10.719	18,706.0	13.1924	139.1	19,507
1975	130.76	41,359	75.3	10.929	19,472.0	13.6594	94.4	20,097
1976	124.01	42,450	74.3	11.644	25,343.0	14.4829	125.0	21,681
1977	116.57	43,532	73.4	12.442	24,676.0	12.8875	128.3	22,882
1978	109.52	44,602	72.5	13.430	27,166.0	13.9070	114.0	25,066
1979	106.39	45,659	71.7	13.785	32,683.0	13.9635	175.7	27,498
1980	103.42	46,700	70.9	14.101	37,512.0	14.7415	192.9	28,151
1981	100.58	47,727	70.26	14.667	45,208.0	14.2719	143.9	30,016
1982	97.88	48,740	69.62	14.941	50,504.0	16.3643	148.8	31,001
1983	96.27	49,739	68.98	15.679	41,998.0	18.6532	140.3	33,148
1984	94.70	50,720	68.34	16.474	44,973.0	14.9588	159.5	34,702

Notes: FC: millions of rai; POP: population in thousands; LABBAGR: percentage of labour force employed in agriculture; PCGDP: millions of 1980 Thai baht; AGREXP: agricultural exports, 1980 constant million baht; AGPR: agricultural production, metric tonnes per hectare; PRLOGS: 1980 constant US$ per cubic metre; ROADS: kilometres.

Moreover the goodness of fit was extremely low. This result seems to suggest that the choice between using deforestation and forest cover as dependent variable does significantly affect the regression results. More research is thus needed on this particular issue.

The poor database might be the cause of the unexpected results, so that it is not appropriate to draw any policy implications from the study until the model can be re-estimated with better data.

15

Government failure and deforestation in Indonesia

Diane Osgood

15.1 Government policy and deforestation

The Indonesian Government turned its attention to the forest sector in the late 1960s when the Suharto government faced severe economic problems and began to see the forest as a ready revenue source. First, it was identified as a source of profits from extracting timber, later as a source of land for agricultural expansion. Prior to 1967, little of Kalimantan's forests had been disturbed, and most commercial forest harvesting took place in the teak forests of Java. All of this changed rapidly after 1967 when the economy was opened to international investments and investors. Generous timber concessions and tax holidays together with greater international investment, created an environment for massive increases in the timber and timber related industries. Between 1970 and 1975, the harvest doubled, with most of the increase in Kalimantan Timor (east). By 1979, gross foreign exchange earnings were US$2.1 billion, and Indonesia was the world's leading exporter of tropical logs with a 41 per cent market share. After 1979 the log harvest began to fall due to several factors, including a weakened world market and government policies.

Government policies have been used to favour forest-based industrialization. Government taxation and export restrictions greatly affect the rate of timber harvest and provide an example of how economic loss as well as natural resources waste can result when incentives permit or encourage inefficient activities. In this case, as forest stocks are depleted, neither the government treasury nor national economy benefits much from the exploitation. One can argue that these policies gave rise to new processing mills in Indonesia and created desperately needed jobs, but this benefit is far outweighed by the social and economic costs incurred. The timber industry is one of the least labour intensive of all major productive sectors in Indonesia. In 1982 it employed less than 0.2 per cent of the Indonesian workforce.

From 1972 to 1985 employment in the timber industry grew by only 1.7 per cent – the "boom" years for the industry. Moreover, employees were often migrant workers from the Philippines and Malaysia. The gains for local employment therefore appear to have been small.

In order to encourage local processing, the Indonesian government raised the log export tax from 10 per cent to 20 per cent. However, most sawn timber and plywood was exempted. Mills were also exempt from income tax for their first five years. Starting in the early 1980s, export controls of logs were progressively enforced; and in 1985 an outright ban was introduced. One of the consequences of these incentives was a jump in the number of mills from 16 in 1977 to over 100 in 1983. With the favourable export trends predicted, the only constraint on Indonesian wood production is the capacity of domestic log production. The government plans for continued increase in mill capacity throughout the 1990s. At 1991 conversion rates and mill capacities, there is a maximum annual demand for logs of 47 million cubic metres per year (Barbier et al. 1992a). Currently mills are only running at 60–70 per cent capacity, implying that current levels of wood demand will rise (assuming no change in utilization). However, processing operations are inefficient under the protected status of mills; Indonesia has the lowest log conversion rate in Asia. Gillis (1990a) found that Indonesian plywood production utilizes 15 per cent more wood than elsewhere in Asia. Such inefficiency and taxation policies lead to costly consequences. For example in 1983 the conversion rate was 2.3:1 for plymills and 1.75:1 for sawmills. That year a cubic metre of plywood exported earned $250, but the export value of the roundwood (log) equivalent was only $109. The logs were worth $100 on the international market, which means the labour added value in plymills was only $9 in export earnings per cubic metre, while government sacrificed $20 in foregone tax revenues on diverted log exports. Furthermore, in 1983 plymill exports worth $109 per cubic metre of processed logs (international prices) cost the equivalent of $133 to produce. This loss was sustainable only because of the government's financial incentives and forgiveness on log export taxes. The folly is even more obvious in the case of sawn timber (plywood). In 1983 a cubic metre of sawn timber fetched only $155. The logs, unprocessed, could be exported for $100. Thus, due to the low conversion rate, Indonesia exported sawn timber worth $100 for only $89. Here the government was also foregoing $20 in export tax in order to lose $11 in export earnings on every cubic metre of logs sawn domestically. Repetto and Gillis (1988) estimated the annual revenue loss from 1988 to be $400 million. In October 1989 export taxes on sawn timber were increased to shift processing activities to plywood. Indonesia is not alone in such policy moves. Gillis (1990a) has observed a trend of export tax structures largely replaced by export bans in tropical forest countries.

Subsidies were also established to curtail deforestation as well. By the mid-1960s, deforestation and erosion arising from fuelwood collection on Java was a serious problem. A domestic kerosene subsidy was in place throughout most of the 1970s and early 1980s and by 1979–80, the subsidy ran up an estimated economic cost of US$16 million. However, kerosene in rural households is not a good substitute for firewood, as it is rarely used for cooking. Official Indonesian statistics (Biro Pusat Statistik 1982) recorded 85 per cent of rural Javanese households still using firewood to cook in 1980.

The main force behind deforestation in Indonesia is the need for land for agriculture to feed a quickly growing population (World Bank 1991, Constantino & Ingram 1990, Pearce et al. 1990). Government strategies such as the first 5 Year Development Plan (*Repelita 1*, 1969–74) which centred around improving agricultural production and expanding agribusiness, and *Repelita 3* (1980–84) which focused on self-sufficiency in basic foods, may have led to forest conversion. In theory, rational farmers will continue to deforest land for frontier agricultural production when it is in their economic interest to do so. That is, land will continue to be cleared up until the point where the marginal cost of clearance and production equals the marginal benefits. Modelling the decision by farmers to move onto new plots over time has received relatively little attention in the literature (see Barbier 1991, Southgate & Pearce 1988, and Schneider et al. 1990 for further discussion).

In Indonesia, one million families are estimated to be engaged in shifting cultivation which results in a reported clearing of 400,000 hectares of forest annually (Setyono et al. 1985 as referenced in Repetto & Gillis 1988). Government policies on agriculture have taken the view that forest land is there to convert, often disregarding the other services and goods provided by the forest. Programmes such as fertilizer subsidies result in direct incentives to expand forest conversion activities. In 1984 and 1985, the World Bank issued reports on Indonesia that were most critical of such subsidy programmes. Despite resistance to structural change (see Robison et al. 1987), some subsidies were removed. One subsidy cut was on pesticides. Between 1985 and 1988, subsidies were cut by 80 per cent of 1985 levels, resulting in a $1.2 million savings, annually.

The second area of government failure is the mismanagement of public forests by inefficient forestry policy. Forestry policy includes the terms of timber harvest, (duration, permissible annual harvest, harvest methods, royalties and licence fees), the utilization of non-wood forest products, and reforestation. In Indonesia, and most other tropical forest countries, timber royalties are assessed on timber removal rather than the stock of trees standing. This "ad valorem" system means that selective cutting is favoured and results in more forest damage and depletion of virgin forests than under other royalty regimes. Repetto & Gillis (1988) recommend a differentiated and constantly updated specific royalty system by which the least valued trees are taxed the least. They recognize that most countries lack both staff and information to operate such a system.

The size and duration of concessions also affects the production capacities and incentives for conservation. Because the concessions are large (well above other Asian countries' averages) and the time span short (20 to 25 years maximum), logging companies have weak incentives to reduce damage or safeguard future productivity. According to FAO guidelines, tropical hardwoods should be allowed to stand for at least 20 to 25 years after a harvest. With such short concessions, most loggers return to reharvest within 5 to 10 years. It is thought that up to 40 per cent of the stand is damaged and may not survive after loggers have entered the forest; furthermore, there are other techniques which could reduce the residual damage by 50 per cent (Whitmore 1984). However, these techniques are both time consuming and require direct incentives to be employed. Concessions of 70 years or more and smaller land size would help reduce destruction (Repetto & Gillis, 1988, Whitmore 1984, Barbier et al. 1992a).

Some government failures are the result of lack of policy and resources. Policies that promote non-wood forest products not only increase the value of the forest area, they are protective *and* productive. However, only a very small proportion of the forestry department budget goes into this area, and non-wood products are not mentioned in the latest Forestry Plan. A larger budget share and national recognition could help the marketing and research necessary to bring these products and their potential into the deserved and desired spotlight.

Another very critical factor in the conversion of Indonesian forests is transmigration. Transmigration was first employed by the Dutch colonial government as a pressure valve with which to relieve over-populous and politically unstable Java. More recent policies have resulted in over 2.5 million people moved between 1971 and 1980. In the early 1980s, cost were around $10,000 per moved household. The government currently projects the resettlement of one million families or more by the year 2000. This does not include the numerous unofficial or "spontaneous" transmigrants. Transmigration is potentially linked to deforestation because the resettled families are given land to farm (and clear if necessary). According to Repetto & Gillis (1988), currently 80 per cent of transmigrants are settled in primary forests. However, many of the farming attempts by both official and unofficial migrants have failed in part because of inadequate assessment of the agricultural capabilities of the soils in the Outer Islands. Despite their low population densities, the Outer Islands do not offer large areas of good, unutilized agricultural land. For the most part, the tropical soils are nutrient-poor, easily leached, and erodible. Most of the nutrients are held in the biomass or the first inch or two of the soil. Repetto (1988b) found for the most part the better agricultural lands in the Outer Islands are already occupied. Hence the farmers are left with a need to move on after a few years, extending into the forest and using "slash and burn" or swidden techniques. In all countries studies by Repetto & Gillis (1988) they found that deforestation by shifting cultivators and timber operations are closely linked. Settlers and cultivators travel along logging roads after the timber harvest and further clear the forest for their own purposes. A World Bank Report (1989b) lists smallholder conversion of land as the number one factor of deforestation in Indonesia. Although one cannot deduce that this is entirely or even largely a result of transmigration, we do know that the numbers of new farmers are very large and can therefore assume a sizable share of the conversion is due the transmigration programme.

15.2 A model of deforestation

This section uses econometric techniques to derive a simple model to clarify the significance and relative importance of the various factors of deforestation in Indonesia. Panel data on 20 regions for the years 1972, 1973, 1975, 1979, 1981, 1982, 1984 and 1988 were used. This approach was also used in a recent study by Constantino & Ingram for the FAO (Constantino & Ingram 1990) although they used data for only five years. The model here differs in focus and purpose. The Constantino–Ingram model was developed to predict deforestation rates, and therefore used variables which are relatively easy to project. The objective here is to discover potential areas of government policy influence on the rates of deforestation. Constantino & Ingram found forest cover to be negatively influenced by population density, and time (a proxy for other variables, particularly roads), and positively influenced by rice productivity and income per capita. In contrast, the model presented in this chapter focuses on individual crops, estate versus other plantations, actual kilometres of roads, transmigration, and indicators of macroeconomic policy.

Designing a multi-variable regression model is considerably constrained by the data available, which are not only quite limited, but quite questionable even when available. The results are undoubtedly affected by the many sporadic jumps in the data and the choice of dependent variable – the logarithm of the ratio of forest area to total land area, which from now on will be referred to as forest coverage. Other studies have criticized using forest coverage as a dependent variable as it does not pick up regional differences in original forest coverage. This criticism is not valid here due to the employment of regional dummies which control for regional variation of original forestation. However, by using forest cover, the distinction between primary and secondary forests which could be important for issues of biological diversity and some ecological functions is lost. However until better deforestation data are available, such detail will be lost from models.

The factors tested for can be broadly classified as follows:

(a) government policies – which may directly affect forest coverage such as road construction, and supply side management (afforestation), transmigration, and log production;

(b) agricultural factors which include both individual crop production figures (maize, cassava, peanuts, soya bean, sweet potatoes and paddy rice) and estate and other plantation production;

(c) macroeconomic factors, which also can be considered as government policy, including exchange rate policy, net capital flows, and external debt.

Although ideally some measure of government taxes and subsidies would be included in such an analysis, the lack of data severely constrains this approach. Education and charcoal production may also be relevant but the regional data are not available. Transmigration is thought to affect forest coverage negatively, as is the construction of roads which grant easier access to forests land for conversion by shifting cultivators. Agricultural crop production is expected to affect the demand of agricultural land; however it excludes the impact of a change in productivity. This constraint also remains due to lack of better data.

By testing for specific crops (maize, sweet potato, soya beans, peanuts, wet and dry rice paddy combined, and cassava in production tonnes) the aim was to highlight areas for future research, and to examine any roles of government policy in the production of that specific crop. Estate and "other" plantation production is included for the same reason. Structural adjustment programmes, usually resulting from high debt servicing commitments, may affect macroeconomic policies by constraining the level of imports and encouraging greater exports (for example, timber) so that the level of debt is included in the model. Following Capistrano & Kiker (1990) an overvalued exchange rate acts as a subsidy to urban consumers on imported goods, while implicitly taxing agricultural, timber, and other export goods produced domestically. Real currency devaluations, as frequently required by the structural adjustment programmes, remove existing distortions and provide incentives for greater domestic production of exportables. Capistrano & Kiker (1990) found that in the 1972–75 period, debt had a negative relationship to deforestation due to the easing on international credit. Net capital inflows are therefore included in the model to capture this possibility.

Timber production is expected to have a negative impact on forest cover. The model uses total log production by province as reported by the FAO (1992a). Data for East Nusa Tenggara was incomplete for 1981–88 and for Irian Jaya for 1979 and figures were interpolated accordingly.

Supply side management of the forests should also affect levels of forest cover. The Indonesian government has had an afforestation project since 1948. In 1980 they enacted a reforestation deposit fee of $4 per cubic metre of tropical timber extracted in Kalimantan and Sumatra. The deposit may be refunded upon verification of adequate replanting. However, survival rates of reforestation plantings were estimated at 72 per cent for Sumatra and 54 per cent for Kalimantan (FAO 1981c).

Data were gathered from the FAO (forest coverage, size of regions, log production, foreign exchange, estate and other plantation production, debt, foreign exchange), from the Annual Pocketbook of Indonesian Statistics – *Buku Saku Statistik Indonesia* – for individual crop production, kilometres of roads, transmigration and reforestation figures, and from Lazard Freres et Cie financial report on Indonesia (net capital flows). All the variables except for debt, foreign exchange and net capital in flows were normalized against the total area of the region.

Dummy variables were employed to capture region specific effects not captured by the other variables described above. Region specific dummies were included in all regressions.

15.3 Model results

A general model was run first, regressing forest coverage on log production, estate and other plantation production, reforestation, transmigration, roads, debt, foreign exchange, capital inflows, and region dummies. The results are reported in column 1 of Table 15.1.

As expected, the coefficients of estate crop production and log production were negative and significantly different from zero at the five per cent confidence level. The coefficients for transmigration, debt, and capital inflow were also negative, but not significantly different from zero. The coefficient of roads was also negative but insignificantly different from zero, casting doubt on Constantino & Ingram's use of time as a proxy for roads. The time trend effects they identify are more likely to be caused by other variables. The coefficient for reforestation was surprisingly negative, however weakly so, and only significantly different from zero at the 10 per cent confidence level, indicating reforestation efforts are not having a serious impact on forest cover. Unreliable data may be a reason for such a result. The foreign exchange coefficient was also positive but insignificantly different from zero. The surprise was the "other" plantation production coefficient which was positive and significantly different from zero at the seven per cent confidence level. This finding suggests that non-estate plantations may not be expanding into the forest and may be relieving pressures on the forest. This may be because of farming techniques which do not result in the need for more land (unlike traditional swidden practices) or improved yields per hectare. In any case, more research in this area is required.

In the second regression, the insignificant variables (foreign exchange, debt, capital inflows, roads, reforestation) were dropped. The results are in line with the previous run. The coefficient of reforestation did not strengthen and was still not significantly different from zero at the 5 per cent confidence level, thus it must be rejected as an influence on forest cover.

In a third run, the more specific crop details were added. The coefficients for

Table 15.1 Non-optimal forest coverage: regression results.

Explanatory variable		(1)	(2)
Lestate	Log of estate crop production (tonnes)	−0.4304 (−2.179)	−5.346 (−3.297)
Lother	Log of non-estate crop production (tonnes)	0.1959 (1.784)	0.1570 (2.033)
Llog	Log of log production (tonnes)	−0.1811 (−4.598)	−0.1775 (−5.439)
Trans	Transmigration figures	−0.0001 (0.60)	
Lroad	Log of kilometres	−0.0412 (−0.261)	
Lrefor	Log of hectares which have been "reforested"	−0.0716 (−1.636)	
Debt[a]	National external debt of Indonesia	−0.0001* (1.088)*	
Forex[b]	Exchange rate of Rupia vs. $US	0.0011* (1.275)*	
Cap[c]	Net capital flows for Indonesia	−0.0001* (−0.630)*	
Lpea	Log of peanut production (tonnes)	−0.6276 (−3.247)	−0.7408 (−4.711)
Lspot	Log of sweet potato production (tonnes)	−0.0059 (−0.056)	
Lmaize	Log of maize production (tonnes)	−0.200 (−0.343)	
Lpaddy	Log of Paddy rice production (tonnes)	−0.0979 (−1.280)	
Lcass	Log of cassava production (tonnes)	−0.0049 (0.081)	
Lsoya	Log of soya bean production (tonnes)	+0.2230 (0.408)	
R-squared		0.5743	0.5938

Notes: the figures used are per region, except those marked * which are national figures; a) Dependent variable used: forest cover; b) T-stats are given in brackets; c)Regional dummies were used, but are not given here.

estate, other, and log production remain significantly different from zero as found above. Of the individual crops, only the peanut production coefficient has a negative and significantly different from zero impact on forest cover. The coefficient for rice paddy is negative but only significantly different from zero at the 20 per cent confidence level; and therefore the hypothesis that it influences forest cover must be rejected. This model indicates that the production of cassava, sweet potatoes, soya beans, and maize does not significantly affect deforestation.

In the last run, the insignificant crop variables were dropped. The results are shown in column 2 of Table 15.1. This model seems to be a "good fit" as shown by the R^2 of 0.5938. The coefficients of estate plantation production, log production and peanut production were all negative and significantly different from zero at less than the five per cent confidence level. The coefficient of "other" plantation production remains a positive and significantly different from zero at the four per cent confidence level. In this final model, the coefficients for transmigration and rice paddy production remain negative but insignificantly different from zero, and must be rejected as significant influences on forest cover.

The model can be summarized as:

$$FOCOV = f (logs, \quad estate, \quad other, \quad peanuts)$$
$$(-) \quad (-) \quad (+) \quad (-)$$

The findings seem robust. There was little variance in the findings when the model was tested for different specifications, nor did the use of time dummies alter the findings. The rôle of population pressures, as measured solely by transmigration figures, are contradictory to expectations. The poor database and lack of figures relative to pre-transmigration populations may be the cause of the unexpected results. The results call for more research into the rôle of the government in estate and other plantations, as well as peanut production.

15.4 Conclusions

The results of the econometric analysis appear only to add ambiguity to the efforts of finding global causes of deforestation because they seem to conflate past findings of agricultural production as negatively related to forest cover; nor do they offer positive insight on the rôle of macroeconomic polices, such as debt and capital flows. In this study, as in others, log production has been identified as a force bringing negative fortune to the forests. This is an intuitively appealing result.

16

An analysis of the causes of deforestation in India

Manab Chakraborty

16.1 Introduction

This chapter develops an econometric model of deforestation for India using largely macro level data. The Indian case is particularly interesting for two reasons: a) since 1878 India has had a trained cadre of state forest managers; and b) concerted and successful efforts in recent years have been made to reduce the rates of deforestation.

16.2 Extent and rate of deforestation

According to FAO (1990b) estimates, India lost 132,000 hectares of forests in 1990 which is approximately 1/24th of the rate of loss in Brazil, or 1/10th of Indonesia. However, such international comparison is spurious. First, both Brazil and Indonesia have respectively eight times and 2.5 times more forests than India. And more importantly, at the present rate of deforestation and competing land-use pressures, India may lose all its forest by AD 2010. In official pronouncements, deforestation ranks among the highest of the environmental problems faced by modern India (DOE 1989). Despite the continuous, often fractious debate on the source and degree of deforestation, it is widely agreed that deforestation has reached an alarming rate and must be arrested.

As early as 1950, the national planners deemed a total forest cover of one-third of India as a desirable target. This objective has been clearly reiterated in the National Forest Policy, 1988: "the national goal should be to have a minimum of one third of the total land area of the country under forest or tree cover". However, the forested area stood at only 19 per cent of India's geographical area in 1989

226

and, what is more, is still declining.

The Forest Department is the key actor in the care and management of the forest land. A notable feature of Indian forest management is the virtual absence of private ownership. In 1980, of the total forest land, the Forest Department owned 93 per cent, parastatals five per cent, private individuals two per cent. Typically, the forests in India are classified for legal purposes as reserve, protected and unclassed. The non-reserve forests are open to collective public use and are generally degraded. The reserve forests impose severe restrictions on public use and are considered to be in "a state of reasonable maintenance" (Lal 1989: 432). Critics have pointed out that a large percentage of the recorded forests is without any vegetation (Agarwal et al. 1982). Micro level studies have shown discrepancies between official estimates and true ground assessments (Bowonder et al. 1985).

In the absence of reliable evidence on actual vegetation changes between 1952 and 1980, it is difficult to generalize the micro level studies to estimate the rate and extent of deforestation for the whole of India. What is clear is that India's best forests are represented by the reserved forests. Changes in reserved forest area partially captures the improvement in vegetation cover, provides an ideal ground for conservation and finally depicts political will to address the problem of deforestation.

In India, there are no ideal, consistent measures of deforestation available. Time series data on actual forest cover are missing. Official statistics are at best indicative. Among the many indicators of state of forest used in the Indian forest literature, one comes across:

(a) ratio of good forest cover to total land surface;

(b) ratio of good forest cover to total forest land;

(c) dense forest canopy (over 40 per cent tree cover per hectare);

(d) percentage of closed forest land to total forest land.

However, none of these is well defined or consistently evaluated nor are data published regularly. Under the circumstances, we shall use reserved forest area as a second best proxy. We are fully aware that change in the reserved area does not fully reflect the extent of deforestation. However, this limitation is not overwhelming for the purpose of this study which is to explain the movement in reserved forest area as a product of influences exercized by exogenous variables (livestock and population pressure, afforestation, rent seeking, etc.).

16.3 Changes in forest area

Table 16.1 shows that India's total forest area during the period 1951–80 has risen considerably. Due to large scale diversion of lands for non-forest uses, forest area – particularly reserved land – showed a decline during the 1950s. However, by 1986,

Table 16.1 Classification of area under forests (thousand hectares).

Category	1950–51	1960–61	1970–71	1979–80	1985–86
Total Forest Area	71,803	69,135	74,961	73,669	75,227
I Forest types					
a) Merchantable	58,460	52,701	47,654	56,493	49,246
b) Inaccessible	13,343	15,289	19,426	16,042	25,981
II Legal status					
a) Reserved	34,405	31,631	31,798	37,252	40,612
b) Protected	11,793	20,355	20,444	22,537	21,509
c) Unclassified	25,570	15,069	12,946	10,558	13,107
II Composition					
a) Coniferous	3,630	4,435	4,312	4,767	
b) Broadleaved					
i) Sal	10,554	11,365	8,072	12,013	71,164
ii) Teak	4,347	9,309	7,100	8,351	
iii) Miscellaneous	53,272	43,152	4,4011	46,672	

Sources: Agarwala 1989, Department of The Environment 1987.

the reserved land had increased by 17 per cent. This increase in reserved area has come about by reclassifying protected and unclassified forests. The unclassified forests are owned communally or by small private owners. Interestingly, the area under both coniferous and broad leaved species has increased.

As noted earlier, the total forest area does not provide a good indication of the quality and extent of tree cover. The Indian Forest Department uses an alternative term – "total forest area" – to indicate all land with tree canopy exceeding 10 per cent per hectare. Obviously, total forest area is less than the area under forests.

Table 16.2 on land utilization statistics reports the changes in area under forests

Table 16.2 Land utilization statistics (million hectares).

Type of land	1950–51	1980–81	Change since 1951	1984–85	Change since 1951
Reporting area	284.3	304.2	+19.9	304.4	+20.1
Area under forests	40.5	67.4	+26.9	67.2	+26.7
Area not available for cultivation	47.5	39.6	–7.9	40.5	–7.0
Other uncultivable land excluding fallow land	49.4	32.2	–17.2	31.1	–18.3
Fallow land	28.1	24.6	–3.5	24.9	–3.2
Net area sown	118.8	140.3	+22.2	140.7	+ 21.9
Area sown more than once	13.1	33.3	+19.9	35.2	+22.1
Total cropped area	131.9	173.3	+41.4	175.9	+44.0

Sources: Agarwala 1989, CSO 1989.

Table 16.3 Summary of changes in land-use 1950–1985 (million hectares).

Addition to:		Supply sources:	
net area sown	21.9	changes in India's territorial area	20.1
area under forest	26.7	reduction in uncultivable land	28.5
Total addition	48.6	Total sources	48.6

between 1950 and 1985. Two things are clear: first, the increase in the area under forests and cultivation have come about through a reduction in uncultivable land and India's territorial expansion. In 1961 Goa and 1971 Sikkim were added to the Indian Union (see Table 16.3). It can be safely said the quality of additional forest land was poor and lacked good vegetation cover. Second, the increase in net area sown due to reclamation of uncultivable land was more than compensated by increases in cropping intensity. The increment in the area under cultivation was not at the expense of forest land.

Lack of remote sensing technology and limited ground checks have fuelled the debate on the real extent of forests (Sharma et al. 1990). For the first time, Landsat imagery for 1985–87 has enabled crown density to be taken into account. The remote sensing assessment places the figure for actual forest cover at 64.01 million ha against the officially recorded area of 75.1 million ha. Compared to the 1987 Landsat estimate, there has been a reduction of 0.19 million ha of forest cover during 1986–1989 or roughly an annual rate of loss of 47,500 ha. This figure of total annual deforestation rate is consistent with the 48,000 ha per year estimated for the period 1983–87 by the World Bank (World Bank 1991b).

It must be emphasized that comparison of all changes between the 1989 and 1987 assessments is fraught with technical problems. In general, the 1989 imagery is more reliable due to the wider scale used for land-use mapping and better multi-spectral scanning resolution. One may observe that the structure of Indian forests is gradually changing thanks to conservation efforts. The area under dense forests (crown density over 40 per cent) has registered a small increase, as shown in Table 16.4.

Indirect evidence of the slowing down of deforestation can be deduced from Table 16.5, which shows forest area lost between 1951 and 1987. The average annual loss of forest land has sharply fallen from 155,000 ha in 1950s to merely 16,000 ha in 1980s. Two factors that have contributed to this sharp turn are:

(a) a halt in the extension of crop cultivation to virgin forests;
(b) the introduction of tighter conservation measures in the form of the 1980 Forest (Conservation) Act which has transferred powers to the central bureaucracy to intervene on any change of forest land-use for development purposes.

A cursory look at Table 16.5 reveals that 61 per cent of the forest area (1951–1980) was lost primarily due to agricultural activities. In India, unlike Brazil or

Table 16.4 Changes in forest cover in assessment years* (million hectares).

Category	1987	1989
Dense forest (crown density 40% & over)	36.14	37.85
Open forest (crown density 10% to less than 40%)	27.66	25.74
Mangrove forest	0.40	0.42
Total	64.20	64.01

Note: * based on Landsat imagery and ground truth verification. *Source: DOE 1989: 9.*

Table 16.5 Forest area lost between 1951 and 1987 (thousand hectares).

Purpose	1951–72	1973–80	1981–87	Cumulative 1951–1987
Agricultural activities	2,433	190		2,623
Submergence due to river valley projects	401	101		502
Industries and townships	125	9		134
Roads & communications	55	6		61
Miscellaneous	388	620		1,008
Total	3,402	926	115	
Average annual loss	155	84	16	

Sources: 1951–72: FAO 1981b; 1973–87: Department of Environment 1987.

Indonesia, there were no policies to encourage transmigration of people to forested areas. The primary source of forest conversion came from typical small peasant farmers. In 1983, 70,063 ha of forest area was under encroachment by illegal users (DOE 1987: 21). Thirty-five per cent of these encroachments occurred in India's most forested state, Madhya Pradesh (hereafter MP).

The source of the other 39 per cent of forest loss lies in non-agricultural demand. Major irrigation schemes and river dam projects have systematically undervalued forest functions, leading to large scale deforestation. Where the true value of forest products is assessed below the market rate, the artificial lowering of costs improve the attractiveness of dam projects (for the controversy on teak forests values in the Narmada Sagar project see Paranjpye (1988)). In the Himalayan mountain regions, road building is associated with shallow landslides and soil loss (Agarwal & Narain 1992: 41). The effect of roads in opening up inaccessible regions to timber logging is considered significant in the densely wooded Bastar district in central India. However, there is little evidence to suggest that logging trails have been followed by agricultural colonization. It has been argued that the major contribution of forests lies in building up soil and binding soil nutrients for agricultural resilience. When the forests go, the most valuable natural loss is topsoil which has knock-on effects on future crop yield and nutrient recycling.

16.4 Sources of deforestation: some hypotheses

The debate on the causes of deforestation in India has produced two diametrically opposite viewpoints. The first school – which is the official viewpoint – has maintained that demand side factors are causing forest depletion. The second school stresses participatory failure.

The official viewpoint

In an agricultural economy the growth in population is translated into agricultural extension. The 1976 National Commission on Agriculture held the lifestyle of rural people and their dependence on forests to be a major source of deforestation. Their report maintained:

> Free supply of forest produce to rural population and their rights and privileges have brought destruction to the forests, and so it is necessary to reverse the process. The rural people have not contributed much towards the maintenance or regeneration of the forests.   (quoted in Khator 1989: 18)

Increased population, modernization and urbanization necessarily put a burden on the country's scarce resources. The burden comes in the form of an increased demand for fuelwood and timber as well as additional land for production of food and housing space. Bowonder et al. (1985: 7) have found that the "major impact of fuelwood extraction has been the rapid depletion of forests in and around the urban centres."

A recurring theme in the deforestation debate is the impact ascribed to the livestock population. Unregulated grazing on common property resources (village woodlot and pastures) provide the main source for fodder. Excessive and continuous grazing has degraded these lands to an extent that their productivity is negligible. Consequently, forests bear the brunt of grazing incidence. The State governments have extended grazing rights to reserved forests. The 1981 Committee for Review of Rights and Concessions in Forest Areas, appointed by the Government of India, has observed (Lal 1989: 94):

> Excessive trampling makes the soil compact and impervious and prevents the circulation of air and water needed for its organic life. Herbaceous plants disappear increasingly and denuded soil is exposed to erosion by wind and water. In the wooded areas, the trees wither, their roots are exposed, injured by hoofs and rot sets in. No wonder, it is remarked that, "the greatest enemy of the forest is the cattle and not the human population".

Several hypotheses can be formulated empirically to verify the official viewpoints.

(a) growth in human and animal population increases deforestation;

231

(b) increases in per capita income at low stages of development furthers the extraction of renewable natural resources;

(c) the higher is fuelwood extraction, the greater is the level of deforestation.

The popular perception

In contrast to the official position, the popular view (Agarwal et al. 1982, Joshi 1983, Guha 1989, Khator 1989) maintains that the above stated factors are not the real causes of deforestation. In particular they point out that the major cause lies in the failure of the forest bureaucracy to secure people's participation. In other words, what is at fault is the monopolistic nature of public ownership. The power of bureaucracy in post-independence India has considerably increased as it was seen to be the only institution equipped with skills to secure rapid development. The bureaucracy consequently was characterized by a) centralized efforts; b) formal process with pre-determined outcomes; and c) profit oriented objectives.

Although in principle forest policy is formulated with several goals in mind, at the time of implementation the goal of revenue earning remains primary. The rent sought by the bureaucracy is threefold:

(a) Personal gains extracted in the form of corruption. Khator (1989: 21) mentions that department officials take a 6–10 per cent commission on minor forest produce from the tribal areas. It is widely believed that for every tree cut, three more trees are cut to gratify the officials, guards and contractors.

(b) Discriminatory pricing of forest produce favours the large industries against the needs of small artisans. An often quoted case is the distortionary price of bamboo supplied by the Forest Department. While the paper and pulp manufacturers paid only Rs 12 per tonne of bamboo, the rural artisans had to pay Rs 1200 per tonne. Subsidies also encourage excessive consumption: "The wood supplied to the plywood millers [in Assam] is highly subsidized. Timber priced at Rs 1485 per cu m in the market is supplied to them for Rs 500 per cu m for tea chests and Rs 740 per cu m for decorative wood" (Agarwal et al. 1982: 46).

(c) Throughout the British period and thereafter, the forests were expected to earn ever more to support the general revenue needs of the government (see Table 16.6). Note that surplus as percentage of gross revenue has exceeded 46 per cent between 1951 and 1972 and reached a peak of 59 per cent during 1955–1959.

This high level of profit generation – both official and private – is considered a prime factor for deforestation. The forest bureaucracy is perceived as the exploiter of the natural resources. One author observes (Joshi 1983):

> Thus, while the forests are dwindling and the inhabitants are getting pauperized, the forest department is earning enormous revenues and is in turn enabling industrialists to earn huge profits.

Table 16.6 Average annual revenue and expenditure of forest departments in India (million rupees).

Average for the period (1)	Gross revenue (2)	Expenditure (3)	Surplus (4)	Surplus as % of gross revenue (5)
1936–37 to 1938–39	25.9	19.4	6.4	24.9
1939–40 to 1941–42	32.7	20.9	11.8	36.2
1942–43 to 1944–45	83.0	49.2	33.8	40.7
1945–46 to 1947–48	149.3	67.9	78.6	53.7
1951–52 to 1953–54	240.1	106.2	133.9	55.8
1955–56 to 1958–59	418.4	171.2	247.2	59.1
1961–62 to 1963–64	693.8	322.3	371.5	53.5
1966–67 to 1968–69	1075.0	566.3	508.7	47.3
1969–70 to 1971–72	1358.7	734.9	623.8	45.9
1980–81 to 1982–83	4725.5	3178.3	1547.2	32.7
1983–84 to 1985–86	8417.0	6056.2	2360.8	28.0

Sources: Agarwala 1989: 302, DOE 1987: 56.

Puri et al. (1983: 523) echo the same feeling:

> Though, it is claimed that scientific forestry is more than a century old in India even after it was more fully articulated in 1894, it has remained clearly one sided. Its character was essentially that of crown forests being regulated in the interest of feudal lords.

Hypotheses that can be formulated to verify the popular perception are as follows (continuing the list started above):

(d) the higher is the benefit/rent to the Forest Departments, the greater is the forest loss;

(e) timber extraction is the main source of deforestation.

Additional factors

India has followed a strong affirmative policy to reforest and allow regeneration of its forests. Since 1950 annual afforestation targets have been set and results reported. This is expected to have a positive impact in reducing deforestation. Another contributory factor in slowing the rate of deforestation is said to be rising agricultural productivity. We have already pointed out that in the last four decades, cultivated area has not expanded in India at the expense of forest area. It remains to be seen whether productivity rises in India will provide ameliorative influences in deflecting land pressures, as in to Latin America (Southgate 1991). To encompass these additional factors in our model, we shall extend our hypothesis:

(f) Regeneration and afforestation are inversely related to deforestation;
(g) Increase in agricultural productivity postpones forest loss.

16.5 Econometric analysis of the causes of Indian deforestation

This section uses econometric techniques to clarify the significance and relative importance of the various causes of deforestation.

Model specification

To capture the cause and effect relationship between the level of reserved forest and the factors that contribute to its variation, a simple model was developed. These factors may be broadly classified as follows:

(a) *demand side pressures* for fuelwood and timber, grazing pressures, urban space – the level of demand for these products varies according to income levels;
(b) *government policies* such as differential prices, supply side management (afforestation measures), bureaucratic control;
(c) *exogenous factors* that directly affect forest use such as population growth, agricultural productivity.

The regression equation attempts to integrate as many of the factors thought to be significant in determining deforestation. Table 16.7 explains each of the terms used in the model.

$$\text{FCOVER} = \text{INPT} + a_1.\text{LIU} + a_2.\text{PCY} + a_3.\text{IAP} + a_4.\text{NRR}$$
$$(-) \qquad (-/+) \qquad (+) \qquad (-)$$

$$+ a_5.\text{UBP} + a_6.\text{RAA} + a_7.\text{IWP} + a_8.\text{FUP}$$
$$(+) \qquad (+) \qquad (-) \qquad (-)$$

The basic model was run without IAP as its t-ratio was statistically insignificant.

NRR is taken as a measure of government failure. Fuelwood and timber products – the two major use values of forests – are included to represent their respective impact on deforestation. Population pressure – both human and livestock – is accounted for. An interesting inclusion is changes in urban density (UBP) which simultaneously captures population increase and reflects occupational changes due to modernization. Unlike other econometric studies we have not included external debt in the model as India's export of forest products and cash crops is minimal. The purpose of including debt is normally to capture the effect of indebtedness on exports of cash crops and forest produce in order to meet debt obligations.

Regeneration/afforestation and agricultural productivity are included since they are thought to reduce pressure on the stock of natural capital, including forests.

Data for regression analysis

The current study bases its analysis on time series data from India. The annual level of reserved forest area (million square kilometres) is used as the dependent variable. The data are taken from various official publications of the central Department of Environment.

In India, the forests are classified under three broad divisions: reserved, protected and unclassified. The protected and unclassified area are open to public use and deemed as denuded. The reserved forests are exclusively under the control of Forest Departments, which impose severe restrictions on public use and are better organized. There are indications that the reserved forests have a high degree of the area under dense forests (over 40 per cent crown cover) (DOE 1987). Ideally one would have preferred to use actual forest cover as the dependent variable. However such figures are not available for the time period (1951–1990) chosen for the study. Actual forest cover by tree density is available only for the period 1987–89 which is very short time span to show any sustained tendencies. The choice of reserved forest area is therefore only second best under the circumstances.

The data for explanatory variables have come from a variety of sources (see Table 16.7). Grazing pressure is reflected by the absolute increase in livestock units (LIU X2) of cattle, buffalo, goat and sheep numbers. These are the four main grazing animals and LIU is expressed in cow equivalents. According to Indian official practice each buffalo is equal to two cows; sheep and goats are set equal to half-cow units. The data come from official livestock surveys and FAO Agricultural Yearbook and monthly bulletins. Per capita net national product (PCY X3, i.e. national income) is

Table 16.7 The regression equation.

Model: FCOVER = INPT + LIU + PCY + NRR + UBP + RAA + IWP + FUP

where:

- Y FCOVER: Reserved forest area (million square kilometres)
- a INPT: Intercept term, set equal to 1
- X2 LIU: Livestock units in millions
- X3 PCY: Net national per capita income (1970–71 prices, Indian Rs)
- X4 NRR: Net Revenue receipts of Forest Department (1970–71 prices, Rs)
- X5 UBP: Urban population as percentage of total population
- X6 RAA: Regenerated and afforested area (million square kilometres)
- X7 IWP: Outturn of timber, pulpwood, roundwood production (thousand cubic metres)
- X8 FUP: Firewood and charcoal production

compiled from National Accounts Statistics and Tata (1990). Urban population (UBP X5) as percentage of population represents population growth and shifts in settlement patterns. Net Revenue Receipts (NRR X4) figures are given in various Statistical Abstracts and deflated to 1970–71 prices according to the time series of wholesale price index as in Tata (1990: 179). Regenerated and afforested area (RAA X6, in million square kilometres) covers achievements for both natural and artificial regeneration. The data are compiled from various editions of Statistical Abstracts. Forest produce is classified under fuelwood and charcoal (FUP) and industrial wood (IWP), the latter includes timber, pulpwood and roundwood. The sources are FAO Forest Products Yearbooks and Statistical Abstracts.

Interpretation of regression analysis

The regression equation is estimated using ordinary least squares (OLS), the results of which are summarized in Table 16.8. The coefficient of multiple determination, or R-square (i.e. measuring the proportion of the total sum of squares explained by the regression) test the goodness of fit of the regression explaining the variation in the dependent variable. For a time series study using aggregate national level data for a heterogenous group of explanatory variables, an R-square of 0.86 and an adjusted R bar square (i.e. reworked to account for the reduction in the degrees of freedom with additional explanatory variables) indicates a good fit of the estimated model in explaining the variation in reserved forest area. The R-square value of 0.86 shows that the seven explanatory variables explain about 86 per cent of the variation in reserved forest area over the period 1952–80. The adjusted R bar square shows that after taking into account the degree of freedom and explanatory variables, the variables still explain 81 per cent of the variation in Y. The F-statistic (i.e. testing the overall significance of the regression with the ratio of explained variance) of 18.46 well exceeds 9.772 which is the minimum value for rejecting the possibility that there is no relationship between the extent of reserved area and the explanatory variables at the 99 per cent confidence interval level. Thus the specified regression models are not rejected as the framework for analyzing changes in reserved forest area.

The statistical significance of the individual parameter estimates are shown by the t-ratios given in Table 16.8 against each estimated coefficient. Using a two-tail test, a 95 per cent confidence interval and 19 degrees of freedom one can reject the null hypothesis that the estimated coefficient equals zero for the parameter estimates with statistics greater that than 2.045 i.e. LIU (X2), PCY (X3), NRR (X4), UBP (X5), RAA (X6) and IWP (X7). FUP (X8) is significant only at the 10 per cent level of significance.

The sign of the parameter estimates for LIU (X2), PCY (X3), NRR (X4) and FUP (X8) is negative, as anticipated. That is, an increase in any of these factors would be

Table 16.8 Ordinary least squares estimation.

	Regressor	Coefficient	Standard error	T-ratio [Prob]
A	435.4	66.48	6.55	[1.00]
X2	−0.91	0.35	−2.63	[.98]
X3	−0.17	0.075	−2.24	[.96]
X4	−0.078	0.021	−3.67	[1.00]
X5	17.07	3.21	5.32	[1.00]
X6	722.08	181.93	3.97	[1.00]
X7	0.0073	0.0016	4.54	[1.00]
X8	−0.003	0.0018	−1.69	[.89]

R-squared	0.86
R-bar-squared	0.81
F-statistic F(7,21)	18.46 [1.0]
S.E. of regression	8.32
Residual sum of squares	1453.5
Mean of dependent variable	349.7
S.D. of dependent variable	19.3
Maximum of log-likelihood	−97.9
DW-statistic	1.9050

Notes: Dependent variable is Y; 29 observations used for estimation from 1952 to 1980; see also Table 16.7.

expected to lead to a decrease in the reserved forest area. For example a unit change in livestock (X2) is expected to decrease reserved forest area by 0.91 per cent, i.e. less than one per cent.

The relationship between net revenue receipts (X4) and reserved forest area is negative and statistically significant. The results confirm that increased proclivity in rent seeking by the Forest Department is likely to reduce reserved forest area.

Urbanization of the population (X5) deflects pressure on land and fuelwood demand. In our model, livestock numbers (X2) grow in tandem with per capita income (PCY). In an agricultural country, rising standards of living in rural areas as reflected in PCY (X3) encourage holding of assets in the form of animals. This leads to an increase in grazing intensity and pressure on reserved forest area. Income per capita (PCY X3) is statistically significant in our model, with an increase of Rs 1 per capita estimated to lead to a decrease of 0.16905 (thousand square kilometres) of reserved forest area. Some comments on the forest regeneration (X6) and industrial wood (X7) and fuel wood extraction (X8) are in order. Fuelwood exploitation bears a negative relationship to reserved forest area and suggests that it is more damaging than industrial extraction. India's remarkable achievement in forest regeneration through natural means as well as a concerted afforestation drive is a major factor in resisting further deterioration of forests. This shows up in the model with forest regeneration (X6) being statistically significant.

16.6 Conclusions

In contrast to the studies carried out in Latin America (Southgate 1991), Thailand (Panayotou & Sungsuwan 1989) and the global context (Kahn & McDonald 1992, Ehui & Hertel 1989, Capistrano & Kiker 1990), the Indian case study stands out in many important respects.

First, frontier agricultural expansion is not a major force in forest loss in India. The major difference between India and Latin American countries is the presence of strong official machinery that deters land invasion and subsequent colonialization.

Secondly many authors (e.g. Southgate 1991, Burgess 1992) have reported the positive influence of increase in agricultural productivity in reducing deforestation. In contrast, we have found that an index of agricultural productivity (which has risen from 71 in 1952 to 98 in 1980, base year 1979–81 = 100) was statistically insignificant in explaining the changes in reserved area. The plausible explanation to us seems that agricultural productivity has significance only where rapid deforestation is taking place, which is not the case in India. Other more significant factors in reducing deforestation are the increase in urban population density, which shows reduced dependence on the agricultural base, and a population movement towards industrialization.

Thirdly, the vested interest in the forest bureaucracy seeks continued expansion of the reserved forest area. However, from time to time, the bureaucracy succumbs to the demands of a political elite which gains dividends by regularizing encroachers on public land. This explains for the variation in the reserved forest area, especially around years when provincial and national elections occur. An indirect effect of the extension of forest area is the chronic low productivity of the Indian managed woodlands. Average production in India is estimated at $0.53\,m^3$/ha per year against the world average of $2\,m^3$/ha (Puri et al. 1983: 524). In the coming years the emphasis will move to better supply management, including stress on higher productivity and removal of distortionary subsidies. In an effort to reduce open access, free for all forest use, the Indian government has launched a number of experiments in social forestry where limited use rights are given to the farmers to take care of their own woodlots and specified areas with reserved forests.

Any revision of property rights raises the important question of who can do what best. This is intimately related to changes in land-use. Forest farming in tropical countries has to address demands for subsistence (fuelwood/fodder); commerce (timber, paper and pulp); and the maintenance of environmental balance. At present the government of India is trying to satisfy all these three varied demands without a stated order of prioritization.

Our discussion suggests that the establishment of property and usufruct rights could deflect fuelwood pressure through social forestry programmes, energy efficiency and fuel switching to other sources. This requires application of common property management regimes for which many communities and voluntary agencies within India have demonstrated their capacity (Chambers et al. 1989). The commercial and industrial needs are best met by allowing the market to function without distortionary subsidies. The environmental and ecological functions of the forests should rest with the government where the concept of total economic value can offer good justification. However, new land-use management demands that the absolute control of the government agencies (Forest Departments own 98 per cent of India's forests) is suitably modified. In other words, government deregulation and privatization of forest land is required to successfully meet the commercial and subsistence needs of the population.

Finally, the Indian case study highlights the continued need to complement econometric models of deforestation with historical analysis. This is all the more important where the availability of data and reliable proxies is limited. Moreover, in a country as diverse as India, no single econometric model can reflect all tendencies affecting forest use. Therefore, an aggregate econometric model can serve only as a complement to regional and micro studies.

IV
The tropical timber trade

17

The timber trade and tropical deforestation in Indonesia

Edward Barbier, Nancy Bockstael,
Joanne Burgess, Ivar Strand

17.1 Introduction

This chapter examines the links between the trade in tropical timber products and deforestation in Indonesia. We review some of the evidence suggesting that timber production is a factor in tropical deforestation, and the rôle of timber trade policy in Indonesia in influencing this process by affecting forest-based industrialization. These issues are of particular concern to Indonesia, as the country has recently banned sawnwood exports to encourage further rapid development of plywood processing.

We develop a partial equilibrium timber trade model of Indonesia to analyze the effects of various policy interventions on the trade and tropical deforestation. The basic timber trade model is developed as a *simultaneous equation* system determining supply and demand in the logging, sawnwood and plywood sectors of Indonesia. The system is linked to a *recursive* relationship determining tropical deforestation, which is estimated separately. We use the model to simulate several policy options, including the impacts of sawnwood export taxes/effective bans, import bans imposed by consumer countries, revenue-raising import taxes, and increased harvesting costs associated with "sustainable management".

The chapter concludes by summarizing the results of the policy analysis, and discusses the policy options open to the Government of Indonesia (GoI) and importing countries.

17.2 The timber trade and tropical deforestation in Indonesia

Southeast Asia currently accounts for around 20 per cent of the world's tropical moist forest (TMF). Indonesia alone has over 50 per cent of the region's TMF and

Table 17.1 Tropical deforestation in Southeast Asia and other regions[a].

	Total forest area $(10^6$ ha)	Undisturbed operable forest $(10^6$ ha)	Annual deforestation $(10^3$ ha)	Total TMF area deforested $(10^6$ ha)
SE Asia insular	167.3	72	1707	117
Indonesia	108.6		1315	
Papua New Guinea	33.5		22	
Malaysia[b]	18.4		255	
Philippines	6.5		110	
Brunei	0.3		5	
Other SE Asia[c]	39.3	18	346	n.a.
Myanmar	31.2		102	
Thailand	8.1		244	
Total SE Asia	206.6	90	2053	>117
Amazonia	613.6	453	4129	100
Central Africa	167.1	107	325	30
Other regions	58.4	< 10	1900	177[d]
World total	1045.7	652	8480	424

Notes: a) unless indicated, 1990 estimates based on Schmidt (1990). *Tropical moist forests* are defined as broadleaf high closed tropical forests, including wetland and mangrove forests but excluding the deciduous dry forests of South Asia, sub-Saharan Africa and sub-tropical South America. In 1988, the FAO estimated the total area of all *tropical closed forests* (including deciduous dry forests) to be 1269.6 million ha; b) Includes forests from Peninsular Malaysia; c) End of 1980 revised estimates based on FAO 1988a; d) May include estimates for Thailand and Myanmar.

over 10 per cent of the world's total (see Table 17.1). The rate of deforestation in Southeast Asia, measured in terms of hectares (ha) lost per year, is also fairly high – approximately 2 million ha annually – and the total area deforested is much higher than in Amazonia and Central Africa. Close to 85 per cent of Southeast Asia's annual deforestation occurs in Indonesia.

The major "cause" of tropical deforestation in Southeast Asia and in Indonesia in particular is generally thought to be the conversion of forests to agricultural land. Tropical timber production, although significant in large areas of "production" forests, is considered to be a less significant factor in overall tropical deforestation. More recently, however, attention has focused on the indirect rôle of timber production in "opening up" inaccessible forest areas, which then encourages other economic uses of the forest resources, such as agricultural cultivation, that lead to deforestation on a wider scale. For example, Amelung & Diehl (1992) identify the major shifts in land-use changes and the causes of deforestation in tropical countries, including Indonesia. As indicated in Table 17.2a, the direct impact of forest activities on deforestation appears to be minimal (i.e. less than 10 per cent of total deforestation), as compared to agriculture. For Indonesia, this is partly because

Table 17.2a Sources of deforestation in tropical countries, 1981–88[a].

	Brazil	Indonesia[b]	Cameroon	All major tropical forest countries
Forestry	2[d]	9	0	2 (10)[e]
Agriculture	89	80	100	(83)[f]
shifting cultivators[c]	13 (23)	59 (67)	92 (95)	na (47)
permanent agriculture:	76	21	8	36
– pastures	40	0	0	17
– permanent crops	4	2	5	3
– arable land	32	19	3	16
Mining and related industries	< 3	< 0.3	0	na
Hydroelectric production	4[b]	0	0	2[b]
Residual[g]	2	11	0	(13)[h]

Notes: a) Percentage shares in deforestation refer to averages for the respective period; b) Data refer to the 1980–90 period; c) Figures in parentheses show the results of the FAO for 1980. These data include also market oriented farmers who produce cash and export crops and engage only partly in shifting cultivation; d) Deforestation due to logging is due to charcoal production; e) The figure in parentheses refers to the estimation of Enquete-Kommission zum Schutz der Erdatmosphäre, *Schutz der tropichen Wälder*, Deutscher Bundestag 11, Wahlperiode, Drucksache 11/7220, 24.05.1990, Bonn, 1990. The calculation includes only Indonesia and Brazil, since these countries account for the largest share in clear cutting by the forestry sector; f) This percentage rate is based on the assumption that the percentage share calculated for shifting cultivators can be taken as an average for the 1981–88 period; g) This includes other industries, housing, infrastructure services and fire loss; h) The residual has been calculated from the data in this column which includes data from different periods.

Table 17.2b Sectoral share in forest degradation and forest modification, 1981–85[a].

Sector	Percentage share in biomass reduction (degradation)				Percentage share in forest modification			
	Brazil	Indonesia	Cameroon	Total[b]	Brazil	Indonesia	Cameroon	Total[b]
Forestry	6	44	10	10	(100)[c]	(100)[c]	98	71
Agriculture[d]	85	49	90	76	0	0	2	26
Others[d]	9	7	0	13	0	0	0	4

Notes: a) For the definition of modification and biomass reduction (degradation) see FAO, *Tropical Forest Resources*, FAO Forestry Paper 30, Rome, 1982; b) Total refers to all major rainforest countries; c) Following FAO statistics, deforestation in virgin forests is 0, since clearing by agriculture and other sectors concentrates on disturbed forests. Even though some clearing occurs in virgin forests, there is reason to assume that the bulk of deforestation is due to forests that have been logged over prior to the clearing of the respective areas; d) These figures have been derived from Table 17.2a and reflect averages for the 1981–88 period or, in the case of Indonesia, the 1980–90 period. *Source: Amelung & Diehl, 1992.*

timber is harvested predominately by selective logging, which does not meet the "strong" definition of deforestation that is commonly used. In contrast, the forestry sector in Indonesia is almost completely responsible for converting virgin forests into productive closed forests or other forms of land-use through forest modification, and has a much more significant rôle in biomass reduction (see Table 17.2b). In short, there are some indications that timber extraction is largely responsible for opening up previously unexploited forest, leading to further forest degradation and outright deforestation.

Other studies indicate the possible acceleration of this process, given the trends in timber production in Indonesia (Burgess 1989). With the exception of Sabah

Table 17.3a Forest areas in major Southeast Asian tropical timber countries (sq km).

	Total land area	Total forest area	Permanent protection forest	Permanent production forest	Conversion production forest	Total production forest	Remaining virgin forest
Thailand	513,115	142,958	n.a.	n.a.	n.a.	n.a.	n.a.
Malaysia							
Peninsular	131,596	63,532	10,679	34,507	9,100	43,607	9,600
Sabah	73,711	44,869	6,000	29,984	4,080	34,064	7,815
Sarawak	123,253	94,384	4,200	32,400	37,784	70,184	50,387
Philippines	300,000	63,830	16,800	44,030	0	44,030	10,420
Papua New Guinea	468,860	359,900	n.a.	n.a.	n.a.	n.a.	n.a
Indonesia	1,930,270	1,439,700	490,410	338,666	305,370	644,036	524,000
Total	3,540,805	2,209,173	548,089	479,587	356,334	835,921	602,222

Source: based on Burgess 1989.

Table 17.3b Production and yield potential in major Southeast Asian tropical timber countries.

	Permanent production forest (million ha)	Sustained yield at 1.5 m^3 per year	Sustained yield at 2.0 m^3 per year	Annual timber harvest (m^3)
Malaysia				
Peninsular	2.85[a]	4,275,000	5,700,000	7,914,328
Sabah	3.00	4,500,000	6,000,000	11,739,262
Sarawak	3.24	4,860,000	6,480,000	11,470,000
Philippines	4.40	6,600,000	8,800,000	3,433,774
Indonesia	33.87	50,805,000	67,740,000	28,500,000
Total	47.36	71,040,000	94,720,000	63,057,364

Note: a) the author does not explain the discrepancy between Tables 17.3a and b over the size of the permanent production forest in Peninsular Malaysia. *Source: based on Burgess 1989.*

Table 17.3c Production and trade of wood products in major Southeast Asian tropical timber countries (1989, '000 m³).

	Products	Production	Imports	Exports	ADC[a]
Indonesia	logs	23,684.0	0.0	0.0	23,684.0
	sawnwood	10,546.0	0.0	2,692.0	7,854.0
	plywood	8,500.0	0.0	8,040.0	460.0
	veneer	53.9	0.0	29.0	24.9
Malaysia	logs	38,900.0	10.0	21,100.0	17,810.0
	sawnwood	7,660.0	197.0	5,134.0	2,723.0
	plywood	1,001.0	15.0	915.0	101.0
	veneer	445.0	5.3	248.7	201.6
Papua New Guinea	logs	1,700.0	0.0	1,260.0	440.0
	sawnwood	118.0	0.0	3.0	115.0
	plywood	18.0	0.0	0.0	18.0
	veneer	0.0	0.0	0.0	0.0
Philippines	logs	2,773.0	393.5	6.0	3,160.5
	sawnwood	975.0	12.0	438.1	548.9
	plywood	341.0	3.0	130.9	213.1
	veneer	75.0	0.0	53.1	21.9
Thailand	logs	1,770.0	1,135.0	0.0	2,905.0
	sawnwood	1,160.0	744.0	30.0	1,874.0
	plywood	185.0	4.0	35.0	154.0
	veneer	60.0	2.0	5.0	57.0
Total SE Asia					
Producers	logs	68,827.0	1,538.5	22,366.0	47,999.5
	sawnwood	20,459.0	953.0	8,297.1	13,114.9
	plywood	10,045.0	22.0	9,120.9	946.1
	veneer	633.9	7.3	335.8	305.4
Total ITTO					
Producers[b]	logs	126,967.0	2,536.7	25,749.2	103,754.5
	sawnwood	48,846.0	1,083.5	9,518.1	40,411.4
	plywood	12,115.6	59.2	9,565.2	2,609.6
	veneer	1,705.7	42.5	555.5	1,192.7
Total World					
Producers[c]	logs	1,677,454	130,230	124,846	1,682,838
	sawnwood	500,685	100,231	99,088	501,828
	wood panels[d]	129,108	28,692	28,443	129,357

Notes: a) ADC: Apparent domestic consumption; b) ITTO: International Tropical Timber Organization; c) All tropical and temperate timber producers, derived from FAO 1991b; d) Includes veneer sheets, plywood, particle board and fibreboard (compressed and non-compressed).

Source: Based on International Tropical Timber Council, Elements for the 1990 Annual Review and Assessment of the World Tropical Timber Situation, *Tenth Session, Quito, Ecuador, 29 May – 6 June, 1991.*

and probably Papua New Guinea, Indonesia has one of the last remaining reserves of virgin forest in Southeast Asia, yet well over one third of the country's tropical forests have been allocated to conversion or permanent production forest (Table 17.3a). As a very large part of the production in Indonesia in the past has come from the conversion forests, which are essentially clear cut and turned over to agriculture, future supplies must depend on the permanent production forests (and any remaining virgin forest reserves allocated to them) – provided that these supplies can be maintained on a sustained basis.

Although Table 17.3b suggests that Indonesia's permanent production forests are not being "mined" to the extent of those in other Southeast Asian countries, there is concern that Indonesia has over-estimated the extent of its production forest, under-reported current timber extraction (e.g. there is substantial illegal felling), or both (Burgess 1989). Moreover several analysts predict a decline in Southeast Asia's (and Indonesia's) share of world production and trade, and some deterioration in the quality of hardwood timber produced from forests in the region (Arnold 1991, Burgess 1989, Sedjo & Lyon 1990). Thus, it is unlikely that Indonesia will be able to maintain its current dominance of world production and trade of timber products, as indicated in Table 17.3c, without further and extensive exploitation of its remaining old growth reserves.

17.3 Indonesian timber trade policies

The tropical timber trade has been subject to severe distortions by export restrictions imposed by log producing countries, including Indonesia. One justification often cited for these policies is that they compensate exporters for the import barriers in developed economy markets by making the price of raw logs higher to the processors in the importing country while reducing the cost disadvantage faced by domestic processors within the timber producing countries. This strategy usually has as its primary aim the creation of more export revenues and employment for the forestry sector, with a secondary objective of reducing harvesting pressure on the forests by increasing value added per log extracted.

Several authors have recently reviewed the rôle of export taxes and bans in encouraging forest-based industrialization and sustainable timber management in tropical forest countries (Gillis 1990b; Vincent & Binkley 1991; Vincent this volume, Chapter 19). Initially, the preference seems to have been for export taxes, through employing escalating rates. For most countries, export tax rates on logs generally ranged between 10 and 20 per cent. Export taxes on sawn timber, veneer and plywood have been negligible. Where sawn timber exports were taxed, rates were typically half that of logs. More recently, the use of export tax structures to

Table 17.4 Export taxes and bans on tropical timber, Southeast Asia, 1989.

Country	Tax rate/export policy	Remarks
Indonesia	20% *ad valorem* on logs.	Tax imposed only on some logs in inaccessible regions. Log export ban since 1985 has made the tax irrelevant for other regions.
	Specific export taxes on sawn timber, ranging from US $250–2400 per m³.	Specific export taxes on sawn timber introduced in 1989; plywood exempt from all export taxes.
Malaysia:		
Peninsular	Log export ban since 1971.	
Sabah	No specific export tax or ban but see remarks.	The Sabah timber royalty has a strong export tax feature: the royalty rate for log exports is almost 10 times the rate for logs used domestically.
Sarawak	15% *ad valorem* of f.o.b. log values	Tax applies to one hardwood species only.
Papua New Guinea	10% of f.o.b. log values.	Tax reported to have been widely evaded through transfer pricing. Log export ban proposed.
Philippines	Log exports restricted to 25% of annual allowable cut since 1979	Ostensibly to control deforestation.

Source: Gillis 1990a.

promote forest-based industrialization has become largely replaced by export bans in tropical forest countries, although export taxes are still being used in certain regions and for specific timber products (see Table 17.4).

Tropical timber export taxes and bans have proved only moderately successful in achieving the desired results in Southeast Asia. For example, although expanded processing capacity was established in Malaysia, the Philippines and Indonesia, it was achieved at high economic costs, both in terms of the direct costs of subsidization as well as the additional costs of wasteful and inefficient processing operations (Barbier 1987, Gillis 1988a and 1990, Vincent & Binkley 1991, Vincent 1992a and 1992b).

In addition, as a long-term forest industrialization strategy, ensuring export sales of processed products through denying processors in other countries access to logs may prove difficult to sustain (Bourke 1988). Importers of Southeast Asian logs, such as Japan, have been known to substitute other raw materials (e.g., cement, steel, plastics, fiberboard, etc.) for timber, and alternative sources of supply, such as sawlogs from temperate and other developing regions (Bourke 1988, Vincent et al. 1990).

In Indonesia, the *ad valorem* export tax on logs was doubled from 10 to 20 per cent in 1978, while most sawnwood and all plywood were exempted. Beginning in 1980, controls on the export of logs were progressively enforced, until an outright ban was introduced in 1985 (Gillis 1988a). The export tax structure created effective rates of protection of 222 per cent for plywood manufacture, and the drop in export revenue to the government from diverting log exports was not compensated by any gain in value-added in sawmilling, resulting in a loss of US$15 per m^3 at world prices. The consequence has been the creation of inefficient processing operations and expanded capacity, with consequences for the rate of timber extraction and forest management. Gillis (1988a) has estimated that during the period 1979–82, due to the inefficient processing operations resulting from this policy, over US$545 million in potential rents was lost to the Indonesian economy, or an average cost of US$136 million annually. Moreover, as the switch from log to processed exports occurred at a time when forest product prices were falling sharply in real terms, the cost to the economy in export earnings was high. Over 1981–84, the net loss in export earnings amounted to US$2.9–3.4 billion, or approximately US$725–850 million annually. Additional losses were also incurred through selling plywood below production cost, which amounted to US$956 million in 1981–84, or US$239 million annually (Fitzgerald 1986).

Although the switch to value-added processing of timber initially slowed down the rate of timber extraction, the inefficiencies and rapidly expanding capacity of domestic processing may have actually increased the rate of deforestation over the medium and long term. For example, by the early 1980s, the major operational inefficiencies in domestic processing due to high rates of effective protection in Indonesia led to the lowest conversion rates in Asia. As a result, for every cubic metre of Indonesian plywood produced, 15 per cent more trees had to be cut relative to plymills elsewhere in Asia that would have processed Indonesian log exports (Gillis 1988a). Thus the protection given to Indonesian mills not only increased total log demand, but the gross operational inefficiencies also ensured that millions more logs may have been harvested than if a more efficient policy – to boost domestic processing capabilities – than forced industrialization through export taxes and bans had been implemented.

More recent analysis of the impacts of the log export ban in Indonesia on the efficiency of timber processing industries confirms that the policy has not increased wood recoveries and thus reduced log consumption compared to pre-ban levels (Constantino 1990). By depressing raw wood prices, the log export ban has led to the substitution of wood for other factor inputs, with substantial wood recovery losses in sawmilling and a slower growth in wood recovery than other factor productivities in plymilling. In both industries, wood consumption has been the main source of output growth over the 1975 to 1987 period, with efficiency gains contributing very little. However, the log export ban has not affected efficiency in the plywood industry so seriously, which could arise from the much newer capital vintages in plywood milling, its export orientation that requires higher production standards, and possibly its access to better quality logs.

Despite the problems with the export restriction policy, it is still being aggressively promoted by the GoI to encourage forest-based industrialization. In Indonesia, not only does the export log ban still remain in place, but in October 1989 export taxes on sawn timber were also increased substantially to prohibit exports and shift processing activities to plywood (see Table 17.4). A secondary objective of the policy is to eliminate the marginal mills, leaving only the competitive ones operational, thus improving overall industrial efficiency. The implications of the policy for wood recovery, log demand and thus tropical deforestation in Indonesia is less clear.

We develop a timber trade model for Indonesia, linked to impacts on deforestation through log demand, in order to compare the sawnwood export ban with other possible trade policy options. Through the various policy option simulations, we investigate the relative merits of the different policy interventions in terms of production, prices, trade and deforestation in Indonesia. However, we first briefly review other models applied to the tropical timber trade in Southeast Asia.

17.4 Other Southeast Asian timber trade models

Various partial equilibrium models of forest product markets and timber supply have been developed for countries in Southeast Asia to examine the implications of policies that restrict log exports.

Vincent (1989) employed a simulation model of the timber trade of Malaysia, other Southeast Asian producers, and the major importers from the region to determine the optimal tariffs on intermediate and final goods for log, sawnwood and plywood products. The results of the model indicated that the large export tariffs imposed by Malaysia and other Southeast Asian exporters reduced domestic prices in those countries, leading to losses in producer surplus. More recently, Vincent (1992a) constructed an economic model of sawlog and sawnwood production, consumption and trade in Peninsular Malaysia during 1973–89 to examine in more detail how the log-export restrictions imposed in 1985 affected the forests products industry in Peninsular Malaysia, and whether the restrictions generated net economic benefits. Although the export restrictions seemed to stimulate growth in the processing industries and employment, the economic costs were high. On an average annual basis, Peninsular Malaysia lost US$6,100 in economic value-added,

US$16,600 in export earnings, and US$34,300 in stumpage value for every sawmill job created by the log-export restrictions. No attempt was made in either of the above models to link the impacts of timber trade interventions on timber harvesting levels and tropical deforestation.

As part of the forest sector policy review of Indonesia conducted by the United Nations Food and Agricultural Organization (FAO) and the GoI, a domestic and international trade model of timber products centred on Indonesia was developed (Constantino 1988a and 1988b, Constantino & Ingram 1990). The simulation model had as its main purpose the determination of supply and demand projections for the Indonesian forestry sector, but it was also used to run policy scenarios on different trade interventions, including the implications of Indonesia's log export restrictions. The impacts of different government harvesting policies were simulated through scenario assumptions concerning the elasticity of the supply of forest land in Indonesia. For example, a small elasticity of supply was interpreted as a deliberate government policy to impose sustained yield constraints; whereas large elasticities reflected the GoI allowing expansion of forest land harvested. Using these different elasticity assumptions, the analysis could then focus on whether, in response to each policy scenario, the GoI should expand the area of forest land harvested, at the risk of greater tropical timber depletion and deforestation, or whether Indonesia would be better off limiting supply through greater harvest restrictions. Detailed results are provided in Constantino (1988a), which are summarized here:

- If international competitors follow conservative harvesting practices, then Indonesia should do the same in order to take advantage of greater employment, foreign exchange earnings and rent capture. On the other hand, if competing countries follow expansionary policies, it is not clear what Indonesia ought to do. Harvest restrictions will lead to a loss in international market share and to declines in employment – but will also increase government revenues.

- Export restrictions on sawnwood and plywood result in a loss in international market share for Indonesia, lower foreign exchange earnings, employment, labour income, royalty revenues for logging and economic rent to forest land. On the other hand, the restrictions lead to lower domestic prices thus benefitting Indonesian consumers, to higher profits in the processing industries and to less forest land harvested. Additionally, imposing harvest restrictions would improve matters, as employment would decline less, even less forest land would be harvested and economic rent to forest land would increase more, thus leading to more revenues if the GoI increased rent capture.

- A 10 per cent currency devaluation by the GoI coupled with restrictive harvesting policies will lead to the conservation of the resource and greater economic rent from forest land, but at the expense of higher domestic prices and less employment, value added, foreign exchange earnings and royalty revenues. This would suggest that some increase in forest land harvested in conjunction with a currency devaluation would be preferred.

– The impacts of a 10 per cent US import tariff on tropical plywood is dissipated somewhat if it is imposed on all exporters and not just Indonesia, and if there is limited substitution between tropical and temperate plywood. For Indonesia, the overall effects are a decline in total plywood exports, a smaller increase in sawnwood exports, and increased domestic consumption and trade diversion to other regions for both products. If harvesting restrictions are imposed, consumer prices will fall more, but employment, capital, foreign exchange earnings and value added will decline less.

The following simulation model is an additional attempt to examine the relationship between policy interventions, tropical timber product trade and deforestation in Indonesia.

17.5 Timber trade and deforestation model of Indonesia

The simulation model employed to examine timber trade and tropical deforestation in Indonesia is a static (single-period), partial equilibrium model of the production, consumption and trade of forest products that is related to the impact of log harvesting on forested area. The model comprises two components: a *simultaneous equation system* determining supply and demand in the logging, sawnwood and plywood sectors of Indonesia, and a *recursive relationship* determining tropical deforestation. Each component was estimated separately over the period 1968–88, before the 1989 tax rises on Indonesian sawnwood exports. The resulting estimated relationships were linked together in the simulation model using 1988 data. The time series and cross-sectional data used in estimating these relationships and constructing the model came from a variety of sources, including World Bank, FAO, IMF and GoI publications, and previous studies on the Indonesia forestry sector cited in this chapter. The model was then used to examine the varying impacts on Indonesia's timber markets and tropical forests of the 1989 sawnwood export tax policy and other policy interventions.

The simultaneous equation system of the supply and demand for Indonesia's roundwood (log), sawnwood and plywood markets was estimated using two-stage least squares employing linear functional relationships (See Table 17.5). As Indonesia essentially prohibits competing foreign timber product imports, supply comes solely from domestic production. In all three markets, demand is assumed to consist of both domestic consumption and foreign demand for Indonesian exports. Separate relationships for domestic production, consumption and import demand for each of the three products were then estimated. All values were based on 1980 real prices.

In the case of the log market, domestic consumption was assumed to be equal to the level of sawnwood and plywood production, multiplied by the respective wood recovery rates plus any local residual demand. Log export demand was considered to be influenced by log export unit values relative to the price of substitutes in world markets and by macro-economic factors affecting final demand in those countries importing Indonesian logs.[1] Due to insufficient data on domestic log prices, round-wood production (harvesting) was related to log export unit values, net of log royal-ties and export taxes, and the costs of harvesting. Domestic plus foreign log demand was therefore assumed to equal total log production. However, because the log export ban in 1985 effectively reduced Indonesian export volumes and unit values to zero, estimation of the log supply and foreign log demand relationships was only pos-sible for the period 1968–84. As a check on the recovery rates used as coefficients in the domestic log consumption equation employed in the simulation model, a regres-sion representing this demand was run for the entire 1968–88 period.

In the sawnwood and plywood markets, domestic consumption was assumed to be a final demand, determined by the respective product prices and Indonesian macroeconomic factors. Again, as domestic price and taxation time series data were scarce, plywood and sawnwood export unit values were used as proxies for domestic prices for these products.[2] Plywood export demand was considered to be influenced by its export unit value relative to the international price of a substitute (i.e. Philippine luan), but in the case of sawnwood export demand, only the export unit value alone proved significant. Both plywood and sawnwood exports were influenced by macro-economic factors affecting final demand in the main import-ing countries for each product.[3] Sawnwood and plywood production in Indonesia were determined by their respective prices (as represented by export unit values), processing capacity and costs – including the price of logs.[4] In the sawnwood market, domestic production and consumption was estimated for the entire 1968–88 period; however, as sawnwood was not exported significantly until after 1973, foreign demand could be estimated only for the 1974–88 period. As plywood pro-duction did not take off in Indonesia until after 1974, the demand and supply rela-tionships for this market were estimated over 1975–88 only.

The separate recursive relationship linking tropical deforestation to log produc-tion (harvesting) was estimated using a logit equation for pooled cross-sectional and time series data across the principal tropical hardwood forest provinces of Indonesia.[5] The relationship estimated the probability of forested relative to non-forested area as a function of log production per square kilometre (km^2), population density and GNP per capita for each province and time period. A dummy variable for 1988 was also employed, as the inventory methods for provincial forestry statistics were modi-fied in this year leading to revisions in the estimates for total forest area.

The results of the estimation of the simultaneous equation system for the three timber product markets and of the deforestation equation are presented in Table 17.5. The twelve endogenous variables of the timber trade supply and demand system are

Table 17.5 Indonesia: timber trade simultaneous equation system and deforestation equation.

1. Roundwood market

$$\text{QLI} = \begin{array}{c} 30868.4 \\ (2.43) \end{array} + \begin{array}{c} 0.172 \times (\text{RPXLI} - \text{XTLI} - \text{ROYLI}) \\ (2.59) \end{array} - \begin{array}{c} 0.433 \times \text{CSTLI} \\ (-1.1) \end{array}$$

$t = 1968–84$
$\text{SE} = 5534.3$
$\text{DW} = 0.76$

$$\text{XLI} = \begin{array}{c} -165246.2 \\ (-2.64) \end{array} - \begin{array}{c} 34562.3 \times (\text{RPXLI}/(\text{PLTENC} \times \text{EXR})) \\ (-2.03) \end{array} - \begin{array}{c} 4.74 \times \text{YPMCL} \\ (-1.51) \end{array} + \begin{array}{c} 1.397 \times \text{POPML} \\ (2.83) \end{array}$$

$t = 1968–84$
$\text{SE} = 6131.6$
$\text{DW} = 0.89$

$$\text{CONLI} = \begin{array}{c} 2513.5 \\ (0.96) \end{array} + \begin{array}{c} 2.09 \times \text{QSWI} \\ (2.08) \end{array} + \begin{array}{c} 1.43 \times \text{QPWI} \\ (1.43) \end{array}$$

$t = 1968–88$
$\text{SE} = 3679.5$
$\text{DW} = 0.86$

2. Sawnwood market

$$\text{QSWI} = \begin{array}{c} 776.3 \\ (1.32) \end{array} + \begin{array}{c} 0.011 \times \text{RPXSWI} \\ (2.00) \end{array} + \begin{array}{c} 0.427 \times \text{CAPSWI} \\ (3.94) \end{array} - \begin{array}{c} 0.021 \times (\text{CSTSWI} + \text{RPXLI} - \text{XTLI}) \\ (-1.41) \end{array}$$

$t = 1968–88$
$\text{SE} = 702.8$
$\text{DW} = 1.53$

$$\text{XSWI} = \begin{array}{c} -2149.3 \\ (-2.24) \end{array} - \begin{array}{c} 0.029 \times (\text{JEXR} \times \text{RPXSWI}) \\ (-1.96) \end{array} + \begin{array}{c} 68.27 \times \text{JAPIND} \\ (5.66) \end{array} - \begin{array}{c} 0.201 \times \text{YPCMSW} \\ (-0.98) \end{array}$$

$t = 1974–88$
$\text{SE} = 315.5$
$\text{DW} = 2.13$

$$\text{CONSWI} = \begin{array}{c} -5914.5 \\ (-0.89) \end{array} - \begin{array}{c} 0.106 \times \text{RPXSWI} \\ (-2.43) \end{array} + \begin{array}{c} 0.000077 \times \text{POPI} \\ (1.19) \end{array} + \begin{array}{c} 34.6 \times \text{GDFI} \\ (2.90) \end{array} - \begin{array}{c} 0.209 \times \text{INDVA} \\ (-1.82) \end{array}$$

$t = 1968–88$
$\text{SE} = 452.8$
$\text{DW} = 1.63$

3. Plywood Market

$$\text{QPWI} = \begin{array}{c} 537.8 \\ (0.73) \end{array} + \begin{array}{c} 0.0042 \times (\text{RPXPWI}) \\ (1.41) \end{array} + \begin{array}{c} 0.706 \times (\text{CAPPWI}) \\ (5.45) \end{array} - \begin{array}{c} 0.008 \times (\text{CSTPWI} + \text{RPXLI} - \text{XTLI}) \\ (-1.48) \end{array}$$

$t = 1975–88$
$\text{SE} = 371.3$
$\text{DW} = 1.75$

$$\text{XPWI} = \begin{array}{c} -43836.4 \\ (-9.45) \end{array} - \begin{array}{c} 1537.9 \times (\text{RPXPWI}/(\text{PHIPW} \times \text{EXR})) \\ (-2.34) \end{array} + \begin{array}{c} 0.943 \times \text{YPCMPW} \\ (4.71) \end{array}$$
$$+ \begin{array}{c} 0.148 \times \text{POPMPW} \\ (8.33) \end{array} - \begin{array}{c} 104.2 \times \text{MFXUV} \\ (-6.65) \end{array}$$

$t = 1975–88$
$\text{SE} = 370.9$
$\text{DW} = 1.97$

$$\text{CONPWI} = \begin{array}{c} -7014.6 \\ (-3.96) \end{array} - \begin{array}{c} 0.0033 \times \text{RPXPWI} \\ (-2.59) \end{array} + \begin{array}{c} 0.000055 \times \text{POPI} \\ (4.16) \end{array}$$

$t = 1975–88$
$\text{SE} = 219.7$
$\text{DW} = 1.75$

Deforestation equation

$$\ln(F/A) - \ln(1 - FA) = \begin{array}{c} 0.875 \\ (4.45) \end{array} - \begin{array}{c} 5.695 \times \text{LOGA} \\ (-1.98) \end{array} - \begin{array}{c} 0.013 \times \text{POPD} \\ (-5.82) \end{array} + \begin{array}{c} 0.00037 \times \text{GDPA} \\ (2.84) \end{array} + \begin{array}{c} 0.997 \times \text{D88} \\ (4.40) \end{array}$$

$R^2 = 0.51$
$\text{SE} = 0.82$
$\text{DW} = 1.57$

Table 17.5 Indonesia: timber trade simultaneous equation system and deforestation equation (cont.)

Dependent variables

QLI	Industrial roundwood production ('000 m³)
XLI	Industrial roundwood exports ('000 m³)
CONLI	Apparent domestic log consumption ('000 m³)
QSWI	Sawnwood and sleeper production ('000 m³)
XSWI	Sawnwood and sleeper exports ('000 m³)
CONSWI	Apparent domestic sawnwood and sleeper consumption ('000 m³)
QPWI	Plywood production ('000 m³)
XPWI	Plywood exports ('000 m³)
CONPWI	Apparent domestic plywood production ('000 m³)
F/A	Forest area (F) per total land area (A), km²

Independent variables

RPXLI	Log export prices (unit values) in domestic currency (Rp/m³)
XTLI	Log export taxes (Rp/m³)
ROYLI	Log royalties and other taxes (Rp/m³)
CSTLI	Logging and log transport costs (Rp/m³)
PLTENC	Average world price of temperate non–coniferous logs (US$/m³)
EXR	Exchange rate, Indonesian Rupiah (Rp)/US$
YPCML	Average GNP per capita of Indonesian log importers (US$)
POPML	Average population of Indonesian log importers
RPXSWI	Sawnwood and sleeper export prices (unit values) in domestic currency(Rp/m³)
CAPSWI	Sawmill capacity ('000 m³)
CSTSWI	Total input sawmill costs (Rp/m³)
JEXR	Exchange rate, Japanese yen (¥)/Rp
JAPIND	Japan – industrial production index 1985 = 100
YPCMSW	Average GNP per capita of Indonesian sawnwood importers (US$)
POPI	Indonesia – population
GDFI	Indonesia – GDP deflator index 1980 = 100
INDVA	Indonesia – industrial value added (Rp billion)
RPXPWI	Plywood export prices (unit values) in domestic currency (Rp/m³)
CAPPWI	Plymill capacity ('000 m³)
CSTPWI	Total input plymill costs (Rp/m³)
PHIPW	Export wholesale Tokyo spot price of Philippine (luan) plywood (US$/m³)
YPCMPW	Average GNP per capita of Indonesian plywood importers (US$)
POPMPW	Average population of Indonesian plywood importers
MFXUV	Average world manufactures export unit values index 1980 = 100
LOGA	Roundwood production per total area ('000 m³/km²)
POPD	Population density (persons per km²)
GDPA	GDP per capita ('000 Rp/km²)
D88	Dummy variable for 1988

Statistical data

t	Time period of the regression
SE	Standard error of the regression
DW	Durbin–Watson statistic
R^2	R-squared statistic

Notes: all values are in constant (1980) prices; t-statistics are indicated in parentheses under the relevant coefficients.

the prices (export unit values), production levels, domestic consumption levels and export levels for roundwood, sawnwood and plywood respectively. The remaining variables used to estimate the supply and demand equations are exogenous.[6] In Table 17.5, t-statistics are displayed in parentheses under each coefficient, and other regression statistics are placed to the right of each estimated equation.

The deforestation regression indicates that increases in population density have a more significant impact than log production in terms of changes in forested area in Indonesia's forested provinces. For every cubic metre (m^3) of timber extracted per km^2, the proportionate change in forested to non-forested area is -0.6 per cent. For every additional person per km^2, the proportionate fall in forest area is -1.3 per cent. In contrast, a rise in incomes of Indonesian Rupiah (Rp) 10,000 per person, approximately US$6 in 1988, would counteract deforestation at a rate of 0.4 per cent.[7]

However, the deforestation equation in Table 17.5 may over-estimate the contribution of log extraction alone to regional changes in forest area in Indonesia. First, there is the usual problem of reliability and accuracy of data. More importantly, however, data limitations prevent distinguishing in the regression between log production from permanent production forests and that from conversion forests. If log production is mainly from conversion forests, then timber extraction is essentially a precursor or by-product to agricultural conversion, which is the principal factor in the resulting deforestation. As indicated in Table 17.3a, close to 50 per cent of Indonesia's total production forests consist of conversion forests, and anecdotal evidence suggests that a good deal of the log production during the period 1973–88 came from the latter forests (Burgess 1989). To the extent that this is the case, then the deforestation equation may be reflecting mainly the impact of agricultural conversion, rather than the initial logging operations, on changes in forested area. This is supported somewhat by the evidence presented in Table 17.2, which shows logging to be much more responsible for "opening up" the forest than is outright forest conversion.

On the other hand, in the 1980s Indonesia's permanent production forest began contributing an increasingly large share to overall log production – and is expected to be the main source of tropical hardwood logs in the future. Although forest regulations for this permanent forest estate are increasingly requiring concessions to harvest timber on a sustained production basis, clear cutting is allowed under certain conditions. Moreover, problems of implementing these regulations and policy failures in forestry management also contribute to over-harvesting and clear-cutting of permanent production forests, as well as illegal felling of non-production forests (Barbier 1987, Burgess 1989, Gray and Hadi 1990, Sedjo 1987, World Bank 1989b). Evidence for the 1980–90 period suggests that logging damages alone contributed *directly* to just under 10 per cent of the annual average rate of deforestation in the Outer Islands of Indonesia (Pearce et al. 1990). Moreover, if timber operations in Indonesia are also "opening up" permanent production and virgin forests to subsequent encroachment and deforestation, then logging may have a major rôle in *indirectly* furthering the deforestation process of these forested areas.

In short, it is difficult to determine the extent to which the deforestation equation in Table 17.5 over-estimates, if at all, the impact of log production on tropical deforestation in Indonesia. The available evidence would suggest that, at the very least, the equation is correct in indicating that this impact is significantly greater than zero and is positive.

The key relationships in the simultaneous equation system of the three main timber markets of Indonesia are for price and quantity. Table 17.6 compares the price elasticity estimates for each of the supply and demand equations of the model with other recent estimates for the Southeast Asian region. In general, the elasticity estimates compare favourably. According to our estimates, the supply of timber products in Indonesia appears to be fairly price inelastic. In the case of the processed products, this may reflect the Indonesian policy of expanding sawnwood and plywood capacity over the 1968–88 period. In our regressions, export demand for sawnwood seems slightly more elastic than domestic demand, whereas for plywood the opposite seems to be the case. A similar result was obtained by Vincent (1989) for all Southeast Asian exporters.

The simultaneous equation system and the deforestation regression are combined to form a simulation model of timber trade–deforestation linkages, using data for 1988. The results are displayed in Table 17.7. Given the Indonesian policy of banning log exports, the export log demand equation from Table 17.5 was not included in the model. It is assumed the total log production in Indonesia supplies domestic log demand. The latter is again related to sawnwood and plywood production, although in the simulation model actual recovery rates for the mid-1980s were used as coefficients rather than those generated by the regression for the entire 1968–88 period.[8]

Table 17.7 compares the results of the base case scenario with the actual values of key price, quantity and deforestation variables for Indonesia in 1988. The model appears to be a reasonably good simulation of these variables. The impact of Indonesia's forest industries and log production on the tropical forest are indicated by effects on forest cover and the annual rate of deforestation. The latter indicator was not simulated by the base scenario; instead, the most recent FAO estimate of annual deforestation in Indonesia was used.

The following sections compare different policy scenarios to the base case, focusing particularly on the sawnwood export tax intervention, the imposition of an import ban, the use of revenue-raising import surcharges and the implementation of improved sustainable timber management practices. There are, of course, a number of important limitations of the use of the simulation model for these policy scenarios.

Table 17.6 Comparative price elasticity estimates for Southeast Asian tropical timber.

	Time period	Short run	Long run
1. This study			
Indonesia – log supply	1968–84		0.20
Indonesia – log export demand[a]	1968–84		−1.51
Indonesia – sawnwood supply	1968–88		0.27
Indonesia – sawnwood domestic demand	1968–88		−0.36
Indonesia – sawnwood export demand	1974–88		−0.68
Indonesia – plywood supply	1975–88		0.31
Indonesia – plywood domestic demand	1975–88		−0.91
Indonesia – plywood export demand[b]	1975–88		−0.46
2. Constantino (1988a)[c]			
Indonesia – sawnwood export demand[d]	See note	−0.08	−0.21
Indonesia – plywood export demand	1975–85	−0.26	−0.58
Importers – elasticity of subst., tropical and temperate sawnwood	1975–85	1.30	2.11
Importers – elasticity of subst., tropical and temperate plywood	1975–85	0.75	1.23
World – elasticity of substitution, source of origin, sawnwood	1979–85	2.50	4.39
World – elasticity of substitution, source of origin, plywood	1979–85	4.56	12.3
3. Vincent (1992a)[e]			
Peninsular Malaysia – log supply	1973–89		1.1
Peninsular Malaysia – sawnwood supply	1973–89		1.7
Peninsular Malaysia – sawnwood domestic demand	1973–89		−0.55
Other tropical exporters – log export supply	1973–89		2.7
Other tropical exporters – sawnwood export supply	1973–89		0.7
Importers – tropical log demand	1973–89		−1.59
Importers – tropical sawnwood demand	1973–89		−5.67
4. Vincent (1989)[f]			
Malaysia – log supply	1960–85		0.64
Malaysia – sawnwood supply	1960–85		1.00
Malaysia – plywood supply	1960–85		1.00
Malaysia – sawnwood domestic demand	1960–85		−0.27
Malaysia – plywood domestic demand	1960–85		−0.94
Other SE Asia exporters – log supply	1960–85		0.46
Other SE Asia exporters – sawnwood supply	1960–85		1.00
Other SE Asia exporters – plywood supply	1960–85		1.00
Other SE Asia exporters – sawnwood domestic demand	1960–85		−0.53
Other SE Asia exporters – plywood domestic demand	1960–85		−0.85
Importers – SE Asian and Malaysian sawnwood demand	1960–85		−1.22
Importers – SE Asian and Malaysian plywood demand	1960–85		−0.46

Notes: a) Export price of Indonesian logs relative to average world price of temperate non-coniferous logs; b) Export price of Indonesian plywood relative to export wholesale spot price of Philippine (luan) plywood; c) Elasticity of substitution between sources of origin is defined as the percentage reduction in the ratio of imports from two different countries if the ratio of import prices of the two countries increases by one per cent; d) Based on Buongiorno (1979) for coniferous sawnwood; e) Elasticities based on Cardellichio et al. (1989). Note that in the latter study, Indonesian sawnwood and plywood supply are both estimated to have elasticities of 0.7, and Indonesian sawnwood and plywood domestic demand have own-price elasticities of −0.92 and −1.55 respectively; f) Elasticities for sawnwood and plywood supply were assigned a value of one in the analysis.

Table 17.7 Indonesia: timber trade and tropical deforestation simulation model.

Key variables	Base case 1988 values	Actual 1988 values
1. Prices (Rp/m³)		
Log border-equivalent price (unit value)	36,714	n.a.[a]
Sawnwood export price (unit value)	333,930	262,013
Plywood export price (unit value)	458,179	464,124
2. Quantities ('000 m³)		
Log production	28,766	29,819
Log domestic consumption	28,480	28,887
Sawnwood production	9,351	10,290
Sawnwood exports	2,923	3,083
Sawnwood domestic consumption	6,427	7,207
Plywood production	7,770	7,733
Plywood exports	6,383	6,372
Plywood domestic consumption	1,387	1,361
3. Deforestation (km²)		
Total forest area	1,401,163	1,401,144
Annual rate of deforestation[b]	13,150	13,150

Notes: a) no data available; b) not calculated in simulation model but based on Schmidt 1990.

First, as discussed above, the linkages of the impacts of log production on tropical deforestation do not take into account changes in the type of forests being exploited (i.e. permanent production forests, conversion forests or new forest areas) or in the production management regime (i.e. selective cutting or clear cutting). The deforestation equation also cannot distinguish between the direct versus indirect impacts of logging on the forest.

Second, although the model is well suited to examining the effects of a particular policy intervention on the *internal diversion* of timber product flows in Indonesia, i.e. between export and domestic markets for sawnwood and plywood, the model does not explicitly indicate the effects of an intervention in terms of *external diversion* of timber products, i.e. between different import markets for sawnwood and plywood. That is, elasticities of substitution by product or by sources of origin in existing importing markets are not explicitly modelled, and the possibility of new import markets opening up as a result of the intervention is not taken into account. In addition, "leakages", or the ability to avoid any restrictions imposed by an intervention, are also not shown.

Finally, the model is capable of only showing "static", or one-period, partial equilibrium effects of a policy intervention. Many of the impacts may have economy-wide, or general equilibrium, impacts that feed back to affect the timber product markets and deforestation. In addition, the impacts may have lag, or dynamic, effects that manifest themselves over a medium or long term horizon. Some of the policy interventions themselves, such as the implementation of sustainable management practices, would realistically involve a longer process of implementation than a single period (i.e. one year).

Keeping these obvious limitations in mind, it is nevertheless useful to use the simulation model to obtain an approximate indication of the relative impacts of different trade and forest sector policy interventions on Indonesia's timber product markets and forest resource base.

17.6 Sawnwood export taxes

As indicated in Table 17.4 and discussed above, in 1989 Indonesia imposed substantial taxes on sawnwood exports in an effort to shift processing activities to plywood. A secondary objective of the policy is to improve competitiveness and overall efficiency of sawmills. However, the implications of the policy for wood recovery, log demand and thus tropical deforestation in Indonesia are less clear.

Table 17.8 displays the results of imposing varying levels of sawnwood export taxes in the simulation model of Indonesia's forest sector. The results compare the changes in key price, quantity and deforestation variables to the 1988 base case scenario.

In the model, the sawnwood export taxes appear to have the effect of reducing export demand for Indonesian sawnwood, thus lowering the export price received

Table 17.8 Indonesia: timber trade and tropical deforestation simulation model policy scenario – sawnwood export tax (% change over base case).

Key variables	10% tax	50% tax	100% tax	250% tax	700% tax
1. Prices (Rp/m³)					
Log border-equivalent price (unit value)	−0.70	−3.35	−6.39	−14.07	−29.03
Sawnwood export price (unit value)	−0.98	−4.72	−9.01	−19.84	−40.93
Plywood export price (unit value)	−0.05	−0.23	−0.44	−0.97	−2.00
2. Quantities ('000 m³)					
Log production	−0.16	−0.77	−1.47	−3.23	−6.66
Log domestic consumption	−0.15	−0.74	−1.42	−3.12	−6.44
Sawnwood production	−0.33	−1.58	−3.01	−6.64	−13.69
Sawnwood exports	−2.26	−10.89	−20.79	−45.79	−94.47
Sawnwood domestic consumption	0.55	2.66	5.07	11.17	23.06
Plywood production	0.01	0.07	0.13	0.29	0.60
Plywood exports	0.01	0.03	0.06	0.13	0.26
Plywood domestic consumption	0.05	0.25	0.48	1.06	2.18
3. Deforestation (km²)					
Total forest area	0.00[a]	0.02	0.03	0.06	0.13
Annual rate of deforestation	−0.3	−1.60	−3.06	−6.73	−13.88

Note: a) negligible increase over the base case forest cover of 44 sq. km.

and the quantity exported. The price effects appear to be outweighed by the quantity effects. Although some sawnwood production is diverted to domestic consumption, it is not sufficient to overcome the fall in export demand. Thus overall sawnwood production falls. The impacts on Indonesia's sawnwood markets are more severe the greater the tax. As indicated in the model, a tax rate of 700 per cent imposed in 1988 would effectively choke off sawnwood export demand. With proposed taxes of US$250–2400 per m^3, actual government policy would be represented by the higher range of sawnwood export taxes shown in Table 17.8.

Significantly, the imposition of a tax on sawnwood exports does not appear to instigate a major shift of processing capacity to plywood – at least not in the one-period duration of the simulation model. Only at extremely high rates of taxation does this occur even slightly, and it appears that increased domestic plywood consumption is the most noticeable effect. Evidence would suggest that there are several structural factors limiting diversion of production between Indonesia's sawnwood and plywood industry in the short term, such as the much newer vintages of capital in plywood milling, and its export orientation that requires higher production and possibly access to better quality logs (Constantino 1990). The simulation model therefore confirms that, if a sawnwood export tax is to shift Indonesian processing capacity towards greater plywood expansion, then it will be a long term process. Unfortunately, the economic costs of reduced sawnwood exports and production appear to be severe from the outset.

Only at high rates of taxation does the sawnwood export tax have a modest impact on deforestation. The inability of the policy to increase plywood output in the short term means that the net reduction in sawnwood production translates into less log extraction. At higher rates of taxation, the fall in log production is much larger, thus resulting in a lower rate of deforestation. However, as noted, the economic costs particularly in terms of lost exchange earnings of a prohibitive sawnwood export tax are severe. Moreover, if the policy is successful over the long run in shifting processing to plywood production, then one would expect log demand, and thus deforestation rates, also to be revived.

In sum, a prohibitive sawnwood export tax in Indonesia appears to be a high cost strategy for shifting processing capacity to plywood production and export, which may only be successful – if at all – over the long term. In addition, the policy does not appear to be an effective approach for reducing timber-related deforestation in Indonesia. Only modest reductions in deforestation may occur, and these will be short-lived if plywood production begins replacing the lost sawnwood output.

17.7 Import ban

Some environmental pressure groups in Western tropical timber importing countries have been urging their governments to ban the import of tropical timber, or at least imports of timber that is "unsustainably" produced. Consumer-led boycotts against tropical timber products have also been instigated. The presumption is that an import ban is the most effective way of ending timber-related deforestation in tropical forest countries. Despite the various economic and legal implications of such a move, as well as questions about its effectiveness, an import ban on tropical timber is a realistic possibility in the near future.

The first column of Table 17.9 indicates the effects of a total import ban, compared to the base case scenario, in the timber trade–deforestation model of Indonesia. In order to simulate an import ban in the model, large price changes were used

Table 17.9 Indonesia: timber trade and tropical deforestation simulation model policy scenario – import ban and revenue raising taxes (% change over base case).

	Total import ban[a]	1% revenue raising import tax[b]	5% revenue raising import tax[d]
1. Prices (Rp/m³)			
Log border-equivalent price (unit value)	–	0.17	−0.82
Sawnwood export price (unit value)	–	−0.11	−0.54
Plywood export price (unit value)	–	−0.21	−1.03
2. Quantities ('000 m³)			
Log production	−28.33	−0.04	−0.19
Log domestic consumption	−27.37	−0.04	−0.18
Sawnwood production	−10.64	−0.03	−0.14
Sawnwood exports	−100.00	−0.23	−1.12
Sawnwood domestic consumption	30.01	0.06	0.30
Plywood production	−43.84	−0.04	−0.22
Plywood exports	−100.00	−0.10	−0.51
Plywood domestic consumption	214.51	0.23	1.12
3. Deforestation (km²)			
Total forest area	–	0.00[c]	0.01
Annual rate of deforestation	–	−0.41	−0.72

Notes: a) Large price changes were used deliberately to constrain sawnwood and plywood exports to zero in this simulation and therefore are no longer endogenously generated by the model. Also, the functional form of the deforestation equation and its estimation using regional panel data imply that the large changes in log production associated with the import ban scenario cannot be used to predict reliably the effects on forest cover and deforestation. Thus both price and deforestation effects are eliminated from this policy scenario simulation; b) A total of US$23.1 million (1980 prices) in revenue would be raised, with US$5.8 million and US$17.3 million from Indonesian sawnwood and plywood exports respectively; c) A negligible increase over the base case forest cover of 53 sq. km; d) A total of US$113.9 million (1980 prices) in revenue would be raised, with US$28.5 million and US$85.4 million from Indonesian sawnwood and plywood exports respectively.

deliberately to constrain sawnwood and plywood exports to zero. Timber product prices are therefore no longer endogenously generated by the model, and as a result, are not shown in Table 17.9. Also, the functional form of the deforestation equation and its estimation using regional panel data imply that the large changes in log production associated with the import ban scenario cannot be used to predict reliably the effects on forest cover and deforestation. Thus the impacts on deforestation are also not reported for this policy scenario simulation.

As shown in Table 17.9, an import ban would have a devastating impact on Indonesia's forest industry in the short term. Although there would be significant diversion of plywood and sawnwood exports to domestic consumption, this would be insufficient to compensate for the loss of exports. Net production in both processing industries would fall. Given its export orientation, the plywood industry would be particularly hurt – reducing its output by over 40 per cent. Net production losses in the sawnwood industry would be closer to 10 per cent. The overall effect is to lower domestic log demand in the short term by around 25–30 per cent.

The policy scenario is of course assuming that the import ban is 100 per cent effective. It is unlikely that all importers of Indonesia's tropical timber products – many of which are also newly industrializing or producer countries with processing capacities – would go along with a Western-imposed ban. In any case, one would expect that over the longer term there would be some diversion of Indonesian plywood and sawnwood exports to either new import markets or existing markets that prove to be less stringent in applying the ban. As discussed above, these effects cannot be captured in our model.

In addition, the long-term implications of an import ban on tropical deforestation are also uncertain. Even if the ban is 100 per cent effective in the short term, any resulting reduction in tropical deforestation is likely to be short-lived. There are several reasons for this.

First, as indicated in the model, diversion of timber products to satisfy domestic demand is likely to continue as Indonesian population and economic activity expands (see Table 17.5). Thus one can expect the wood processing industries to recover somewhat through re-orientation to meet local consumption of timber products. Moreover, past evidence suggests that domestic-oriented processing in Indonesia is less efficient than export-oriented processing, implying poorer wood recovery rates and greater log demand (Constantino 1990). Pressure on the tropical hardwood forests may therefore increase after the initial "shock" of a ban.

However, it is unlikely that the domestic market in Indonesia will generate the same demand for higher valued timber products as the international market. Instead, domestic demand is likely to be strongest for high-volume but lower valued wood products. As a consequence, it may be difficult for Indonesia to justify holding as large a proportion of its tropical forests as permanent production forests if the expected economic returns from sustainable management decline as a result of the ban. More of the resource may be shifted to conversion forests, and timber production will become residual to satisfying the growing demand for agricultural land. In short, without the timber trade providing increased value added in the form of external demand for higher valued products, there may be less reason for Indonesia "holding on" to these forests as opposed to converting them to an alternative use, such as agriculture (Barbier 1992, Vincent 1990b).

To summarize, a total import ban would cause a major diversion of Indonesian timber products to meet domestic demand. Although in the short term net production of wood products, and thus log demand, would fall, this situation would not necessarily be sustained over the long run. Even if this is not the case, the ban may be ineffective in permanently reducing tropical deforestation because, in the first place, timber production is not the main source of deforestation in Indonesia, and secondly, as the value of holding on to the forest for timber production decreases, the incentives to convert more of the resource to agriculture will increase.

17.8 Revenue-raising import tax

International co-operation and compensation to assist developing countries in achieving sustainable management of timber production forests have also been discussed in recent years. For example, the International Tropical Timber Council (ITTC) unanimously adopted a "Year 2000 Target" that encourages "ITTO members to progress towards achieving sustainable management of tropical forests and trade in tropical forest timber from sustainably managed sources by the year 2000" (Decision 3(X) ITTC 1990). One suggestion for raising additional funds for this strategy is for importing countries to impose a small import surcharge on tropical timber, which is then directed to tropical forest countries that have shown "demonstrable progress" towards achieving the Year 2000 Target.

A recent study has indicated that a one to three per cent surcharge on the tropical timber imports (excluding plywood and based on 1986 trade levels) of the EEC, Japan and USA would raise approximately US$31.4 to 94.1 million with little additional distortionary effects (NEI 1989). Buongiorno & Manurung (1992) also examine the scope for a five per cent revenue-raising import levy on tropical timber by the Union pour le Commerce des Bois Tropicaux (UCBT) or the EEC.[9] The results indicate that tropical timber exporters would lose around US$44.8 million in additional revenues. Thus, if the funds raised by the tax were rebated to exporting countries, they could be made better off by over US$40 million.

The final two columns of Table 17.9 show the impacts of a one and five per cent revenue-raising import tax on Indonesian tropical timber exports in the simulation model. In the case of a one per cent surcharge, a total of US$23.1 million (1980 prices) in revenue would be raised, with US$5.8 million and US$17.3 million from Indonesian sawnwood and plywood exports respectively. For the five per cent surcharge, a total of US$113.9 million (1980 prices) in revenue would be raised, with US$28.5 million and US$85.4 million from Indonesian sawnwood and plywood exports respectively. When compared to the revenue estimates from the NEI (1989) study, the above figures would suggest that, in the case of Indonesia, applying the import surcharge to plywood would significantly raise the total amount of financing appropriated.

The results shown in Table 17.9 confirm that a small import surcharge would have very little distortionary effects on Indonesia's timber product flows and prices. There would also be a negligible direct impact on deforestation. However, around the five per cent level, the impacts of the surcharge on exports in particular would become more noticeable. Thus from the standpoint of minimizing additional distortionary effects, the policy scenario confirms that an import surcharge of less than five per cent would be optimal.

A more pertinent issue is whether it is worth imposing an import surcharge to raise revenue for sustainable management of tropical forests. In the simulation model, the same result, and thus the same amount of revenue, could be achieved if Indonesia raised the money for its own "sustainable management" initiatives by imposing a revenue-raising export surcharge of one to five per cent. The NEI study argues that the imposition of an export levy by producing counties themselves has the advantage of directly addressing the forest management systems of these countries, as well as avoiding the transaction costs involved in international transfers, but prevents obvious problems of monitoring and evaluating success in achieving sustainable management. The counter-argument is that an import surcharge not only has problems of transaction costs and administration, but that it is also possibly discriminatory if it is limited only to the *tropical* timber trade. Moreover, the import-surcharge-cum-international-transfer mechanism would still require the co-operation of producer countries, as well as raising similar problems of monitoring and evaluation of progress towards sustainable management and expenditures. Finally, there is the issue of whether the amount of funds raised through any trade surcharge would be adequate for the task, and whether it would be a more appropriate avenue for raising additional large-scale funding *outside* the timber trade altogether (Barbier et al. 1992c).[10]

In sum, it is possible to raise revenue through an import surcharge of less than five per cent with minimal distortionary effects on Indonesia's timber trade and production. The key issue is whether this is the most efficient and appropriate means of raising financing for improved sustainable management of production forests.

17.9 Sustainable timber management

An alternative to the above trade interventions would be a more direct policy initiative by Indonesia to improve "sustainable" management of its remaining production forests. As discussed above, several studies have pointed to considerable problems of policy failures in Indonesian forestry management that are contributing to over-harvesting and clear-cutting of permanent production forests, as well as illegal felling of non-production forests (Barbier 1987, Burgess 1989, Gray and Hadi 1990, Sedjo 1987, World Bank 1989b). Correcting these policy failures would improve sustainable management of Indonesia's remaining tropical forests, and thus reduce timber-related deforestation, but would also mean higher harvesting costs per cubic metre of wood extracted.

Table 17.10 indicates the most likely effects of such a "sustainable" management policy scenario, which is represented by an increase in harvesting costs. As there is insufficient data to determine the extent to which costs would rise if such a policy were to be implemented in Indonesia, we have arbitrarily chosen a 25 per cent and 50 per cent harvest cost increase for comparison.

A surprising feature of the scenario is that although log prices are affected significantly by the increased harvest costs, any resulting impacts on the rest of Indonesia's forestry sector seem to be somewhat dissipated. There appear to be several factors at work. First, the log supply for Indonesia is very inelastic with respect to price in our model (see Tables 17.5 and 17.6). This is not surprising given that

Table 17.10 Indonesia: timber trade and tropical deforestation simulation model policy scenario – sustainable timber management (% change over base case).

Key variables	25% rise in harvest costs	50% rise in harvest costs
1. Prices (Rp/m^3)		
Log border-equivalent price (unit value)	41.59	83.06
Sawnwood export price (unit value)	4.04	8.09
Plywood export price (unit value)	2.86	5.72
2. Quantities ('000 m^3)		
Log production	−0.94	−1.87
Log domestic consumption	−1.37	−2.73
Sawnwood production	−1.89	−3.77
Sawnwood exports	−1.03	−2.05
Sawnwood domestic consumption	−2.28	−4.55
Plywood production	−0.87	−1.73
Plywood exports	−0.38	−0.75
Plywood domestic consumption	−3.12	−6.24
3. Deforestation (km^2)		
Total forest area	0.02	0.04
Annual rate of deforestation	−2.28	−4.23

Indonesian policy has been to devote log production solely to supplying domestic processing capacity, which of course has expanded considerably over the 1968–88 period. Second, increased log costs are only one component of the total factor costs of Indonesia's processing industries, and have increasingly become the least important component in recent years (Constantino 1990). Finally, Indonesia's sawnwood and plywood exports seem to be the least affected by the increased harvest costs, which would suggest that external demand factors exert an important counter-acting influence.

There is reason to believe that the impacts of the sustainable management scenario on reducing timber-related deforestation may be underestimated in Table 17.10. As discussed above, the deforestation equation in the model can only record changes in forest cover due to changes in overall log production rates. It cannot distinguish between *qualitative* changes in the management of log production that may also affect timber-related deforestation more indirectly, such as controlling residual stand damage, improved replanting and reforestation, reduced high-grading of stands, limiting trespass, improving the incentives to control stand abandonment and encroachment, and so forth.[11] The latter factors may be more important in timber-related deforestation than log extraction alone. In addition, some sustainable management techniques, such as utilising lesser known species and imprproved harvesting techniques, may actually increase the amount of logs produced from a given timber stand, thus simultaneously limiting harvest costs and reducing the pressure to exploit the remaining tropical forest.

Finally, it is unlikely that the implementation of any sustainable management techniques would occur in a single year, as implied by the simulation scenario. A longer lead time for implementation would mean more time for the Indonesian forest industries to absorb the increased costs, but would also imply a longer period before the effects on reducing deforestation would be fully felt.

In sum, Indonesia's timber industries may not be badly affected by the higher harvest costs associated with implementing sustainable forestry management policies. Although there would be some reduction in tropical deforestation due to reduced log production, this direct effect may be less significant than the improvement in forest management and protection resulting from qualitative changes in timber stand management practices and ownership.

17.10 Conclusion

The simulation model developed to examine the relationship between Indonesia's timber trade and tropical deforestation has provided some important insights into this linkage as well as the scope for the use of trade policy instruments to limit deforestation. As discussed throughout, there are obvious constraints on the use of such models to examine such dynamic, complex and pervasive effects. Nevertheless, the model and the policy simulations would suggest extreme caution in the use of broad trade policy interventions as a means to affect timber-related deforestation in Indonesia, and in some cases even as an economic tool to develop further Indonesia's timber processing capacity.

First, it is clear that timber production is not the major cause of tropical deforestation in Indonesia. Even where timber production is a factor in deforestation, most timber-related deforestation, including subsequent deforestation by agricultural encroachment, may have more to do with the management and regulation of the timber stand than with the amount of logs extracted from the stand *per se*.

Secondly, extreme trade interventions, such as the current GoI policy of prohibitive sawnwood export taxes and the imposition of a total import ban on tropical timber products, may have an initial "shock" impact on tropical deforestation in Indonesia. However, this may be short-lived as dynamic factors in the economy – notably the shifting of processing capacity or the transfer of permanent production forests to conversion forests – may take hold. Both forms of trade intervention clearly impose high economic costs on Indonesia's forestry industries. Surprisingly, a high sawnwood export tax would appear to make little headway in the short run in achieving its stated objective of spurring development of Indonesia's plywood sector.

Thirdly, an import surcharge on tropical timber imports would have minimal distortionary impacts on Indonesia's timber trade – provided that it was imposed at a level of less than five per cent. A more pertinent issue is whether this policy is the correct means for raising funds for sustainable management of production forests, and whether the financing raised would be sufficient for the task.

Finally, improvements in sustainable timber forest management and regulation by Indonesia could raise log harvesting costs, but there may not be such significant impacts on Indonesia's processed products and trade. The reduction in timber-related deforestation in Indonesia resulting from the policy may not be fully captured by the model.

In sum, there seems little scope for the use of trade policy interventions as a means to reducing tropical deforestation in Indonesia. If there is concern over timber-related deforestation, a more appropriate approach may be to deal more directly with the problem by improving sustainable production forest management and regulation at the timber stand level. However, a key factor is whether the GoI has sufficient incentive to ensure that better forest management policies are implemented and enforced. By ensuring access of sustainably managed timber to import markets and by providing financial assistance for Indonesian policy efforts, the major tropical timber consumer countries could go a long way toward encouraging the appropriate incentives for action by the Indonesian government.

Notes

1. The main importers of Indonesian logs over 1968–84 were China, Hong Kong, Japan, Korea and Singapore.

2. Given that Indonesian sawnwood and plywood are tradeables, the assumption that domestic and export prices of these product are highly correlated over time seems reasonable. This is confirmed by the limited time series data available on both export and domestic prices.

3. The main importers of Indonesian plywood over 1975–88 were China, Hong Kong, Japan, Singapore and the USA. The main importers of Indonesian sawnwood over 1974–88 were China, Hong Kong, Italy, Japan, Korea, Malaysia, the Netherlands, Singapore, Thailand and the UK. Given the predominance of Japan and its main Southeast Asian newly industrializing competitors among these importers, the Japanese/Indonesian exchange rate and Japanese industrial activity proved to have strong explanatory power in the demand for Indonesian sawnwood exports (see Table 17.5).

4. However, because the price of logs was represented by log export unit values, it is only included in the estimation for the period before the ban, 1968–84.

5. Twenty provinces were selected, spread over the Indonesian islands of Sumatra, Kalimantan, Sulawesi, Maluku, Irian Jaya, and Nusa Tenggara. Data were available for the years 1973, 1979, 1981, 1982, 1984 and 1988.

6. As some of the instrumental variables were highly correlated, the list of instruments included in each two-stage regression was sometimes modified.

7. Indonesia's real GNP per capita in 1988 (1980 prices) was approximately US$230. A rise in income reflects economic development and may be associated with reduced pressure on deforestation in a number of ways: a) through an increasing concentration of the population in urban centres which reduces the direct pressure of rural populations on tropical forests; b) through changing the composition of the economy from being based on primary extractive industries (i.e. timber, agriculture, etc) to industrial processing and service based industries; and c) through improved efficiency and management of resource use which may accompany economic development, again reducing the indirect pressure of natural resource consumption on tropical forests.

8. In the simulation model, the log domestic demand equation is therefore CONLI = 77.95 + 1.5 × QSWI + 1.85 × QPWI. A comparison with the regression for CONLI of Table 17.5 indicates that, by the mid-1980s, recovery rates for sawmills in Indonesia had improved slightly, whereas they had deteriorated somewhat for plymills. This seems reasonable intuitively, given that by the 1980s the expansion in sawmill capacity as the GoI attempted to improve efficiency and quality, whereas plymill capacity was deliberately encouraged to expand.

9. The UCBT countries are Belgium, Denmark, France, Germany, Luxembourg, the Netherlands and the United Kingdom.

10. One problem is that the scale of financing required may be well beyond what can be raised from either an import or export surcharge levied on the trade. For ex-

ample, Agenda 21 of the UN Conference on Environment and Development has estimated that, internationally, over US$1.5 billion annually will be required by tropical forest countries to reduce deforestation (ITTC 1992).

11. *Trespass* in the forestry context refers to losses due to logging theft, which could also be extended to include losses due to graft. *High-grading* refers to the removal of high-valued timber and leaving a degraded timber stand.

18

Deforestation:
the rôle of the international
trade in tropical timber

Edward Barbier, Joanne Burgess,
Josh Bishop, Bruce Aylward

18.1 Introduction

In recent years, concern about the destruction of tropical forests has led to increased interest in the rôle of the international timber trade in promoting deforestation in the tropics, and in ways to restrict or reform the trade as a means to protect and preserve tropical forests. This chapter, based on Barbier et al. (1993), assesses the strength of the link between the trade and deforestation and the likely effectiveness of policy interventions in slowing or halting deforestation.

18.2 Tropical timber production and trade: a brief overview

A large proportion of the timber harvest in tropical countries is consumed domestically and does not enter the international trade. Only 17 per cent of total tropical timber production is used for industrial purposes; the remainder is consumed for fuelwood and other non-industrial uses. Out of the total volume of industrial timber produced by tropical countries, approximately 31 per cent is exported in round or product form. In other words, only about 6 per cent of total tropical roundwood production enters the international trade (Bourke 1992). In addition, tropical timber plays a minor rôle in the global timber market; tropical countries account for approximately 15 per cent of the total volume of global timber production, and 11 per cent of the value of global exports (Tables 18.1a & b).

Tropical Asia and Oceania dominate the tropical forest products trade – accounting for half of the industrial roundwood, sawnwood and wood-based panels produced by tropical countries and representing over 85 per cent of total exports of tropical forest products. Whereas tropical Central and South America produce

Table 18.1a The volume of world and tropical timber production and trade, 1990.

	Production	Exports	Tropical production as a % of world production	Tropical exports as a % of world exports
Total roundwood (mn m³)				
World	3450.4	120.5		
Tropical	1397	30	40.5	24.9
Industrial roundwood (mn m³)				
World	1654.2	118.2		
Tropical	275	29.1	16.6	24.6
Saw and veneer logs (mn m³)				
World	979.8	66.8		
Tropical	149	27	15.2	40.4
Sawnwood and sleepers (mn m³)				
World	485.9	88.9		
Tropical	59	9	12.1	10.1
Wood based panels (mn m³)				
World	124.9	31.2		
Tropical	19.4	13	15.5	41.7
Wood pulp (mn tonnes)				
World	154.4	25		
Tropical	7.2	0.7	4.7	2.8
Paper and paper board (mn tonnes)				
World	238.2	55.2		
Tropical	15.5	2.2	6.5	4.0

Note: tropical countries are all developing countries excluding China, Chile, Argentina, Turkey and S. Korea.

some 36 per cent of tropical forest products, the vast majority of this output is consumed domestically. Tropical Africa produces the smallest regional share (14 per cent) of tropical forest products and exports roughly 10 per cent of its production. The major exporting countries are Indonesia, Malaysia, Côte d'Ivoire, Brazil, Gabon and Congo, each of which had net forest product exports worth over US$100 million in 1990 (Table 18.2).

Whereas the production of tropical roundwood has been steadily increasing over the past few decades, timber consumption in tropical countries has been growing at an even faster rate due to population and income growth. The rapid expansion of timber demand in many tropical countries has led to increased domestic consumption, reduced timber exports and increased timber imports. As a result, many tropical timber producing countries are becoming net timber importers. The forest products trade generates positive net foreign exchange earnings in only 7 of 14 Af-

Table 18.1b The value of world and tropical timber production and trade, 1990.

	Exports (US$ billion)	Tropical as a % of world exports
Saw and veneer logs		
World	6.57	
Tropical	2.28	34.7
Sawnwood		
World	16.99	
Tropical	2.15	12.7
Wood based panels		
World	10.14	
Tropical	4.15	40.9
Wood pulp		
World	15.82	
Tropical	0.86	5.4
Paper and paper board		
World	45.27	
Tropical	1.5	3.3
Other		
World	2.68	
Tropical	0.16	6.0
Total		
World	97.47	
Tropical	11.1	11.4

Note: tropical countries are all developing countries excluding China, Chile, Argentina, Turkey and S. Korea. *Source: Bourke 1992.*

rican exporting countries, 3 of 14 Central and Southern American exporting countries, and 3 of 9 exporting countries from Asia and Oceania (Table 18.2).

Although exports of industrial roundwood, sawnwood and sleepers continue to dominate tropical timber exports, producer countries are increasingly exporting more highly processed products (e.g. wood-based panels, wood pulp and paper and paperboard) in an effort to capture a higher proportion of the value added in timber processing. On average, since the 1970s, the proportion of industrial roundwood and sawnwood exported relative to total production has steadily decreased, while the corresponding proportion of wood-based panels has doubled. This pattern is especially apparent in the Asia region where the percentage of wood-based panels exported relative to total production has risen from 40 per cent in 1960–70s to 90 per cent in 1990 (Table 18.3b; data in Table 18.3a refer to coniferous and non-coniferous timber products).

The real price of tropical logs followed a rising trend during the 1970s; prices briefly declined between 1979 and 1985, only to pick up again thereafter.[1] The real

Table 18.2 Forest products trade balance and percentage of total trade in tropical countries, 1990.

	Forest product imports (10^3 US\$)	Forest product exports (10^3 US\$)	Forest product net exports (10^3 US\$)	Total imports (10^6US\$)	Total exports (10^6 US\$)	Forest products as a % of total imports (%)	Forest products as a % of total exports (%)
Tropical Africa							
Cameroon	35,412	99,833	64,421	1,300	1,200	2.7	8.3
Central African Republic	468	29,994	29,526	170	130	0.3	23.1
Congo	4,500	106,087	101,587	570	1,130	0.8	9.4
Côte d'Ivoire	27,200	236,147	208,947	2,100	2,600	1.3	9.1
Gabon	3,655	136,774	133,119	760	2,471	0.5	5.5
Ghana	5,129	76,526	71,397	1,199	739	0.4	10.4
Kenya	23,594	4,054	0	2,124	1,033	1.1	0.4
Madagascar	8,546	534	0	480	335	1.8	0.2
Malawi	8,058	1,993	0	576	412	1.4	0.5
Nigeria	33,083	1,680	0	5,688	13,671	0.6	0.0
Sierra Leone	1,028	146	0	146	138	0.7	0.1
Tanzania	15,700	1,539	0	935	300	1.7	0.5
Zaire	3,666	17,032	13,366	888	999	0.4	1.7
Zimbabwe	5,765	4,169	0	n.a.	n.a.	n.a.	n.a.
Tropical C and S America							
Costa Rica	40,020	21,895	0	2,026	1,457	2.0	1.5
El Salvador	21,800	2,725	0	1,200	550	1.8	0.5
Guatemala	69,410	18,326	0	1,626	1,211	4.3	1.5
Honduras	137,921	31,061	0	1,028	916	13.4	3.4
Mexico	403,605	13,884	0	28,063	26,714	1.4	0.1
Nicaragua	10,566	2,569	0	750	379	1.4	0.7
Panama	76,979	3,988	0	1,539	321	5.0	1.2
Trinidad & Tobago	54,396	458	0	1,262	2,080	4.3	0.0
Bolivia	4,060	22,160	18,100	716	923	0.6	2.4
Brazil	299,402	1,750,981	1,451,579	22,459	31,243	1.3	5.6
Colombia	104,056	20,060	0	5,590	6,766	1.9	0.3
Ecuador	157,834	24,373	0	1,862	2,714	8.5	0.9
Paraguay	13,055	24,971	11,916	1,113	959	1.2	2.6
Peru	104,914	2,558	0	3,230	3,277	3.2	0.1
Tropical Asia and Oceania							
Hong Kong	1,752,273	705,535	0	82,495	29,002	2.1	2.4
India	290,967	16,337	0	23,692	17,967	1.2	0.1
Indonesia	330,157	3,069,199	2,739,042	21,837	25,553	1.5	12.0
Malaysia	483,372	3,040,884	2,557,512	29,251	29,409	1.7	10.3
Philippines	173,662	123,119	0	13,080	8,681	1.3	1.4
Singapore	747,548	663,302	0	60,647	52,627	1.2	1.3
Sri Lanka	28,771	600	0	2,689	1,984	1.1	0.0
Thailand	1,002,371	101,551	0	33,129	23,002	3.0	0.4
Papua New Guinea	5,504	115,500	109,996	1,288	1,140	0.4	10.1

Sources: FAO 1992b and World Bank 1992c

Table 18.3a Production and trade in timber products by tropical countries, 1990 (thousand cubic metres).

	Production	Exports	Imports	ADC
All tropical countries				
Industrial roundwood	257587	28,705	4,318	233,200
Sawnwood	72584	8,719	4,841	68,706
Wood-based panels	18483	12,818	1,891	7,556
Tropical Africa				
Industrial roundwood	41687	3,959	40	37,768
Sawnwood	6598	814	235	6,019
Wood-based panels	1108	243	66	931
Tropical C. and S. America				
Industrial roundwood	95697	172	95	95,620
Sawnwood	26641	1,032	1,557	27,166
Wood-based panels	4289	833	296	3,752
Tropical Asia and Oceania				
Industrial roundwood	120203	24,574	4,183	99,812
Sawnwood	39345	6,873	3,049	35,521
Wood-based panels	13086	11,742	1,825	3,169

Note: ADC: apparent domestic consumption.   *Source: FAO 1992b.*

Table 18.3b Export of timber products as a percentage of production in tropical countries (%).

	1961	1970	1980	1990
All tropical countries				
Industrial roundwood	15.6	27.1	18.2	11.1
Sawnwood	15.3	17.3	16.2	12.0
Wood-based panels	34.0	33.0	32.5	69.4
Tropical Africa				
Industrial roundwood	23.8	23.3	16.3	9.5
Sawnwood	32.1	28.5	12.7	12.3
Wood-based panels	34.0	33.0	28.6	21.9
Tropical C. and S. America				
Industrial roundwood	1.6	1.0	0.2	0.2
Sawnwood	14.7	13.6	6.9	3.9
Wood-based panels	34.0	33.0	15.3	19.4
Tropical Asia and Oceania				
Industrial roundwood	22.2	44.2	33.4	20.4
Sawnwood	11.9	18.5	24.5	17.5
Wood-based panels	40.5	39.9	49.4	89.7

Note: tropical countries are taken here to be countries with the majority of their land mass lying between the tropics. This table includes data on coniferous and non-coniferous timber products.   *Source: FAO 1992b.*

price of tropical sawnwood followed a similar trend, although with larger fluctuations and a steeper decline in the mid-1980s. The real price of other industrial timber products, including wood based panels, pulp, and paper and paper board, have been increasing since the early 1970s and have sustained this rise throughout the 1980s. The real price increases may reflect increasing product scarcity due to declining forest inventories and increasing demand for tropical timber products, whereas the recent downturn in some real prices may reflect the depressed state of the global economy during the late 1970s and early 1980s.

A number of studies have attempted to forecast future trends in supply and demand in the forest sector on a global and a regional level (Sedjo & Lyon 1990, Kallio et al. 1987, FAO 1990d, Cardellichio et al. 1989, ECE/FAO 1986, ECE/FAO 1989, ECE/FAO 1990, USDA Forest Service 1990). A comprehensive review of many of these studies is provided by Arnold (1991). The results of the previous studies are generally supported by the projections of supply, demand and trade in hardwood timber products recently undertaken by the Center for International Trade in Forest Products (Perez-Garcia & Lippke 1993). Baseline short term projections to the year 2000 reveal a number of significant trends:

- Decreasing commercial inventory of tropical timber is beginning to constrain harvests significantly, particularly in Malaysia. There is no comparable shortage of temperate hardwoods, which supply large consuming country markets in the US, Europe and other non-Asian markets. Even with increased sawlog production in Indonesia and Brazil, two tropical hardwood suppliers with a large inventory, the tropical share of all hardwoods will begin to decline.
- The combination of expected strong economic growth in tropical timber producing countries, more modest demand growth in other consuming countries and declining supply of tropical hardwood logs will produce a substantial shift away from export to domestic markets by the major tropical hardwood suppliers. Declining log exports will be offset only partially by increased product exports.
- Declining tropical hardwood inventory will lead to steadily rising sawlog prices, reaching levels 60 to 80 per cent above 1990 levels by the year 2000, in real terms. Most of this price increase is anticipated to occur in Southeast Asia. Prices in countries with adequate temperate hardwood sources will remain more stable.
- While product prices will increase with rising log prices, the availability of other supply sources in the more developed consuming countries, in conjunction with lower demand growth, will constrain product price increases which in turn will squeeze profits for processors of tropical timber. The trend in the developed consuming countries is therefore for a reduction in log imports and the processing of imported tropical logs, and increased product imports with more substitution away from tropical products.

Long-term projections to 2040 suggest that the commodity in short supply will continue to be tropical hardwood logs, not processing capacity. With the available tropical hardwood inventory in several countries declining rapidly by 2000, either harvest levels will be reduced quickly to more sustainable levels or they will drop even more abruptly just a few years later with the depletion of the inventory. Long term projections also suggest that tropical sawlog prices will continue their upward trend and that product price increases will not keep pace.

Although only a small proportion of tropical timber harvested enters the international trade, the industrial forest sector and timber trade may be important to tropical countries for a number of reasons. First, the industrial forest sector makes a direct contribution to the economy of producer countries – for example, in 1989 the industrial forest sector accounted for three to six per cent of total gross domestic product (GDP) in Malaysia and Indonesia, and two to three per cent of total GDP in Côte d'Ivoire, Gabon, Ghana, Brazil and Costa Rica. In most other tropical timber producing countries, however, the industrial forest sector represented less than two per cent of total GDP in 1989 (World Bank 1992c). Second, wood-related industries are an important source of employment generation in the manufacturing sector of many developing countries, although the level of employment in wood-related industries is modest compared to the size of economically active populations in developing countries. Third, forest product exports are a valuable source of foreign exchange for a few countries. For example, forest product exports account for over 10 per cent of the total value of exports from Central African Republic, Ghana, Indonesia, Malaysia and Papua New Guinea. However, across the tropics as a whole, exports of forest products do not generate substantial foreign exchange earnings compared to total export earnings (Table 18.2). Finally, there are numerous wider economic and social benefits associated with timber production and trade that are also important for tropical developing countries, such as rural infrastructure development and the provision of other social amenities.

Given the small and declining volume of tropical timber production that actually enters the international trade there appears to be little direct linkage between the trade and tropical deforestation. As discussed later in this chapter, domestic and trade policy distortions in producer countries may further reduce the impact of the timber trade on deforestation in tropical countries. Although the value of the tropical timber trade is not particularly significant for most exporting countries, it is an important source of foreign exchange earnings and wider socio-economic benefits for a few key producers. In addition, the economic benefits associated with the timber trade can help to reinforce incentives for tropical countries to manage and conserve their forest resources. The following section looks at the relationship between timber production for the trade and tropical deforestation in more detail.

18.3 Analysis of the links between timber production, timber trade and tropical deforestation

A number of studies have attempted to assess the relative importance of various economic activities, including timber extraction, in causing *tropical deforestation*. However, most of the work is extremely tentative and constrained to qualitative analysis. For example, Binswanger (1991) and Mahar (1989b) highlight the rôle of subsidies and tax breaks, particularly for cattle ranching, in encouraging land clearing in the Brazilian Amazon. More recent analysis by Schneider et al. (1990) and Reis & Margulis (1991) emphasize the rôle of agricultural rents, population pressures and road building in encouraging small-scale frontier settlement in this region. The study by Schneider et al. (1990) also identifies the importance of logging – log production from the Amazon region increased from 4.5 million m^3 in 1975 to over 24.5 million m^3 in 1987 – in forest exploitation, primarily through opening access to previously inaccessible lands. Commercial timber extraction has been encouraged by both a range of public policies, increasing domestic demand and strong international demand for tropical hardwoods.

A few studies have used statistical analysis to explore the linkages among factors thought to cause tropical deforestation and forest land clearance. There are a number of important "caveats" that need to be borne in mind when reviewing these statistical studies of the factors thought to contribute to deforestation. Firstly, all of the analyses suffer from problems of reliability and accuracy of data. Data limitations make it difficult to distinguish between production and conversion forests across the tropics, which may undermine an estimated relationship between log production and changes in forest area. If log production is mainly from conversion forests, then timber extracted is essentially a precursor or by-product of agricultural conversion, which is the principal factor in the resulting deforestation. In addition, due to the level of aggregation required to undertake such studies, the analyses are not sensitive to the different types of forests and different patterns of wood use. Finally, although these studies are relevant to the relationship between timber production and tropical deforestation, most were not designed explicitly to explore this relationship. Nevertheless, statistical approaches can provide interesting insights to the existence and relative importance of relationships between deforestation and factors thought to contribute to forest conversion.

Building on the lessons learned in previous studies, a regression analysis to examine the relationship between timber production and forest clearance in the tropics was developed by Barbier et al. (1993). The analysis covers 53 tropical countries and examines the relationship between a range of variables (i.e. timber production, agricultural yield, population density, income growth and tropical forest stock) and forest clearance. The statistical model (presented in Table 18.4) is fairly robust given the complexity of the problem being analyzed, the aggregate

Table 18.4 The linkages between timber production and tropical deforestation.

Dependent variable: Five-year change in closed forest area (logarithm forest area ('000 ha) 1985 − logarithm forest area ('000 ha) 1980

Explanatory variables	Estimated coefficient (t statistic)
constant	−0.1009 (−4.188)
X1 (logarithm of closed forest area as a percentage of total area 1980 (log of forest area 1000 ha(*100)/total area 1000 ha))	0.01253 (1.609)
X2 (population density 1980 (total population/total land area (ha))	−0.0474 (−2.695)
X3 (industrial roundwood production per capita 1980 (m^3/total population))	−0.0849 (−2.3029)
X4 (real GNP per capita 1980 (US$ GNP/total population))	0.000195 (1.8098)
X5 (agricultural yield 1980 (cereal production 1000 mt/cereal production area 1000 ha))	0.02301 (1.1298)
X6 (dummy Latin America)	−0.06809 (−2.4086)

Estimated elasticities
    X1 = 0.0125
    X2 = −0.0285
    X3 = −0.0186
    X4 = 0.1870
    X5 = 0.0339
    X6 = −0.0216
    R^2: 0.268
F Statistic: 2.8089
Number of observations: 53

Source: Barbier et al. 1993.

level of the analysis, the cross-sectional nature of the model and data limitations. The model supports the hypothesis that industrial roundwood production is positively associated with forest clearance in the tropics for the 1980–85 period − i.e. increasing levels of industrial roundwood production per capita leads to higher rates of forest loss.[2]

Previous statistical analyses provide only limited evidence of the linkages between tropical timber production, trade and tropical deforestation. This suggests that we need to be cautious when making statements about the rôle of the trade in tropical deforestation. Further research is required at a regional and country level to provide more conclusive results for policy makers. Other factors pertaining to the timber trade also suggest that caution is required when developing appropriate policy interventions to encourage sustainable forest management. For example, the following section considers market conditions and nature of demand for tropical timber and its implications for trade policy.

18.4 Prices, substitution and demand for tropical timber products

In order to discuss trade policy options there is a need to understand the market conditions for tropical timber products and consumers' response to price changes. An important issue concerning trade in tropical timber products is the degree to which demand for these products in consumer markets is affected by changes in price and, in particular, by the availability of substitute products derived from temperate timber or softwood. A related issue is the extent to which non-timber products compete with tropical timber products in various end uses; another is the degree to which tropical timber products from one country or region substitute for products from another country or region. This section discusses what is currently known about each of these demand and substitution effects by focusing on the results of various empirical estimates of the corresponding *price elasticity* measures. Until recently, there were few studies of these effects, and most current studies generally focus only on the *own-price* elasticity of demand for tropical timber products.

Past empirical studies suggest that the own price elasticities of demand for tropical timber products are on the low side (see NEI 1989 for a review). In other words, consumers do not respond dramatically to price changes. Similar results are obtained in more recent analyses of specific regions and countries, notably Southeast Asia (see Ch. 17, Table 17.6). For Southeast Asia in general, it appears that the elasticities of demand for log exports are higher than for plywood and sawnwood exports.[3] It is not surprising, therefore, that these countries have moved to restrict log exports in favour of expanding exports of processed products, notably plywood.

However, what holds true for specific regions or countries may not be applicable to the global market in tropical timber products. A recent analysis of the international demand and supply for non-coniferous wood products indicates that the long run elasticities of demand are generally low (see Table 18.5). Only the demand for plywood has an elasticity greater than one. Although not all of the products analyzed were of tropical origin, in contrast to the Southeast Asian region, the demand elasticities for logs in the world market appear to be lower than for sawnwood, which in turn are less than that of plywood.

One explanation for the very low elasticity of global demand for non-coniferous logs is that over the 1968–88 period many major tropical timber producers implemented policies to restrict log exports (see Section 18.5, below). Although the fall in log exports by these producers was compensated for by increased exports by other producers, the overall effect was that the total supply of world exports grew slowly, relative to demand. On the import side, this would mean that real prices of non-coniferous logs would rise faster than the quantity imported, which would translate into a low long-run elasticity of import demand. Such a scenario suggests that over the 1968–88 period there were not many substitutes for non-coniferous (mainly tropical) logs. On the other hand, the higher elasticities for non-coniferous sawnwood and, in particular, plywood imports would suggest that there may have been more substitutes available for these products in importing markets. This is not surprising, given that many of the major importing countries of these products also produce their own processed products, especially plywood, from either the non-coniferous logs they import or from timber (largely coniferous) which they produce themselves.

Table 18.5 also indicates that over 90 per cent of the adjustment of tropical timber product imports to changing prices occurs within four to seven years. Relatively slow adjustment of log imports to price changes probably reflects the longer lead time required to adjust primary and secondary processing capacity in importing countries. The degree of substitution between tropical and temperate products in consumer markets illustrates the extent to which the markets for these two type of products are inter-related, or whether there are essentially two different markets for two distinct commodities. This has important implications for exporters of tropical timber products. As noted in Section 18.2, global markets for temperate (mainly softwood) products are currently much larger than for similar products produced by tropical countries. If there is a high elasticity of substitution between these two types of products, then an expansion of exports of tropical timber prod-

Table 18.5 Long term price elasticities of demand and supply across major tropical timber importers and exporters.

Estimation period: 1968–88	Elasticity	Adjustment lag (years)
Import demand		
Non-coniferous logs	−0.16	6.7
Non-coniferous sawnwood	−0.74	4.6
Plywood	−1.14	4.1
Export supply		
Non-coniferous logs	0.70	5.8
Non-coniferous sawnwood	1.02	7.6
Plywood	0.20	6.0

Note: although not all of this trade was of tropical origin, 83 per cent of the world imports of non-coniferous logs were from tropical sources, 62 per cent for non-coniferous sawnwood and 75 percent for plywood. *Source: Buongiorno & Manurung (1992).*

ucts could lead to expanded market share and increased export earnings – at the expense of the temperate product market. On the other hand, a high elasticity of substitution would also mean that any increase in the export price of tropical timber exports would lead importing countries to substitute with temperate wood products.

The results for certain markets presented in Table 17.6 indicate that the elasticities of substitution between temperate and tropical wood products are very low.[4] This suggests that there are two distinct markets, and tropical producers of these products would have difficulty in penetrating the larger temperate market. This hypothesis is supported by market trends, which show that the market share of tropical plywood in the global Japanese and US plywood markets has remained essentially constant over time – in spite of the rapid rise in exports from tropical countries, especially Indonesia (Constantino 1988a).

Other studies have examined the degree of substitution of temperate for tropical logs. For example, Vincent et al. (1990) sought to quantify the extent to which Japanese consumption of imported sawlogs has shifted over the past two decades away from tropical and towards temperate logs. Their analysis suggests that temperate softwoods from different regions are still closer substitutes than softwoods and tropical hardwoods. On the other hand, imports of tropical sawlogs were further reduced by technical changes in Japan, which outweighed the price effects. This impact probably reflects strategic efforts by the Japanese sawnwood industry to broaden its source of logs, due to the imposition of export restrictions by tropical timber producers.

Diversification of sources of supply, including substituting softwoods for hardwoods, may also be occurring in the South Korean market. A recent study of the market over 1970–90 found that the consumption of imported tropical hardwood logs is responsive to the price of domestically produced softwood logs (Youn & Yum 1992). However, consumption of softwood logs is not affected by the price of imported tropical logs. This would suggest that in the Korean market there are strong possibilities for substitution of domestically produced softwood for tropical logs, but not the other way around.

Of course, tropical timber products may also be substituted by other, non-wood products in end uses. Although there is increasing anecdotal evidence of this occurring in many consumer markets, particularly in construction and furniture industries, estimating the magnitude or scale of this effect has proven more difficult. For example, the Netherlands Economic Institute (NEI 1989) has reviewed studies on the substitution of wood by other construction materials and concluded that evidence of this effect is not statistically reliable. However, NEI also conducted its own survey which suggested that this substitution effect may be significant. For example, plywood is believed to face severe competition from solid synthetic panels, with price strongly influencing the choice of product in the construction industry.

A further important issue is whether there is significant substitution between tropical timber products from different importing regions or countries. Table 17.6 indicates that the short and long run estimates of elasticity of substitution by origin for tropical sawnwood and plywood in certain importing countries are very high, especially for plywood.[5] This would suggest that importing countries can substitute between sources of origin with relative ease. Moreover, the elasticities indicate how certain tropical product exporters – especially Indonesia – have been able to increase their market share of world tropical sawnwood and plywood exports very rapidly in recent years. For example, because of its large timber resource base, abundant labour supply and a policy of restricting log exports, Indonesia has been able to expand plywood processing capacity, production and exports fairly quickly.[6] As a result, Indonesia has been able to undercut the price of its competitors and take over import markets where substitution between different product sources appears to be high (Constantino 1988a, Barbier et al. 1993).

To summarize, empirical studies suggest that demand for tropical timber products is relatively price insensitive and that substitution between tropical and temperate timber products in importing markets has not been very significant. However, in response to export log bans by tropical producers, some importers are increasingly diversifying the source of their supply. Substitution of non-wood products for timber may be occurring, but the evidence is largely anecdotal. Substitution between tropical timber products originating from different countries or regions does appear to be very high, particularly for plywood. This would suggest that importers can substitute between sources of origin with relative ease but also that tropical exporters can easily capture market share through price competition. If accurate, the empirical evidence on the demand for tropical timber indicates that producers as a group may enjoy significant market power. Co-ordinated price increases by all tropical producers would not lead to a significant decline in total demand; on the other hand, the short-term gains are also great for individual large producers attempting to undercut rivals and increase their market share, particularly for value-added products.

18.5 Trade policy measures and sustainable forest management

The evidence presented in Sections 18.2 and 18.3 on the linkages between the trade in tropical timber products and tropical deforestation suggests that the trade is not the major source of the problem. In fact, the main linkage may be the other way around: the impact of increasing scarcity of tropical timber resources, and thus tropical hardwood logs, on production and export (Barbier et al. 1993). Moreover, as discussed in the previous section, there is little evidence that tropical timber products are being "squeezed out" of consumer markets by non-tropical products or through price discrimination. If anything, tropical timber producers appear to be competing among themselves for dominant shares of a strong market.

Nevertheless, there is a genuine cause for concern over the excessive exploitation and rapid depletion of tropical production forests in many regions, including the indirect impacts of "unsustainable" harvesting practices on the loss of non-timber forest values and the incentives to convert forest land to other uses (for example, agriculture). A particularly significant impact may be the rôle of "unsustainable" timber production in opening up forest areas to subsequent agricultural encroachment and conversion. However, these effects do not in themselves warrant major interventions in the timber trade in order to improve sustainable forest management. To the contrary, they suggest that the proper focus for policy intervention ought to be at the concession and forest level.

On the other hand, the timber trade can lead to greater net returns for forestry investments and sustainable management of production forests, making this option more attractive than alternative uses of forest land such as agriculture. Unfortunately, in many producer countries the widespread prevalence of market and policy failures have distorted the incentives for sustainable management. See for example Barbier et al. (1993), Barbier et al. (1991), Gillis (1990b), Pearce et al. (1990), Repetto (1990), Repetto & Gillis (1988), Vincent & Binkley (1991). Failures in concession and pricing systems have produced counter-productive incentives that lead to the "mining" of production forests. Domestic market and policy failures have also had a major influence on the conversion of forest land to agriculture and other uses.

Thus the key factor in reducing timber-related tropical deforestation is ensuring proper economic *incentives* for efficient and sustainable management of tropical production forests. Appropriate forest management policies and regulations within producer countries ought to provide these incentives so that the *long run* income-generating potential of harvesting timber is maximized, and any significant external environmental costs associated with timber harvesting are "internalized". The starting point for sustainable production forest management is therefore to tackle the problem at its source by enabling producer countries to improve forest sector policies.

In particular, Hyde et al. (1991) suggest that *royalty, contract and concessional* arrangements are key incentive mechanisms determining levels of trespass, high-grading and loss of some non-timber values. Proper forest management regulations and controls on encroachment can also exacerbate forest degradation problems (Cruz & Repetto 1992, Poore et al. 1989). In addition, the failure of *stumpage prices* (i.e., the prices of harvested logs at the stand) to take into account the additional social costs associated with the loss of non-timber values (e.g. watershed protection, carbon storage, non-timber forest products) can generate excessive net returns – or *economic rents* – from timber harvesting, and thus lead to socially inefficient harvesting levels.[7] Finally, over the long term, if stumpage values are "underpriced", they will also fail to reflect increasing scarcity as old growth forests are depleted. As a result, efficient development of forest industries will be undermined, particularly the transition from dependence on old-growth to secondary-growth forests and the coordination of processing capacity with timber stocks (Vincent and Binkley 1991).

Timber trade policy in producer and consumer countries has, if anything, magnified the problems created by poor forestry policy and regulations in tropical forest countries. Although log export restrictions in producer countries have stimulated growth and employment in domestic processing, they may have led to serious problems of processing over-capacity and inefficiency. See, for example, Barbier (1987), Constantino (1990), Repetto & Gillis (1988), Vincent (1992a and 1992b). To the extent that this is the case, log prices are artificially depressed and recovery rates fall, thus increasing pressure on timber resources. For larger exporters, such as Indonesia and Malaysia, the expansion in processing capacity from export restrictions has contributed to their ability to capture a large share of international markets. This has proved more difficult for smaller exporting countries. Nevertheless the economic costs for all countries is extremely high; where export restrictions lead to over-capacity and inefficiencies in processing, logging pressure on the tropical forest resource may actually increase over the medium and long term (Barbier et al. 1993).

Although import tariffs on tropical forest products are generally low and declining in major developed consumer markets, non-tariff barriers may be significant and increasing (Bourke 1988 and 1993). Tariff barriers in the smaller developing country markets generally remain much higher, although these also appear to be declining. In some cases the increase in non-tariff barriers in major importing markets may reflect backsliding on previous commitments to reduce tariffs, in order to maintain some degree of protection for domestic industry – particularly plywood processing. In many developed countries, however, these trends also appear to reflect growing pressure from media and advocacy groups to reduce consumption of tropical timber which, rightly or wrongly, is perceived to be an environmentally "unsound" product.

Restrictions on imports depress the global demand for tropical timber products and can feed back to reduce stumpage values in producer countries, thus discouraging the incentives for more efficient processing and better forest management. Moreover, producer countries will continue to argue that they need to compensate their domestic processing through subsidies and export restrictions. The potential impact of any new environmental initiatives to reduce tropical timber imports is similar to that of traditional import barriers. Total trade with the largest and most profitable markets would be reduced, forcing producers to cut back production and look for alternative markets. In the short-term the rate of tropical deforestation might be reduced but the long-term effects could be perverse. If new markets cannot be developed at home or abroad, the income received by tropical timber exporters would presumably fall, along with the economic incentive to invest in forest management. The influence of developed country consumers over tropical forest management would also be reduced by their actions, as the relative importance of these markets declines.

In short, restrictions in trade are not helping to reduce timber-related deforestation in developing countries. In contrast, by adding value to forestry operations, the trade in tropical timber products *could* act as an incentive to sustainable production forest management – provided that the appropriate domestic forest management policies and regulations are also implemented by producer countries. Unfortunately, many proposed trade policy interventions to "save the tropical forests" – such as bans, taxes and quantitative restrictions – will actually work to *restrict* the trade in tropical timber products. Such interventions will reduce rather than increase the incentives for sustainable timber management – and may actually increase overall tropical deforestation. Trade interventions that are implemented *unilaterally*, i.e. by a single trading bloc or nation, have the additional problem of being discriminatory under GATT rules, particularly if they are applied to tropical timber products alone.

However, a recent study suggests that there is a rôle for additional tropical timber trade policies in fostering *trade-related incentives* for sustainable management (Barbier et al. 1993). Trade policies will be the most effective if they: are employed in conjunction with and complement improved domestic policies and regulations for sustainable forest management within producer countries; improve rather than restrict access to markets for tropical timber products so as to ensure maximum value added for sustainably produced tropical timber exports; and assist producer countries in obtaining the additional financial resources required to implement comprehensive, national plans for sustainable management of tropical production forests.

The first set of policies would require producer countries to undertake substantial reviews of their forest sector policies to determine the implications of their existing domestic forestry policies and regulations on timber-related deforestation and the extent to which their timber export policies may also be affecting deforestation, either directly or through exacerbating problems caused by poor domestic forestry policies and regulations. Producer countries ought to correct those policy distortions that work against sustainable timber production objectives, as such distortions seriously undermine efficient and sustainable forest sector development and worsen deforestation.

The second set of policies is aimed at removing any remaining barriers to tropical timber imports into consumer markets, particularly for those producer countries that demonstrate a commitment to forest sector policy reform. For example, the removal of specific tariff and non-tariff barriers on imports could take place on a case-by-case basis, depending on demonstrable progress by each exporting country. This could occur through normal bilateral trade negotiations or through multilateral agreements and organizations. In addition, consumer countries should actively promote, through information and market intelligence campaigns, the use of tropical timber imports from exporting countries that are implementing "sustainable management" policies.

The final set of policies – additional financial assistance – is not straightforward. It raises complicated issues concerning the need for international compensation, the scale of resource transfers required and the possible mechanisms for implementation. We discuss these issues in the next section.

18.6 Measures to raise revenues for sustainable forest management

The main rationale for providing financing for assisting tropical forest countries in moving towards sustainable forest management is that there is an important principle of *international compensation* at stake. There are essentially three reasons for this argument:

- It is often suggested that timber exporting countries receive an insufficient share of the returns from tropical timber product exports – at least to incur the additional harvesting costs and other economic impacts of sustainable timber management.
- Implementation of the forestry policies and regulations required to ensure the proper enforcement and monitoring of sustainable management of production forests will impose substantial additional costs on producer countries that they will find difficult to afford.
- To the extent that all nations benefit from the global external benefits resulting from sustainably managing large tracts of tropical forest lands, then the international community should compensate producing nations for the loss of potential income that they would incur by reducing tropical deforestation, timber sales and conversion of forest land to other uses.

The first point is difficult to substantiate. As will be discussed further below, the issue may have less to do with whether the unequal distribution of revenues along the chain of trade *reduces incentives* for sustainable management at the forest level, than with whether any excess revenues along the chain can be tapped for *additional funds* to assist the move to sustainable management of tropical production forests.

The second and third points are much more relevant to the argument for international compensation. It is now generally accepted, as well as enshrined in the Forest Principles accord of the 1992 UNCED Conference, that compensating tropical forest countries for their rôle in maintaining a resource that has value on a *global level* is a fundamental basis of multilateral policy action. However, the second and third points also suggest that compensation is needed by tropical timber producing countries for the income they may forego in protecting their forests and for the additional costs incurred in implementing sustainable management practices for their production forests. This has to be demonstrated empirically.

A recent policy simulation was conducted using a global forest sector trade model to indicate the additional economic impacts to tropical forest countries of "setting aside" some of their forest resource base (Perez-Garcia & Lippke 1993). Essentially, this was simulated by a reduced timber supply scenario where the inventory of commercial tropical hardwood resources is reduced by 10 per cent – which is equivalent to land being taken out of production forest and permanently protected. The predictable result is to accelerate the underlying trend of increasing timber scarcity and higher sawlog prices in tropical forest regions, notably in Malaysia and Indonesia. The model indicates that such reductions in supply would result in a loss of wealth for tropical timber producing countries. Over the long run, permanent set-asides would mean that the remaining production forest inventory could not support as high a level of sustainable harvest as under base case projections.

There are also indications that the additional costs required to implement sustainable forestry management policies and regulations are significant. Based on broad estimates made for ITTO and UNCED, additional funds required by all producer countries to implement sustainable management of their tropical forest resource could be anywhere in the range of *US$ 0.3 to 1.5 billion annually* (Barbier et al. 1993).

Although these figures would suggest the need for additional financial assistance for producer countries, the real issue is whether the financing ought to be raised from the tropical timber trade or from other sources. There are essentially three policy options available: re-direction of existing revenue from the trade; appropriation of additional revenue from the trade; and additional funding from sources external to the trade.

A recent study for ITTO conducted by the Oxford Forestry Institute (OFI) has argued the case for a *tax transfer* of revenue from the trade between consumer and producer countries (OFI 1991). The main rationale behind the transfer is that, firstly, revenues from stumpage (royalties) and other taxes accruing to producer country governments from tropical timber production and trade are often low in relation to the consumer value of products; i.e., producer country governments capture a low proportion of the total economic rent earned through the trade. Secondly, revenues from taxes on imported tropical timber products accruing to consumer governments, such as value-added tax (VAT), are relatively large. Consequently, a relatively modest reduction in the rate of taxation at the consumer end of the chain would allow a reasonably large increase in the stumpage value of the resource *without* affecting the end price of the resource.

Table 18.6 illustrates the current situation and how a tax transfer might work. In the current scenario, producer country governments are assumed to collect a royalty of US$9 per m³, and consumer country governments impose a VAT on final products of 15 per cent. However, in the tax transfer scenario, VAT could be reduced to 7.5 per cent and the royalty increased to US$30 per m³ – and the final price would be left unchanged. Depending on the product, tax revenues of the consumer country governments would fall by 13 to 18 per cent. These losses are offset by important incentives. Due to the increased royalties, producer country government revenues rise dramatically, by 29 to 82 per cent, which would provide additional resources for sustainable management expenditures. The increased royalty would also raise the sale price of logs within producer countries, which could help to control over-exploitation of timber and inefficient use of logs in processing. Despite the increase in the price of logs, within the forestry sector of producer countries profits would still rise modestly. A slight rise in profits from the trade would also occur in the consumer countries.

In short, the main implications of a tax transfer would be to change significantly the *distribution* of economic rents from the timber trade as an incentive for sustainable management in producer countries. The main advantage would be that additional funds for this purpose could be raised *with little or no* effect on final product prices. As indicated in Table 18.7, just under US$1.5 billion in additional funds could be raised by producer countries through this means – closer to the "upper bound" of the estimated financing for sustainable management required by these countries.

However, consumer countries are likely to be concerned about the fiscal and political implications of a tax transfer scheme. First, as Table 18.7 shows, consumer country governments would have to forego nearly US$3.7 billion in tax revenues from the trade – more than 2.5 times what producer country governments gain in increased revenues. This implies a substantial net loss in revenue "captured" from the trade. Second, exempting tropical timber products from VAT or other taxes could prove politically problematic in that it could set the precedent for other goods being exempted from taxes on "environmental" grounds. Forest industries in temperate forest countries and their governments could also press for similar treatment on the same grounds – the need for additional investment for sustainable management or for compensation for past investment.

Finally, consumer countries would insist on an internationally agreed monitoring and enforcement system that would ensure that: a) producer countries did re-

Table 18.6 Effects of a tax transfer on the distribution of value added in the timber trade

Log exports[a]	Share of total value (%)				Change relative to current share (%)
	Current		Tax transfer		
Producer country	9.2		11.1		21
operating cost		5.7		5.8	3
profits		0.9		1.1	19
government revenue		2.6		4.2	59
Consumer country	90.8		88.9		0
operating cost		40.1		41.3	3
profits		26.2		27.5	5
government revenue		24.5		20.1	0
Totals	100.0		100.0		
		100.0		100.0	

Timber exports[b]	Current		Tax transfer		current share (%)
Producer country	10.5		12.6		19
operating cost		7.5		7.7	3
profits		1.0		1.2	20
government revenue		2.0		3.6	82
Consumer country	89.5		87.4		0
operating cost		40.4		41.5	3
profits		24.1		25.1	4
government revenue		25.1		20.8	0
Totals	100.0		100.0		
		100.0		100.0	

Product exports[c]	Current		Tax transfer		current share (%)
Producer country	35.3		37.1		5
operating cost		20.1		19.3	0
profits		8.4		9.0	7
government revenue		6.9		8.8	29
Consumer country	64.7		62.9		0
operating cost		19.3		19.8	2
profits		22.2		22.8	3
government revenue		23.2		20.2	0
Totals	100.0		100.0		
		100.1		100.0	

Notes: a) Tropical logs exported by producer countries and all processing in consumer countries; b) Primary processing of timber in producer countries and secondary processing in consumer countries; c) Export of final product with all processing done in the producer country.

Source: based on Figures 4.1, 4.2 and 4.3 in OFI (1991).

Table 18.7 Revenue transfers available from value added in the tropical timber trade, 1990 (US$ million).

	Log exports	Sawnwood exports	Processed product exports	Total
FOB export value[a]	2,280	2,150	4,150	8,580
Final end use value[b]	46,588	29,189	12,858	88,635
Tax transfer scenario[c]				
Consumer country				
Current government revenue	11,414	7,326	2,983	21,724
Post-transfer revenue	9,364	6,071	2,597	18,033
Revenue lost/transferred	2,050	1,255	386	3,691
Producer country				
Current government revenue	1,211	584	887	2,682
Post-transfer revenue	1,957	1,051	1,132	4,139
Revenue gained/transferred	745	467	244	1,457

Notes: a) Based on Bourke (1992); b) Based on ratios of FOB value to final product value in OFI (1991); c) Based on Table 18.6.

spond by raising royalty fees; and b) the additional revenue raised was spent on sustainable forest management expenditures. On the other hand, producer countries would consider too much external supervision to be "interference" with their internal affairs and sovereign rights over their resource base. Carefully negotiated bilateral and multilateral agreements would be necessary to avoid confrontations and to make the system workable.

One possible variant on the tax transfer scheme would be a revenue transfer scheme. Rather than lower their VAT or other taxes on the tropical timber trade, consumer country governments could instead transfer directly some proportion of the revenue raised through these taxes to producer countries. Although the consumer country governments would still forego substantial revenues, they could ensure more direct control, and thus leverage, over the allocation of funds to sustainable management, for example indicating which producer countries are to receive them. In addition, consumer country governments would most likely have to forego less revenue if it were directly transferred to producer countries to help them meet their US$0.3–1.5 billion target then under the tax transfer scheme (see Table 18.7). As indicated in the table, due to "leakages" to other sectors in the tropical timber trade, a tax transfer scheme would require consumer country governments to forego around 2.5 times more tax revenue in order to allow producer country governments to achieve their target.

However, a revenue transfer scheme would still require an internationally agreed monitoring and enforcement system. Producer country governments would probably be more concerned about the direct control and conditions that consumer country governments could assert over a revenue transfer scheme. Producers would insist on full involvement in the establishment, implementation and monitoring of any such scheme.

In recent years there has also been renewed interest in the use of trade instruments in appropriating additional tax revenue from the trade for sustainable forest management.

One study by the Netherlands Economic Institute (NEI 1989) has indicated that a surcharge of one to three per cent on the tropical timber imports of the EEC, Japan and USA would raise approximately US$ 31.4 to 94.1 million with little additional distortionary effects. If endorsed by a multilateral agreement, such as the ITTO, the import surcharge would be within GATT rules through sub-article XX(h). A differentiated surcharge could also be imposed so that imports of processed tropical hardwood products face less discrimination than logs, thus reducing existing distortions from escalating tariffs. The funds raised would most likely be transferred to the ITTO for distribution, possibly through specific projects and programmes. Other forms of collecting additional funds were found by the study to be less desirable.

For example, imposition of an export levy by producing countries themselves has the advantage of directly addressing the forest management systems of those countries, but presents obvious problems of monitoring and evaluating success in achieving sustainable management. If the funds were transferred to ITTO, transaction costs could be high, the same rate would need to be implemented in all producer countries simultaneously, and less funds would be raised than through an import surcharge. A parafiscal tax on all timber sold in consumer countries has the advantages of taxing all kinds of timber equally, including non-tropical products, and of generating large revenues at a low rate of taxation. However, such a tax has complications for external trade policies (e.g., temperate forest countries could claim unfair discrimination) and "harmonization" of internal tax rates within trading blocs (e.g., the EEC's Internal Market Strategy). Finally, a voluntary surcharge collected by the tropical timber trade itself could raise funds transferable to ITTO without requiring additional national legislation. Unfortunately, compliance and effectiveness may be diminished because of the current overcapacity and low net profit margins faced by many international traders and wood processing industries; the lack of effective control measures to ensure collection with equal efficiency in all consumer countries; and the possibility of entry and exit by traders in the industry to avoid payment (NEI 1989).

Buongiorno and Manurung (1992) also examine the scope for a five per cent revenue-raising import levy on tropical timber by the Union pour le Commerce des Bois Tropicaux (UCBT) of the EEC. The results indicate that tropical timber exporters would lose around US$44.8 million in trade earnings, with Indonesia and Malaysia suffering the worse, but importing countries would earn US$87.7 million in additional revenues. Thus, if the funds raised by the tax were rebated to exporting countries, they could be made better off by over US$40 million. Moreover, a tax by UCBT countries would have little effect on total world trade of tropical timber products. The model shows that tropical timber exporting countries would compensate for any declines in imports by UCBT countries through increasing exports to non–UCBT countries.

One of the major concerns of producer countries is that any revenue-raising import surcharge, even at very low levels, would be distortionary. In particular, if the tax was levied by all importers, then there could be a more significant impact on total world trade in tropical timber products. Moreover, such a tax would discriminate in favour of temperate forest-based industries of the developed market economies. For example, Buongiorno and Manurung suggest that producer countries may prefer a tax on exports rather than an import surcharge by UCBT countries. This would give producers more direct control over the proceeds of the tax. In addition, an export tax would affect all import markets rather than just one, thus spreading the costs of sustainable management to all producers and consumers.

Finally, there is the issue of whether the amount of funds raised through any trade surcharge would be adequate for the task, and whether there are more appropriate avenues for raising additional large-scale funding *outside* the timber trade altogether. The studies undertaken so far suggest that the amount of *net* funds raised from a trade surcharge of one to five per cent may fall short of the approximate target of US$0.3–1.5 billion required annually by producer countries as additional resources for sustainable forest management.

There is a strong rationale for additional funds to be made available to producer countries for sustainable forest management from sources *outside* the tropical timber trade. Comprehensive international agreements, targeted financial aid flows and compensation mechanisms to deal with the overall problem of tropical deforestation may ultimately eliminate the need to consider intervention in the timber trade. Given that commercial logging is not the primary cause of tropical deforestation, such approaches would avoid unnecessary, and possibly inappropriate, discrimination against the timber trade. On the other hand, a comprehensive tropical forest agreement seems much more difficult to negotiate and raises its own problems of workability and effectiveness (see Barrett (1990) for a discussion of the difficulties involved in securing international environmental agreements).

It is unlikely in the current global economic climate that there will be a concerted international effort to substantially increase bilateral or multilateral aid flows for sustainable production forest management. Nevertheless, there still remains the possibility of designing new sources of financial assistance that are separate from existing developing country aid budgets. The Forestry Principles signed at UNCED are effectively a step in that direction, and international commitments through the Tropical Forestry Action Plan (TFAP) and Global Environment Facility (GEF) continue to reinforce the global interest in forestry and biodiversity protection. The case could be made for more comprehensive international agreements to raise revenues for sustainable management of tropical forests, including production forests. The main focal points of such agreements should be the development of alternative revenue-raising mechanisms other than trade interventions or reliance on existing aid budgets.

For example, Amelung (1991) has explored some of the conditions necessary to ensure efficient international compensation payments, arguing that payments are a better solution to controlling tropical deforestation than trade barriers as they offer the chance to "internalize" global externalities and improve incentives for management. Moreover, rather than focus exclusively on interventions in the timber trade to finance compensation, Amelung (1991) opts for the establishment of an international rainforest fund, as proposed by UNEP, in order to avoid free-riding among non-tropical countries.

Similar arguments are put forward by Sedjo et al. (1991), although their preference is for the establishment of a global system of marketable forest protection and management obligations (FPMOs). Under a voluntary global forestry agreement, FPMOs would be distributed to all signatories, probably through criteria based on a mix of GNP levels and forest area. Holders of FPMOs must either a) fulfil the obligation "on the ground" or b) induce another agent to assume the obligation, presumably in exchange for a payment. Thus countries with large obligations and small forests would have "excess" obligations and hence be forced to meet these externally. Countries with large forests and small obligations could meet them internally but would then have an "excess" of forests. They could then be the object of negotiations with countries that had "excess" obligations.

The advantages of such a system are that, while not requiring compliance, it provides incentives for nations to comply out of their own self-interests, as well as achieving the objectives of international compensation; it would ensure that non-tropical forest countries (i.e., mainly industrialized countries) would bear a substantial portion of the costs; and it would not excessively limit production of traditional and commercial forest products. The disadvantages would be the high degree of monitoring required, the need for a "clearing house" for international trade in permits and the difficulties of negotiating a comprehensive international agreement to establish the system. On the last point, the authors argue for the establishment of a "transition" system with only a few major industrialized and tropical forest countries involved, and initially limited to mainly bilateral agreements with little trading. Although negotiations to establish such a scheme are difficult and arduous, they may be the best hope in the medium and long term of ensuring a global commitment to control tropical deforestation, as well as the sustainable management of production forests.

18.7 Conclusion

The trade in tropical timber products is not a major cause of tropical deforestation. Nevertheless, there is genuine cause for concern over the excessive exploitation and rapid depletion of tropical production forests in many regions, including: the indirect impacts of "unsustainable" harvesting practices on the loss of non-timber forest values and the incentives to convert forest land to other uses (e.g. agriculture); the impact of market, policy and trade distortions on the incentives for tropical timber management and trade; and the consequent failure of many producer countries to ensure efficient and sustainable development of forest-based industries, and, as a result, the failure to match expansion in processing capacity with the economic availability of timber stocks.

The key factor in reducing timber-related tropical deforestation is ensuring proper economic *incentives* for efficient and sustainable management of tropical production forests. Unfortunately, domestic economic and forestry policies in producer countries have tended to distort the incentives for sustainable management of timber resources and have failed to curb the wider environmental problems associated with timber extraction. Trade policy distortions in producer and consumer countries have, if anything, exacerbated this situation.

By adding value to forestry operations, the trade in tropical timber products *could* act as an incentive to sustainable management of production forests – provided that the appropriate domestic forest management policies and regulations are also implemented by producer countries. However, many proposed trade policy interventions – such as bans, taxes and quantitative restrictions – will actually work to *restrict* the trade in tropical timber products. Such interventions will reduce rather than increase the incentives for sustainable timber management – and may increase overall tropical deforestation. The first priority, therefore, is not more trade intervention but to obtain agreement from producer countries to remove forestry sector and trade policy distortions that discriminate against sustainable forest management and from consumer countries to eliminate remaining import barriers to tropical timber products.

Tropical forest countries will need additional financial assistance in implementing sustainable management. Although the additional funds ought to come from sources external to the trade, in the current global economic climate more official development aid for this purpose is unlikely. The use of a *tax* or *revenue transfer scheme* to redistribute the existing revenue from the trade back to producer countries is the most promising and immediate option. If necessary, it could be supplemented by a *revenue-raising trade surcharge* of less than five per cent – but this approach should be used cautiously to avoid any distortionary or discriminatory effects. However, over the long run, the global community ought to be looking toward the establishment of more innovative mechanisms, such as a *tropical forest fund* or a global system of *marketable forest protection and management obligations*, as the best means to combatting the overall problem of tropical deforestation, including the development of sustainable production forest management.

Notes

1. The real prices are derived from current export unit values and deflated with an index of export unit values of manufactured goods from the UN *Monthly Bulletin of Statistics*. The forest products price index includes both coniferous and non-coniferous forest products traded internationally.

2. In this version of the model, the variable used to capture the correlation between roundwood production and forest clearance was scaled using information on *population size*. The model was also run using the variable industrial non-coniferous roundwood production *per total area* in 1980. The coefficient on this alternative explanatory variable was also negative (−0.000167) and there was little change to the estimated coefficients of the other independent variables. However, the explanatory power of this alternative variable was slightly less significant (t statistic −1.68) and the robustness of the model in general was slightly reduced through the inclusion of this variable.

3. In Ch. 17, Table 17.6, the elasticity of demand by importers of tropical sawnwood of −5.67 calculated by Cardellichio et al. (1989) and used by Vincent (1992a) for Malaysia seems way out of line with other estimates. The earlier analysis by Vincent (1989) gives a more realistic estimate of −1.22 for Malaysian and SE Asian sawnwood import demand.

4. For sawnwood, the importing countries were China, Italy, Japan, the United Kingdom and the United States. For plywood, they were Japan, the Netherlands and the United States.

5. For sawnwood, the data were imports from Brazil, Côte d'Ivoire, Indonesia, Malaysia, the Philippines and Singapore into Italy, the United Kingdom and the United States. For plywood, they were imports from Brazil, China, Indonesia, Malaysia, the Philippines and South Korea into the United States.

6. However, as discussed in Chapter 17, these policies have also resulted in processing inefficiencies, over-capacity, artificially low log prices, and, consequently, greater pressure on timber resources in Indonesia.

7. The question of to whom this rent is allocated is not strictly an economic issue. Hyde et al. (1991) strongly suggest that the allocation of rent is merely a matter of distributive preference. Each country must make its own judgement over the equitable distribution of the nation's natural wealth. Such concerns over equitable distribution are, however, likely to be influenced by expectations concerning the ensuing productive use of such economic surplus. More importantly, if a government leaves all the profits to the concessionaires without seeking to appropriate a reasonable return as the owner of the resource, the government leaves a powerful

incentive in place for concessionaires to "mine" the forests. This is particularly true in cases where – as mentioned above – concessions policy and regulation are not strictly controlled and enforced.

19

The tropical timber trade and sustainable development

Jeffrey Vincent

19.1 Introduction

The tropical timber trade appears to have promoted neither sustained forest management nor sustained forest-based industrialization. The boom-and-bust export pattern is often blamed on demand by developed countries, high import barriers, and low international wood prices. In fact, it is rooted in tropical countries' own policies related to timber concessions and wood-processing industries. These policies suppress timber scarcity signals and must be revised if the trade is to promote sustained economic growth. Even if this is done, the trade may not promote sustained-yield forestry in individual countries.

The history of the tropical timber trade is discouraging both to foresters and environmentalists interested in sustained management of tropical forests and to policy-makers interested in sustained industrialization in the forest sector. Since the end of the Second World War, one tropical country after another has followed a boom-and-bust export pattern (Gillis 1988b, Repetto & Gillis 1988, Vincent & Binkley 1991). High initial export earnings are followed by depletion of old-growth forests, a lack of management of second-growth forests (Poore et al. 1989), and a collapse of domestic processing industries. Logging and processing industries enjoy profits during the boom, but the economic activity is not sustained.

This pattern emerged in West Africa in the 1950s and 1960s. It became even more apparent in the 1970s and 1980s as the trade shifted toward Southeast Asia and expanded in volume. In Southeast Asia today, several countries have already gone bust (for example, Thailand and the Philippines), others will shortly (for example, the state of Sabah in Malaysia), and in most remaining countries the boom is either cresting or waning (Gillis 1988b, Scott 1989).

Is the boom-and-bust pattern inevitable? If so, is the tropical timber trade inherently incompatible with sustainable development? International timber prices

Table 19.1 Production and trade volumes for wood products in 1989 in developing countries.

Products	Production	Exports	Imports
Solid wood (1,000 m³)			
Industrial roundwood[a]	306,256	34,199	12,767
Sawnwood	89,012	13,400	11,691
Wood-based panels[b]	21,200	12,050	3,016
Fibre (1,000 tonnes)			
Wood pulp			
Paper and paperboard	20,923	2,548	6,904

Notes: a) logs and pulpwood; b) mainly veneer and plywood, but also includes particle board and fibreboard. Excludes China, which is a major producer but lies in the temperate zone. *Source: FAO 1991a*

reflect the commercial value of tropical wood, not the diverse values of tropical forests as sources of biological diversity, clean water, and non-timber forest produce (for broader discussions of tropical forestry economics, see Repetto 1987, Peters et al. 1989, Barbier et al. 1991). Nevertheless, can the trade indirectly protect these non-market values by generating incentives to maintain permanent forest areas? Although the timber trade provides opportunities for a tropical country to enhance its overall economic performance, the trade does not necessarily create incentives for sustained forest management or for sustained industrialization within the forest sector.

Policies in tropical countries have generally reduced the economic benefits that those countries can reap from the trade and have reduced the likelihood that the trade can promote sustainable development of the forest sector.

19.2 Misconceptions about the tropical timber trade

The inability of tropical timber-exporting countries to break out of the boom-and-bust pattern is often attributed to three factors: developed countries' exploitation of tropical countries' timber resources, high import barriers by developed countries against processed tropical timber products (the International Tropical Timber Organization was established in part to address this problem), and low prices for tropical timber in international markets (Westoby 1978b, Guppy 1984). Consumption in developed countries allegedly drives the boom. Import barriers allegedly inhibit the development of processing industries in tropical countries, reducing those countries' export earnings and the value of their forests as a source of raw materials. Low prices allegedly reflect market manipulation by developed countries and reduce the financial viability of forest management.

None of these three factors holds up well when trade statistics are examined (see Table 19.1). In 1989, developing countries (excluding China), which are mainly tropical, exported 11 per cent of their harvest of industrial roundwood. They exported 23 per cent of their output of solid-wood processed products, and smaller percentages of their output of fibre products. Taken together, these figures indicate that only about a third of the industrial roundwood harvested in developing countries entered international trade in any form. Moreover, many of the exports were to other developing countries.

In 1989, developing countries (excluding China) imported an only slightly smaller value of wood products than they exported: $11.5 billion versus $12.7 billion (FAO 1991b). There is a significant international flow of tropical solid-wood products, which are mainly hardwood (non-coniferous), from developing to developed countries, but there is also a significant flow of temperate fibre products, which are mainly softwood (coniferous), in the opposite direction. The North uses the South's resources, but the South also uses the North's resources.

Global consumption of wood products is rising (FAO 1988b), but at a diminishing rate (Sedjo and Lyon 1990).

Economic models of the global forest sector predict that the rate will continue to diminish (Cardellichio et al. 1989, Dykstra & Kallio 1987a). Developing, not developed, countries account for most of the increase in consumption that has occurred recently or that has been forecast. The timber trade is responsible for a shrinking share of consumption of tropical timber products.

In regard to the second factor – high import barriers trapping timber-exporting countries in the boom-and-bust cycle – import tariffs against processed wood products certainly exist in developed countries. Like most tariffs, however, these tariffs have been brought down markedly by the various rounds of the General Agreement on Tariffs and Trade (GATT) negotiations (Olechowski 1987). Moreover, developed countries' tariffs on wood products are generally comparable to or lower than their tariffs on most products (Bourke 1988; although non-tariff import barriers against wood products also exist, these appear to be no more significant than in the other traded goods), and they are generally lower than corresponding import tariffs in developing countries (see Table 19.2). In many cases, they are lower than developing countries' own export taxes on wood products (Bourke 1988). For example, in 1990 Peninsular Malaysia announced export taxes on sawnwood equivalent to 11 to 22 per cent of the export prices of sawnwood.

Tropical timber-exporting countries have suffered some degree of economic harm from trade barriers on wood products imposed by trading partners and by themselves. Yet, import barriers have not prevented Peninsular Malaysia from becoming the world's largest exporter of hardwood sawnwood, tropical or temperate, or Indonesia from becoming the largest exporter of plywood. Studies indicate that import barriers on wood products in developed countries have modestly

Table 19.2 Import tariff rates in the early 1980s (%).

Products	EC	US	Japan	LDCs[a]
Wood				
"In the rough"	0.0	0.0	0.0	14.4–34.1
Primary	1.9	5.6	7.4	16.2–57.8
Secondary	1.5	1.7	4.8	24.1–73.1
Other				
Machinery, appliances	4.4	3.2	n.a.	n.a.
Textiles, textile articles	5.6	14.7	n.a.	n.a.
Footwear, headgear, other	6.6	12.2	n.a.	n.a.
All items	2.4	2.9	n.a.	n.a.

Notes: Rates for wood products in the European Community (EC), United States (US), and Japan are for imports from developing countries; all other rates are for imports from all sources; n.a.: not applicable; LDC: less developed countries. a) Range of the averages for developing countries in Africa, America, and Asia.

Sources: Olechowski 1987, Bourke 1988.

decreased the trade volume for most wood products, but that much of the increase in trade that would result from their removal would be captured by exports from temperate countries (Olechowski 1987, Drykstra & Kallio 1987b).

The third alleged factor, low prices, holds up better. Since the end of the Second World War, average export prices for tropical hardwood logs and sawnwood have been substantially lower than corresponding prices for temperate hardwood products (see Table 19.3). There is a simpler explanation than market manipulation: although some tropical timber exports compete on the basis of quality with fine temperate hardwood products, most compete on the basis of price with commodity products made from temperate hardwoods and softwoods. If this is the case, then the international prices of tropical hardwood products would be expected to fall between those of temperate softwoods and hardwoods. This is precisely what Table 19.3 shows.

Table 19.3 Export prices for tropical hardwood products relative to those for temperate products: averages for 1945 through 1988.

	Corresponding temperate product	
Tropical hardwood product	Hardwood	Softwood
Logs (from Asia)	0.58	0.98
Logs (from Africa)	0.88	1.50
Sawnwood	0.73	1.52

Sources: based on analysis of data in various issues of the FAO *Yearbook of Forest Products*. Time series data on the export unit value of a tropical product were regressed on corresponding data for a temperate product. Data for logs during 1945–1948, and all products during 1951–1955 were missing.

Although speciality woods such as mahogany, teak, and ebony are the best known tropical woods, most tropical woods have commodity end uses for which there are many substitutes. Econometric evidence that imports of tropical timber products are highly sensitive to price, which is consistent with tropical timber products having readily available substitutes, has been reported. Tropical timber became significant in international trade after the Second World War, when "lauan" plywood and other commodity uses were developed. Tropical logs and plywood are traded primarily from Southeast Asia to East Asia. The logs are processed into plywood, which is used in concrete forms and for structural purposes, and into sawnwood, which is used in joinery products and for decorative purposes (Nectoux & Kuroda 1989). Temperate softwoods provide potential substitutes in the plywood market, whereas both temperate softwoods and hardwoods provide substitutes in the sawnwood market. Tropical sawnwood is imported primarily by western Europe, where it is used mainly in joinery products (Nectoux & Dudley 1987). Temperate hardwoods and, increasingly, temperate softwoods provide competition there.

Competition with temperate timbers has inhibited increases in the international prices of tropical timber products. Since the late 1940s, the export unit value for tropical logs exported from Asia has risen at a nominal rate of 4.1 per cent per year, whereas the export unit value for tropical sawnwood (from all sources) has risen at 3.6 per cent per year. (Calculation based on analysis of data in various issues of FAO *Yearbook of forest products*. The natural logarithms of export unit values during 1945–1989 were regressed on a constant and a time trend. Data were missing for logs during 1945 and for both products during 1951–1955. The cited rates are significant at the one per cent level). These rates barely match general price inflation during this period.

Prices of tropical timber are unlikely to increase substantially in the future because the world has many alterative sources of roundwood for making commodity wood products. In 1989, developing countries (excluding China) produced less than a fifth of the world's industrial roundwood (FAO 1991b). At the global level, the forces that lead to rising scarcity of timber (tropical and temperate combined) appear to be diminishing, not intensifying. As noted earlier, increases in global roundwood consumption are slowing. Roundwood supplies are increasing in many temperate countries both directly, because of increasing areas of plantations and second-growth forests, and indirectly, because of the development of technologies for making reconstituted wood products from mixed species and low-quality timber (Sedjo & Lyon 1990).

Economists measure timber scarcity by examining changes over time in stumpage value, which is the surplus that remains after deducting logging costs from log prices. Stumpage values at the global level have risen in real (inflation-adjusted) terms during the 20th century, which indicates rising scarcity. The rate of increase has steadily diminished, however, which indicates that the causes of the scarcity are slackening (Sedjo & Lyon 1990, Cardellichio et al. 1989). Forecasting models predict that global stumpage values and roundwood prices will increase more slowly in the future than they have in the past (Binkley & Vincent 1988).

19.3 Boom-and-bust as sustainable development

Competition with temperate timbers has limited the impact that physical depletion of timber stocks in tropical forests has had on international timber prices. This observation provides a clue to the causes of the boom-and-bust pattern.

By communicating information about timber scarcity, stumpage values govern the supply and demand adjustments that occur as a country's forest sector develops (Vincent & Binkley 1991). Imagine a tropical country that has not exploited its old-growth forests and does not have an opportunity to import or export wood products. Harvests (Figure 19.1a) would be high early in development, as the country converts forestland to agriculture and uses timber as a source of capital for

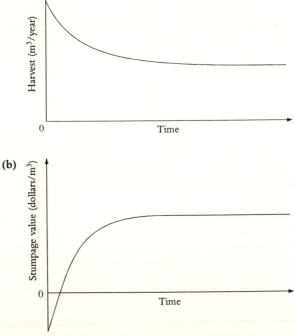

Figure 19.1 Transition from mining old-growth timber to sustained-yield forestry in a closed economy: a) harvest and b) stumpage values.
Source: adapted from Vincent & Binkley 1991: Fig. 1.

industrialization. Stumpage values (Figure 19.1b) would be low because timber is abundant. As timber is depleted, however, rising scarcity would cause the stumpage value to rise, dampening timber demand and stimulating investments in forest management (because their returns increase). These adjustments would promote a transition to a sustained-yield state in which harvests equal growth and stumpage values are constant (the forest earns a return solely through timber growth).

The gradual transition depicted in Figure 19.1 would not necessarily occur in a small tropical country that is open to timber trade. Such a country would face international prices for roundwood and processed products. Because the country is small, these prices would reflect the global economic scarcity of timber, not the physical scarcity of timber within the country's forests. If competition with temperate timber prevented international prices for tropical timber from rising substantially over time, then stumpage values within the country would not increase substantially either.

The country's forests would not be earning a return from either rising stumpage values or net timber growth, which by definition is nil in old-growth forests. From a purely financial standpoint, the country should harvest all its forests as quickly as possible, because it has the alternative of cashing in the stumpage value of the standing timber at the prevailing international price and investing the capital in opportunities that do earn a positive rate of return. Hence, in the face of slowly rising international timber prices, the tropical timber trade will tend to lead to boom-and bust logging in small countries unless policies directly constrain the rate of harvest.

Does this mean that the tropical timber trade is inconsistent with sustainable development? The answer depends on how the phrase is defined. If only the forest sector itself is evaluated, and according to a traditional forestry definition of sustainability; for example, harvesting forests to produce an even flow of timber over time, then a timber boom is obviously not sustainable (it may be viewed as sustainable if forest regrowth provides timber for another boom several decades later). If sustainable development is defined as sustainable macroeconomic growth, then a timber boom can be an integral phase of a sustainable development process. Stumpage value that is invested efficiently can provide fuel for an economy's take-off into sustainable economic growth (Hartwick 1977). By investing timber capital in other industries or public services (for example, infrastructure and education), a country can sustain economic growth after the forest sector goes bust.

Although sustainable development is usually defined at an aggregate, not a sectoral, level, it tends to be defined more broadly than as sustained growth in economic output (WCED 1987, Pearce et al. 1990). Boom-and-bust logging might be considered unsustainable because it creates environmental costs or social problems that reduce human welfare, if not economic growth, either now or in the future. To the extent that stumpage values fail to reflect these non-market costs, timber prices are indeed low. Whether the costs are so great that a small tropical country should refrain from boom-and-bust logging is an empirical question.

19.4 The first policy failure: timber concession policies

Unfortunately, government policies in tropical countries have increased the prevalence of the boom-and-bust pattern. They have created the possibility that, far from being part of an optimal development process, boom-and-bust, forest-sector development has generated economic losses due to excessively rapid harvests, insufficient investments in forest management, and inefficient wood processing. Two sets of policies, which have occurred in almost every tropical timber-exporting country, are most responsible for these problems. They have in common the effect of suppressing both timber scarcity signals and the necessary responses to these signals.

The first set of policies relates to tropical timber concessions. In most tropical countries, forests are government-owned. Harvesting is carried out by private parties who receive timber concessions. For example, concessionaires in Southeast Asia are generally domestic individuals or firms, not multinational corporations from developed countries (see Gillis 1987 and 1988b). Unfortunately, timber concession policies fail to combine forest tenure with capture of stumpage value. This prevents either the government or concessionaires from detecting and responding to rising stumpage values.

Although governments are the owners, the fees they levy on timber extracted by concessionaires bear no relation to stumpage values (Gillis 1980, Page et al. 1976, Ruzicka 1979, Vincent 1990a). These fees are set administratively and arbitrarily. They are generally a fraction of stumpage values, and they do not mimic the trajectory depicted in Figure 19.1. Because they are much lower than stumpage values, the value of forests as a source of government revenue is artificially reduced, inducing governments to favour the conversion of forests to uses that yield greater tax revenue. The reduction in revenue also means that potential funds for forest management are reduced. The lack of funds for forest management in many tropical countries results not so much from low international timber prices as from the failure of governments to capture the existing stumpage value. Indeed, the availability of substitutes for tropical timber suggests that international prices could not be raised even if tropical timber exporters followed the advice of Guppy (1984) and formed a cartel. Moreover, studies indicate that a cartel could hurt tropical timber exporting countries if importing countries retaliated with import tariffs (Vincent 1989), and that a cartel would not necessarily increase the long-run area of forest (Rauscher 1990, Vincent & Newmark 1992).

Although concessionaires capture most of the stumpage value, they have little incentive to invest in forest management because their concession contracts are typically short and of uncertain duration. The uncertainty stems from the allocation of concessions as part of a political patronage process in many countries (Pura 1990, Hendrix 1990, Sesser 1991). Higher international timber prices would simply increase concessionaires' windfall profits, with little impact on investments in forest management.

What is needed is clear: to combine secure forest tenure with sufficient capture of stumpage value by the party holding the tenure rights (Hyde et al. 1991, Paris & Ruzicka 1991). The stumpage value provides the financial incentive for forest management, and the tenure provides confidence in this incentive. Obviously, there are two broad approaches to combining the two. One would be to maintain government ownership but to increase the amount of stumpage value captured by the government to a level sufficient for financing forest management. The other would be to allow concessionaires to continue to capture the lion's share of stumpage value but to restructure concession contracts so that concessionaires had rights comparable to those of a private owner. This would involve lengthening contracts and making them renewable and transferable, so that they would have asset value. In the extreme, forests might be privatized.

The two options have quite different implications for the distribution of the wealth that flows from harvesting a country's tropical forests and for the efficiency with which this wealth is invested. For this and other reasons, the two options may not be equally appropriate in all countries. One or the other is necessary, however, to link timber scarcity signals and forest management responses.

19.5 The second policy failure: log-export restrictions

The second set of policies relates to wood processing. Policy makers in developing countries have repeatedly assumed that the export of raw materials is wasteful and that export revenue and jobs are foregone whenever natural resources are exported in unprocessed form. In the forest sector, this line of thinking has led many countries to attempt to stimulate domestic wood processing by restricting log exports.

Such policies fail to recognize that a country endowed with a natural resource does not necessarily have a comparative advantage in processing that resource (Roemer 1979). The raw material provided by the natural resource is just one of the inputs needed to manufacture the processed product. If a country does not have a comparative advantage in wood processing, then the promotion of wood-processing industries drains labour, capital, and other non-wood factors of production from more efficient sectors of the economy. Although log-export restrictions might succeed in building up a large domestic wood-processing industry, the net impact on economic growth may be negative.

The net impact may be negative even if the loss of output in other economic sectors is ignored. Within the forest sector, there is a trade-off between the value of wood (log price) and the value added to wood. By reducing foreign demand, log-export restrictions depress the domestic price of logs, causing the value of wood itself to decline. This provides the domestic processing industry with a cost advantage in cheaper raw materials. Hence, processing capacity expands. When wood is processed, value is indeed added: payments must be made to employees, managers, machinery suppliers, investors, and others who provide the inputs used in processing. However, although value is added to wood, it is done at the cost of reducing the value of wood itself.

Empirical evidence generally indicates that the value added to wood does not offset the loss in the value of wood when processing expands because of log-export restrictions. In the case of Indonesia, studies have estimated that both export earnings and economic rents (surpluses) were less during the 1980s because of log-export restrictions than they would have been otherwise, in spite of rapid growth by the plywood industry (Gillis 1988b, Lindsay 1989). In the case of Peninsular Malaysia, one study has estimated that for every $2200-per-year sawmill job generated by log-export restrictions during 1973 through 1989, the region gave up $6100 in economic value added (value added to wood, minus the reduction in the value of wood) and $16,600 in export earnings (because of foregone revenue from log exports) (Vincent 1992a); the cited figures are at 1989 price levels in Malaysia, converted to US dollars). Log-export restrictions stimulated expansion of processing capacity and created jobs in Indonesia and Peninsular Malaysia, but at an extraordinary cost.

The most deleterious consequence of log-export restrictions is that they interfere with the price signals that balance timber demand and timber supply. Restrictions are typically imposed after complaints by domestic processing industries about rising domestic log prices. Industries blame the rising prices on log exports, which are often destined for countries whose markets are perceived to be closed to imports of processed wood products. Given their faith in the good of the value added, and a sense that log importers' trade practices are unfair, tropical country governments willingly acquiesce to industry demands that log exports be curtailed.

The industry–government perspective tends to focus on foreign demand for logs as an explanation for price increases. The supply side must also be considered. Log prices rise in a tropical country when timber is becoming more scarce, because of timber depletion either within the country (if it is relatively large) or at the global level (if it is small). For the timber industry to develop along a sustainable path, domestic processing industries must respond to this scarcity signal. The market is signalling that additional capacity expansion is not profitable and that the industry must increase its processing efficiency to remain internationally competitive.

Log-export restrictions artificially suppress this signal. They create an illusion that wood is still abundant. The restrictions are often moderate initially, for example, low levels of export taxes or export quotas on just a few species. As depletion proceeds, they are typically escalated to maintain the illusion. Although protection is sometimes justified for infant industries, tropical countries tend not to wean their wood-processing industries off cheap wood.

Consequently, too much processing capacity develops, too much forest is converted to other uses, and too little management occurs of the remaining forest. Too much capacity develops because mills pay an artificially low price for roundwood. Too much forest is converted because log-export restrictions reduce the stumpage value of timber and hence the value of forests relative to alterative land-uses. The log-export restrictions in Peninsular Malaysia reduced stumpage values per cubic metre by 31 per cent (Vincent 1992a). Too little management, whether active management such as enrichment planting or passive management such as careful logging, occurs because the lower stumpage values reduce the returns to management activities.

For all these reasons, log-export restrictions promote boom-and-bust development. Tropical timber boycotts, which have been advocated by environmental organizations and debated in legislatures in many developed countries, would have similar effects because they too reduce timber prices and therefore stumpage values (Vincent 1990b). Although permitting log exports does not guarantee that stumpage values will be high enough to justify financially the retention of forests or investments in forest management (Leslie 1987), it does improve the chances.

19.6 Conclusions

The boom-and-bust nature of tropical countries' participation in the international timber trade has not resulted from developed countries' consumption of tropical timber or from developed countries' import barriers. It is related to the fact that international prices for tropical roundwood have not risen rapidly over time, but this is more likely due to the fact that the world is not running out of wood, than to market manipulation by developed countries. Tropical countries' own policies, particularly those related to timber concessions and wood processing industries, have exacerbated the tendency toward boom-and-bust development. The tropical timber trade can promote sustainable development of a tropical country's economy, but not necessarily even-flow harvesting of its forests. Tropical countries must relax log-export restrictions and link capture of stumpage value to forest tenure if they are to reap maximum benefits from the trade.

References

Abell, T. M. B. 1988. The application of land systems mapping to the management of Indonesian forests. *Journal of World Forest Resource Management* **3**, 111–27.

Agarwal, A., R. Chopra, K. Sharma 1982. *The state of India's environment*. New Delhi: Centre for Science and Environment.

Agarwal, A. & S. Narain 1992. *Floods, flood plains and environmental myths*. New Delhi: Centre for Science and Environment.

Agarwala, V. P. 1989. *Forests in India: environmental and production frontiers*. New Delhi: Oxford and IBH.

Aiken, S. R. & C. H. Leigh 1992. *Vanishing rainforest: the ecological transition in Malaysia*. Oxford: Clarendon.

Albion, R. G. 1926. *Forests and sea power: the timber problem of the Royal Navy 1652–1862*. Cambridge, Mass.: Harvard University Press.

Allegretti, M. H. 1988. Extractive reserves: an alternative for reconciling development and environmental conservation in Amazonia. In *Alternatives to deforestation*, A. B. Anderson (ed.), 252–64. New York: Columbia University Press.

Allen, J. C. & D. F. Barnes 1985. The causes of deforestation in developing countries. *Annals of the Association of American Geographers* **75**(2), 163–84.

Amelung, T. 1991. Tropical deforestation as an international economic problem. Paper presented at Egon-Sohmen Foundation Conference on Economic Evolution and Environmental Concerns, Linz, Austria (30–31 August).

Amelung, T. & M. Diehl 1992. *Deforestation of tropical rainforests: economic causes and impact on development*. Tugingen: Mohr.

Anderson, A. B. (ed.) 1990. *Alternatives to deforestation: steps toward sustainable use of the Amazon rain forest*. New York: Columbia University Press.

Arensberg, W., M. Higgins, R. Asenjo, F. Ortiz, H. Clark 1989. *Environment and natural resources strategy in Chile*. Santiago: USAID.

Arnold, M. 1991. *Forestry expansion: a study of technical, economic and ecological factors*. Paper No. 3, Oxford Forestry Institute.

Bann, C. 1993. *The private sector and global warming mitigation*. Unpublished paper, Centre for Social and Economic Research on the Global Environment, University College London and University of East Anglia.

Banuri, T. & F. A. Marglin (eds) 1993. *Who will save the forests?* London: Zed Press.

Barbier, E. B. 1987. Natural resources policy and economic framework. In *Natural resources and environmental management in Indonesia*, J. Tarrant (ed.). Jakarta: USAID.

Barbier, E. B. 1989a. Sustainable agriculture on marginal land: a policy framework. *Environment* **31**(9), 12–17, 36–40.

Barbier, E. B. 1989b. *Economics, natural resource scarcity and development*. London: Earthscan.

Barbier, E. B. 1991. Tropical deforestation. In *Blueprint 2: greening the world economy*, D. W. Pearce (ed.), 138–66. London: Earthscan.

Barbier, E. B. 1992. *Economic aspects of tropical deforestation in south east Asia*. Paper presented at the Political Ecology of South East Asia's Forests Workshop, Centre for South East Asian Studies, School of Oriental and African Studies, London University.

Barbier, E. B., B. Aylward, J. C. Burgess, J. T. Bishop 1992a. *Environmental effects of trade in the forest sector*. Joint Session of Trade and Environment Experts, OECD, Paris.

Barbier, E. B., N. Bockstael, J. C. Burgess, I. Strand 1992b. Timber trade and tropical deforestation – Indonesia. In *The economic linkages between the international trade in tropical timber and the sustainable management of tropical forests*, E. B. Barbier, J. C. Burgess, J. T. Bishop, B. Aylward, C. Bann (eds), International Tropical Timber Organization (ITTO).

Barbier, E. B., J. C. Burgess, B. Aylward, J. T. Bishop 1992c. *Timber trade, trade policies and environmental degradation*. LEEC Discussion Paper DP 92–01, London Environmental Economics Centre.

Barbier, E. B., J. C. Burgess, J. Bishop, B. Aylward, C. Bann 1993. *The economic linkages between the international trade in tropical timber and the sustainable management of tropical forests*. International Tropical Timber Organization (ITTO), Final Report.

Barbier, E. B., J. C. Burgess, A. Markandya 1991. The economics of tropical deforestation. *Ambio* **20**(2), 55–8.

Barnes, R. F. W. 1991. Deforestation trends in tropical Africa. *African Journal of Ecology* **28**(3), 161–73.

Basa, V. F. 1991. *Results of a multidate assessment of deforestation using high resolution satellite data in the Philippines*. Manila: National Mapping and Resource Information Centre.

Battjees, J. 1988. *A survey of the secondary vegetation in the surroundings of Araracuara, Amazonas, Colombia*. Amsterdam: Hugo de Vries Laboratory, University of Amsterdam.

Belsley, D. A., E. Kuh, R. E. Welsch 1980. *Regression diagnostics: identifying influential data and sources of collinearity*. New York: John Wiley.

Bertrand, A. 1983. Deforestation en la Zone de Foret en Cote d'Ivoire. *Bois et Forêts des Tropiques* **202**, 3–17.

Bhumibhamon, S. 1986. *The environmental and socio-economic aspects of tropical deforestation: a case study of Thailand*. Bangkok: Faculty of Forestry, Kaesetsart University.

Bigman, D. 1990. A plan to end LDC debt and save the environment too. *Challenge* (July–August).

Bilsborrow, R. E. 1987. Population pressures and agricultural development in developing countries: a conceptual framework and recent evidence. *World Development* **15**(2), 183–203.

Bilsborrow, R. E. & P. F. DeLargy 1991. Land-use, migration and natural resource deterioration in the Third World: the cases of Guatemala and Sudan. In *Resources, environment and population: present knowledge, future options*, K. Davis & M. Bernstam (eds), 125–47. New York: Population Council and Oxford University Press.

Bilsborrow, R. E. & M. E. Geores 1992. *Rural population dynamics and agricultural development: issues and consequences observed in Latin America*. Ithaca, N.Y.: Cornell International Institute for Food, Agriculture and Development, Cornell University.

Bilsborrow, R. E. & P. W. Stupp 1988. *The effects of population growth on agriculture in Guate-*

mala. Occasional paper, 88–24. Carolina Population Centre, University of North Carolina at Chapel Hill.

Binkley, C. S. & J. R. Vincent 1988. Timber prices in the US South: past trends and outlook for the future. *Southern Journal of Applied Forestry* **12**, 15–18.

Binswanger, H. 1991. Brazilian policies that encourage deforestation in the Amazon. *World Development* **19**(7), 821–9.

Birdsall, N. 1988. Economic approaches to population growth. In *Handbook of development economics*, H. Chenery & T. N. Srinivasan (eds), 477–542. The Netherlands: North-Holland.

Biro Pusat Statistik: see Indonesian Department of Information.

Blum, E. 1993. Making biodiversity conservation profitable: a case study of the Merck INBio Agreement. *Environment* **35**(4), 16.

Booth, W. 1989. Monitoring the fate of the forests from space. *Science* **243** (17 March), 1428–9.

Boserup, E. 1965. *The conditions of agricultural growth: the economics of agrarian change under population pressure*. London: Allen & Unwin.

Boserup, E. 1990. *Economic and demographic relationships in development*. Baltimore and London: The Johns Hopkins University Press.

Bourke, I. J. 1988. *Trade in forest products: a study of the barriers faced by the developing countries*. FAO, Rome, Forestry Paper 83.

Bourke, I. J. 1992. *Restrictions on trade in tropical timber*. Rwanda: African Forestry and Wildlife Commission.

Bourke, I. J. 1993. Tariff rates on timber products in selected countries. In *The economic linkages between the international trade in tropical timber and the sustainable management of tropical forests*, E. B. Barbier, J. C. Burgess, J. T. Bishop, B. Aylward, C. Bann (eds), Annex G. London: Environmental Economics Centre.

Bowonder, B., S. S. R. Prasad, N. Y. M. Unni 1985. *Deforestation and fuelwood use in urban centres*. Hyderabad: Administrative Staff College.

Broemeling, L. D. & H. Tsurumi 1987. *Econometrics and structural change*. New York: Marcel Dekker.

Bromley, R. 1981. The colonization of humid tropical areas in Ecuador. *Singapore Journal of Tropical Geography* **2**(1), 15–26.

Browder, J. O. 1989. Development alternatives for tropical rainforests. In *Environment and the poor: development strategies for a common agenda*, H. J. Leonard (ed.), 111–34. New Brunswick, NJ., and Oxford: Transaction Books.

Browder, J. O. 1992. Limits to extractivism. *BioScience* **42**(3), 174–82.

Brown, K. 1994. Medicinal plants, indigenous medicine and conservation of biodiversity in Ghana. In *Intellectual property rights and biodiversity conservation: a multidisciplinary analysis of the values of medicinal plants*, T. M. Swanson (ed.). Cambridge: Cambridge University Press.

Brown, K. & W. N. Adger 1993. *Forests for international offsets: economic and political issues of carbon sequestration*. CSERGE Working Paper GEC 93–15. University of East Anglia and University College London.

Brown, K. & D. W. Pearce 1994. The economic value of carbon storage in tropical forests. In *The economics of project appraisal and the environment*, J. Weiss (ed.), 102–23. Cheltenham: Edward Elgar.

Bruenig, E. F. 1987. The forest ecosystem: tropical and boreal. *Ambio* **16**(2–3), 68–79.

Bunker, S. 1984. Modes of extraction, unequal exchange, and the progressive underdevelopment of an extreme periphery: the Brazilian Amazon. *American Journal of Sociology*

89(5), 1017–64.

Buongiorno, J. 1979. Income and price elasticities for sawnwood and wood-based panels: a pooled cross-section and time series analysis. *Forest Science* **9**(2), 141–8.

Buongiorno, J. & T. Manurung 1992. *Predicted effects of an import tax in the European Community on the international trade in tropical timbers*. Unpublished paper, Department of Forestry, University of Wisconsin, Madison.

Burgess, J. C. 1991. *Economic analyses of frontier agricultural expansion and tropical deforestation*. MSc thesis. Environmental and Resource Economics, University College London.

Burgess, J. C. 1992. *Economic analysis of the causes of tropical deforestation*. Discussion Paper 92–03, London Environmental Economics Centre.

Burgess, P. F. 1989. Asia. In *No timber without trees: sustainability in the tropical forest*, D. Poore, P. F. Burgess, J. Palmer, S. Rietbergen, T. Synnott (eds). London: Earthscan.

Capistrano, A. D. 1990. *Macroeconomic influences on tropical forest depletion: a cross country analysis*. PhD thesis, University of Florida.

Capistrano, A. D. & C. F. Kiker 1990. *Global economic influences on tropical closed broadleaved forest depletion, 1967–1985*. Paper presented at International Society for Ecological Economics Conference, Washington DC USA.

Capra, F. 1982. *The turning point: science, society and the rising culture*. London: Flamingo.

Cardellichio, P. A., Y. C. Youn, D. N. Adams, R. W. Joo, J. T. Chemelik 1989. *A preliminary analysis of timber and timber products production, consumption, trade and prices in the Pacific Rim until 2000*. CINTRAFOR Working Paper 22. Centre for International Trade in Forest Products, University of Washington, Seattle.

Case, A. 1991. Spatial patterns in household demand. *Econometrica* **59** (4).

Chambers, R., N. C. Saxena, T. Shah 1989. *To the hands of the poor*. London: Intermediate Technology.

Cheng, W. 1989. Testing the food-first hypothesis: a cross-national study of dependency, sectoral growth and food intake in less developed countries. *World Development* **17**(1), 17–27.

Chin, S. C. 1989. Managing Malaysia's forests for sustained production. *Wallaceana* 56&57, 1–11.

Clark, C. 1990. *Mathematical bioeconomics*. New York: John Wiley.

Clarke, J. & D. W. Rhind 1991. *Population data and global environmental change*. Programme on Human Dimensions of Global Environmental Change, Working Group on Demographic Data. International Social Science Council, The Hague and Paris.

Cliff, A. D. & J. K. Ord 1973. *Spatial autocorrelation*. London: Pion.

Cliff, A. D. & J. K. Ord 1981. *Spatial processes*. London: Pion.

Cohen, J. & P. Cohen 1975. *Applied multiple regression*. Hillsdale, NJ: Erlbaum.

Constantino, L. F. 1988a. *Analysis of the international and domestic demand for Indonesian wood products*. Unpublished paper, Department of Rural Economy, University of Alberta.

Constantino, L. F. 1988b. *Demand, supply and development issues in the Indonesian forest sector*. Unpublished paper, Department of Rural Economy, University of Alberta.

Constantino, L. F. 1990. *On the efficiency of Indonesia's sawmilling and plymilling industries*. D. G. of Forest Utilization, Ministry of Forestry, Government of Indonesia and FAO, Jakarta.

Constantino, L. F. & D. Ingram 1990. *Supply-demand projections for the Indonesian forestry sector*. D. G. of Forest Utilization, Ministry of Forestry, Government of Indonesia and FAO, Jakarta.

Cruz, M. C. J. 1988. *More people than trees: the Philippine case*. Los Baños: Institute of Environmental Science and Management, University of the Philippines.

Cruz, M. C. J., C. Meyer, R. Repetto, R. Woodward 1992. *Population growth, poverty, and environmental stress: frontier migration in the Philippines and Costa Rica.* World Resources Institute, Washington DC.

Cruz, M. C. J. & I. Zosa-Feranil 1987. *Policy implications of population pressure in the Philippine uplands.* Los Baños: Department of Environmental Studies and the Population Institute, University of the Philippines.

Cruz, M. C. J., I. Zosa-Feranil, C. L. Goce 1988. Population pressure and migration: implications for upland development in the Philippines. *Journal of Philippine Development* **26**(1), 15–46.

Cruz, W. D. & M. C. J. Cruz 1990. Population pressure and deforestation in the Philippines. *ASEAN Economic Bulletin* **7**(2), 200–212.

Cruz, W. D. & R. Repetto 1992. *The environmental effects of stabilisation and structural adjustment programs: the Philippines case.* Washington DC: World Resources Institute.

CSO 1989. *Statistical abstracts of India, 1961 to 1989.* New Delhi: Central Statistical Office (CSO).

Cuddington, J. T. 1987. Macroeconomic determinants of capital flight: an econometric investigation. In *Capital flight and Third World debt*, D. R. Lessard & J. Williamson (eds). Washington DC: Institute for International Economics.

Davis, K. 1963. The theory of change and response in modern demographic history. *Population Index* **29**(4), 345–66.

Davis, K. & M. Bernstam (eds) 1991. *Resources, environment and population: present knowledge, future options.* New York: Population Council and Oxford University Press.

de Milde, R. 1991. *A preliminary estimate of the forested area of peninsular Malaysia.* Kuala Lumpur: FAO.

de Soto, H. 1989. *The other path: the invisible revolution in the Third World.* New York: Harper & Row.

de Vries, M. G. 1986. *The IMF in a changing world 1945–85.* Washington DC: International Monetary Fund.

Deacon, R. T. & P. Murphy 1992. *The structure of an environmental transaction: the debt-for-nature swap.* Allied Social Science Associations Meetings, Annaheim, California.

Denevan, W. M. 1980. Latin America. In *World systems of traditional resource management*, G. A. Klee (ed.), 217–44. New York: Halsted Press.

Detwiler, R. & C. Hall 1988. Tropical forests and the global carbon cycle. *Science* **239**(1, January), 42–7.

Dixon, R., K. Andrasko, F. Sussman, M.-7X Trexler, T. Vinson 1993. Forest sector carbon offset projects: near-term opportunities to mitigate greenhouse gas emissions. *Water, Air and Soil Pollution* **70**, 561–77.

DOE 1987. *India's forests.* Department of Environment (DOE), Government of India, Delhi.

DOE 1989. *The state of forest report 1989.* Department of Environment (DOE), Government of India, Delhi.

Dollar, D. 1991. Outward-orientated developing economics really do grow more rapidly: evidence from 95 LDCS, 1976–85. *Economic Development and Cultural Change* **40**(3), 523–44.

Dornbusch, R. 1987. *Dollars, debts, and deficits.* Leuven, Cambridge Mass. and London: Leuven University Press and MIT Press.

Dourojeanni, M. J. 1988. *Si el Arbol de la Quina Hablara.* Lima: Fundacion Peruana para la Conservacion de la Naturaleza.

Dourojeanni, M. J. 1991. *Amazonia Peruana: Que Hacer?* Iquitos, Peru: Centao de Estudios Tecologicas de la Amazonia.

Dykstra, D. P. & M. Kallio 1987a. Base scenario. In *The global forest sector: an analytical per-*

spective, M. Kallio, D. P. Dykstra, C. S. Binkley (eds), Chapter 28. Chichester, England: John Wiley.

Dykstra, D. P. & M. Kallio 1987b. Scenario variations. In *The global forest sector: an analytical perspective*, M. Kallio, D. P. Dykstra, C. S. Binkley (eds), Chapter 29. Chichester, England: John Wiley.

ECE/FAO 1986. *European timber trends and prospects to the year 2000 and beyond*. ECE/TIM/30, United Nations, New York.

ECE/FAO 1989. *Outlook for the forest and forest products sector of the USSR*. ECE/TIM/48, United Nations, New York.

ECE/FAO 1990. *Timber trends and prospects for North America*. ECE/TIM/53, United Nations, New York.

ECE/FAO 1992. *The forest resources of the temperate zones*. New York: United Nations.

Eckholm, E. P. 1980. The other energy crisis: firewood. In *Energy in the developing world: the real energy crisis*, V. Smil & W. E. Knowland (eds). Oxford: Oxford University Press.

The Ecologist (1993) *Whose common future? Reclaiming the commons*. London: Earthscan.

Edwards, S. 1989. Structural adjustment policies in highly indebted countries. In *Developing country debt and the world economy*, Sachs, J. (ed.), Chicago: University of Chicago Press.

Egler, E. G. 1961. A zona bragantina do estado do para. *Revistal Brasileira de Geografia* **23**, 527–55.

Ehui, S. K. & T. W. Hertel 1989. Deforestation and agricultural productivity. *Journal of American Agricultural Economics* **71**(3), 703–11.

Evans, P. 1979. *Dependent development: the alliance of state, local and multinational capital in Brazil*. Princeton: Princeton University Press.

Fankhauser, S. 1993. *Evaluating the social cost of greenhouse gas emissions*. Unpublished paper, Centre for Social and Economic Research on the Global Environment, University College London and University of East Anglia.

FAO 1977. *FAO production yearbook*. Rome: FAO.

FAO 1981a. *Tropical forest resources assessment project*. Rome: FAO.

FAO 1981b. *Forest resources of tropical Africa, tropical Asia, and America Tropical*. Rome: FAO.

FAO 1981c. *Tropical forest assessment project: forest resources of tropical Asia*. Rome: FAO.

FAO 1983. *Quarterly Bulletin of Statistics* **34**(6). Rome: FAO.

FAO 1984. *FAO production yearbook*. Rome: FAO.

FAO 1985a. *Tropical forestry action plan*. Committee on Forest Development in the Tropics, Rome.

FAO 1985b. *Quarterly Bulletin of Statistics* 1985: **36**(8). Rome: FAO.

FAO 1988a. *An interim report on the state of forest resources in the developing countries*. FAO: MISC/ 88/7. Forest Resources Division, Forestry Department. Rome.

FAO 1988b. *Forest products: world outlook projections*. Forestry Paper 84, FAO, Rome.

FAO 1989a. *Production yearbook, volume 42*. Rome: FAO.

FAO 1989b. *Trade yearbook, volume 42*. Rome: FAO.

FAO 1990a. *Interim report on forest resources assessment 1990 project*. Committee on Forestry, Tenth Session, 24–28 September, Rome.

FAO 1990b. *An interim report on the state of forest resources in the developing countries*. Forest Resources Division, Rome.

FAO 1990c. *Quarterly Bulletin of Statistics* 1990: **34**(4). Rome: FAO.

FAO 1990d. *Forest products prices 1969–88*. Forestry Paper 95, FAO, Rome.

FAO 1991a. *Second interim report on the state of tropical forests*. Forest Resources Assessment 1990 Project, Tenth World Forestry Congress, Paris.

REFERENCES

FAO 1991b. *Yearbook of forest products: 1978–1989*. Rome: FAO.

FAO 1992a. *Third interim report on the state of tropical forests*. Rome: FAO.

FAO 1992b. *The forest resources of the tropical zone by main ecological regions*. Forest Resources Assessment Project, FAO, Rome.

FAO 1993. *Summary of the final report of the forest resources assessment 1990 for the tropical world*. Eleventh Session of the Committee on Forestry, FAO, Rome.

FAO various years. *Forest production yearbooks*. Rome: FAO.

Fearnside, P. M. 1985. *Human use systems and the causes of deforestation in the Brazilian Amazon*. Paper presented at UNU International Conference on Climatic, Biotic and Human Interactions in the Humid Tropics (25 February–1 March), San Jose dos Campos, Sao Paulo, Brazil.

Fearnside, P. M. 1986a. Brazil's Amazon forest and the global carbon problem: reply to Lugo and Brown. *Interciencia* **11**(2), 58–64.

Fearnside, P. M. 1986b. Deforestation in the Amazon: how fast is it occurring? *Ambio* **15**, 82–8.

Fearnside, P. M. 1987. Causes of deforestation in the Brazilian Amazon. In *The geophysiology of Amazonia*, R. E. Dickinson (ed.). New York: John Wiley.

Fearnside, P. M. 1993. Forests or fields? A response to the theory that tropical forest conservation poses a threat to the poor. *Land Use Policy* **10**(2), 108–21.

Feeny, D. 1984. *Agricultural expansion and forest depletion in Thailand, 1900–1975*. Discussion Paper 458, Yale University, Economic Growth Centre.

Fishlow, A. 1985. Coping with the creeping crisis of debt. In *Politics and the economics of external debt crisis*, Wionczek (ed.). Boulder, Co.: Westview Press.

Fitzgerald, B. 1986. *An analysis of Indonesian trade policies: countertrade, downstream processing, import restrictions and the deletion program*. CPD Discussion Paper 1986–22, World Bank, Washington DC.

Forrester, J. W. 1969. Planning under the dynamic influences of complex social systems. In *Perspectives of planning*, E. Jantsch (ed.), 235–54. Paris: Organization for Economic Co-operation and Development.

Fortmann, L. & J. W. Bruce (eds) 1988. *Whose trees? Proprietary dimensions of forestry*. Boulder, Co.: Westview Press.

Galloway, J. N. 1989. Atmospheric acidification projections for the future. *Ambio* **18**, 161–6.

Gámez, R., A. Piva, A. Sittenfield, E. Leon, J. Jimenez, G. Mirabelli 1993. Costa Rica's conservation program and National Biodiversity Institute (INBio). In Reid et al. (eds) (1993), 53–68.

Gandolfo, G. 1987. *International economics II: international monetary theory and open-economy macroeconomics*. Berlin: Springer.

Gasques, J. G. & C. Yokomizo 1986. Resultados de 20 Anos de Incentivos Fiscais na Agropecuario da Amazonia. *XIV Encontro National de Economia* ANPEC 2, 47–84.

Gastil, R. 1989. *Freedom in the world*. New York: Freedom House.

Geary, R. C. 1954. The contiguity ratio and statistical mapping. *The Incorporated Statician* 5.

GEF 1992. *Memorandum of understanding on Norwegian funding of pilot demonstration projects for joint implementation arrangements under the climate convention*. GEF, World Bank, Washington DC.

Gentry, A. H. 1989. Northwest South America (Colombia, Ecuador and Peru). In *Floristic inventory of tropical countries*, D. G. Campbell & H. D. Hammond (eds), 391–400. New York: New York Botanical Garden.

315

Geores, M. E. & R. E. Bilsborrow 1991. *Deforestation and internal migration in selected developing countries.* Annual Meeting of the Southern Demographic Association, Jacksonville, Fla.

George, S. 1988. *A fate worse than debt.* Harmondsworth: Penguin.

Georgiou, S. 1993. *Environmental benefit estimation: a clearer view across the field?* Unpublished paper, Centre for Social and Economic Research on the Global Environment, University College London and University of East Anglia.

Gillis, M. 1980. *Fiscal and financial issues in tropical hardwood concessions.* Development and Discussion Paper 110, Harvard Institute for International Development, Cambridge, Mass.

Gillis, M. 1987. Multinational enterprises and environmental and resource management issues in the Indonesian tropical forest sector. In *Multinational corporations, environment and the Third World*, C. S. Pearson (ed.), Chapter 3. Durham, N.C.: Duke University Press.

Gillis, M. 1988a. Indonesia: public policies, resource management and the tropical forest. In *Public policies and the misuse of forest resources*, R. Repetto & M. Gillis (eds), Chapter 2 & Chapter 18. Cambridge: Cambridge University Press.

Gillis, M. 1988b. The logging industry in tropical Asia. In *People of the tropical rainforest*, J. S. Denslow & C. Padoch (eds), Chapter 12. Berkeley, California: University of California Press.

Gillis, M. 1990a. *Value added taxation in developing countries.* Washington DC: World Bank.

Gillis, M. 1990b. *Forest incentive policies.* World Bank Forest Policy Paper, World Bank, Washington DC.

Government of Bolivia 1990. *Los Recursos Forestales en Bolivia.* Sucre, Bolivia: Department of Natural Resources.

Grainger, A. 1986. *The future role of the tropical rainforests in the world forest economy.* PhD thesis, University of Oxford.

Grainger, A. 1993. *Controlling tropical deforestation.* London: Earthscan.

Gray, J. A. & S. Hadi 1990. *Fiscal policies and pricing in Indonesian forestry.* D. G. of Forest Utilization, Ministry of Forestry, Government of Indonesia and FAO, Jakarta.

Greenwood, M. J. 1975. Simultaneity bias in migration models: an empirical examination. *Demography* **12**(3), 519–36.

Grepperud, S. 1992. *Population–environment linkages: the case of Ethiopia.* Unpublished paper, Department of Economics, University of Oslo.

Guha, R. C. 1989. *Unquiet woods.* New Delhi: Oxford University Press.

Guppy, N. 1984. Tropical deforestation: a global review. *Foreign Affairs* **62**(4), 928–65.

Hafner, J. 1990. Forces and policy issues affecting forest use in northeast Thailand 1900–1985. In *Keepers of the forest. Land management alternatives in southeast Asia*, M. Poffemberg (ed.), 69–94. West Hartford, Conn.: Kumarian Press.

Hansen, S. 1990. Macroeconomic policies: incidence on the environment. In *Development research: the environmental challenge*, J. T. Winpenny (ed.), 27–37. London: Overseas Development Institute.

Hartwick, J. M. 1977. Intergenerational equity and the investing of rents from exhaustible resources. *American Economic Review* **67**, 972–4.

Hayami, Y. & V. Ruttan 1985. *Agricultural development: an international perspective.* Baltimore: The John Hopkins University Press.

Hecht, S. B. 1985. Environment, development and politics: capital accumulation and the livestock sector in Eastern Amazonia. *World Development* **13**(6), 663–84.

Hecketsweiler, P. 1990. *La Conservacion des Ecosystems Forestiers du Congo.* Gland: IUCN.

Hendrix, K. 1990. Vanishing forest tells a way of life. *Los Angeles Times* 18 March. P. A1.

REFERENCES

Hepper, H. N. 1989. West Africa. In *Floristic inventory of tropical countries*, D. G. Campbell & H. D. Hammond (eds), 189–97. New York: New York Botanical Garden.

Higgins, G. M., A. H. Kassam, L. Naiken, G. Fischer, M. Shah 1982. *Potential population supporting capacities of lands in the developing world*. Rome: Food and Agriculture Organization, and Laxenberg, Austria: International Institute for Applied Systems Analysis.

Hirsch, P. 1987. Deforestation and development in Thailand. *Singapore Journal of Tropical Geography* **8**(2), 129–38.

Hirsch, P. 1990a. Forest, forest reserve, and forest land in Thailand. *The Geographical Journal* **156**(2), 166–74.

Hirsch, P. 1990b. *Development dilemmas in rural Thailand*. Oxford: Oxford University Press.

Hoeller, P. 1991. Macroeconomic implications of reducing greenhouse gas emissions: a survey of empirical studies. OECD *Economic Studies* **16**, 45–78.

Houghton, R. A. 1993. The rôle of the world's forest in global warming. In *World forest for the future: their use and conservation*, K. Ramakrishna & G. M. Woodwell (eds), 21–58. New Haven and London: Yale University Press.

Houghton, R. A., R. D. Boone, J. M. Melillo, C. A. Palm, G. M. Woodwell, N. Myers, B. Moore, D. L. Skole 1985. Net flux of carbon dioxide from tropical forests in 1980. *Nature* **316** (15 August), 617–20.

Hsiao, C. 1986. *Analysis of panel data*. New York: Cambridge University Press.

Hyde, W. F., D. H. Newman, R. Sedjo 1991. *Forest economics and policy analysis: an overview*. World Bank Discussion Paper 134, World Bank, Washington DC.

Hyde, W. F. & J. E. Seve 1991. Malawi: a rapid economic appraisal of smallholder response to severe deforestation. In *Pre-proceedings of working group, S6.03–03 and S6.10–00 meetings at the 10th world congress*, R. Haynes, P. Harou, J. Mikowski (eds), Paris, France: International Union of Forest Research Organizations.

Ibrahim, F. 1984. *Ecological imbalance in the republic of Sudan – with reference to desertification in Dafur*. Bayreuth, Germany: Bayreuther Geowissenscaftliche Arbeigen.

Indonesian Department of Information 1982. *Biro purat statistik*. Government of Indonesia, Jakarta.

Instituto Nacional de Pesquisas Espaciais (INPE) 1990. Estado do deforestamento de floresta amazônica, brasileira em 1989. San José dos Campos, INPE, Brazil, Unpublished paper.

Instituto Nacional de Pesquisas Espaciais (INPE) 1992. *Deforestation in Brazilian Amazonia*. San Jose dos Campos: Instituto Nacional de Pesquisas Espaciais.

Intergovernmental Panel on Climate Change (IPCC) 1990. *Climate change: the scientific assessment*. Cambridge: Cambridge University Press.

Intergovernmental Panel on Climate Change (IPCC) 1991. *Emissions scenarios for the IPCC –* an update: assumptions, methodology, and results. Detailed backup documentation, Intergovernmental Panel on Climate Change, Working Group I.

International Tropical Timber Council (ITTC) 1992. *Report on preparations for the 1992 United Nations conference on environment and development*. (XII)/8, 12th Session, 6–14 May, ITTC, Yaounde, Cameroon.

International Tropical Timber Organization (ITTO) 1990. *The promotion of sustainable forest management: a case study in Sarawak, Malaysia*. International Tropical Timber Council, Eighth Session, 16–23 May 1990, Denpasar, Bali, Indonesia.

Ives, J. D. & B. Messerli 1989. *The Himalayan dilemma: reconciling development and conservation*. London and New York: United Nations University and Routledge.

Jacobs, M. 1988. *The tropical rain forest*. Berlin: Springer.

Jenkins, M. D. 1987. *Madagascar: an environmental profile*. Cambridge: WCMC.

Johnson, B. 1991. *Responding to tropical deforestation: an eruption of crisis and array of solutions*.

London: Conservation Foundation; Washington DC: World Wildlife Fund.

Johnston, J. 1984. *Econometric methods*. New York: McGraw-Hill.

Joly, L. G. 1989. The conversion of rainforests to pastures in Panama. In *The human ecology of tropical land settlement in Latin America*, D. A. Schumann & W. L. Partridge (eds), 86–130. Boulder, Co.: Westview Press.

Jones, J. J. 1990. *Colonisation and environment: land settlement projects in Central America*. Tokyo: United Nations University Press.

Jones, W. B. J. & H. V. Richter 1981. *Population mobility and development: Southeast Asia and the Pacific*. Canberra: Development Studies Centre.

Jonkers, W. B. J. & R. Glastra (eds) 1989. *Colombia, Indonesia and Ivory Coast: selected abstracts 1986–88*. Ede, Netherlands: Tropenbos.

Joshi, G. 1983. Forests and forest policy in India. *Social Scientist* (January), 43–52.

Kahn, J. R. & J. A. McDonald 1992. *Third World debt and tropical deforestation*. Unpublished paper, Department of Economics, SUNY-Binghampton, New York.

Kallio, M., D. Dykstra, C. Binkley 1987. *The global forestry sector: an analytical perspective*. New York: John Wiley.

Katila, M. 1992a. *Modelling deforestation in Thailand: the causes of deforestation and deforestation projections for 1990–2010*. Unpublished paper, Finnish Forestry Institute, Helsinki.

Katila, M. 1992b. *Modelling deforestation in Thailand*. Thai Forestry Sector Master Plan, 16H–4071-VIPT - 9. THA/88/R51. Royal Forest Department, Ministry of Agriculture and Cooperatives, Bangkok.

Kazmer, D. R. 1977. The agriculture development on the frontier: the case of Siberia under Nicholas II. *American Economic Review* **67**(1), 429–32.

Kemf, E. 1990. *Month of pure light: the regreening of Vietnam*. London: The Women's Press.

Keyfitz, N. 1991. Population and development within the ecosphere: one view of the literature *Population Index* **57**(1): 5–52.

Khator, R. 1989. *Forests, the people and the government*. New Delhi: National Book Organization.

Killick, T. 1990. Notes on macroeconomic challenge. In *Development research: the environmental challenge*, J. T. Winpenny (ed.), 38–42. London: Overseas Development Institute.

Klankamsorn, B. & P. Adisornprasert 1983. *Study on the changes of forest land-use in Northeastern Thailand using LANDSAT imageries*. Department of Forestry, Bangkok.

Klankamsorn, B. & T. Charuppat 1991. *Deforestation in Thailand*. Bangkok: Royal Forest Department.

Kummer, D. M. 1991. *Deforestation in post-war Philippines*. Unpublished paper, Department of Geography, Clark University, Worcester Mass.

Kummer, D. M. 1992. *Deforestation in the post-war Philippines*. Chicago: University of Chicago Press.

Kuusela, K. 1987. Forest products – world situation. *Ambio* **16**(2–3), 80–85.

Lal, J. B. 1989. *India's forests: myth and reality*. Dehradun: Natraj Publishers.

Lanly, J. P. 1982. *Tropical forest resources*. Rome: FAO.

Lanly, J. P. 1983. Assessment of the forest resources of the Tropics. *Forestry Abstracts* **44**(3), 287–313.

Lanly, J. P. 1988. *An interim report on the state of tropical forest resources in the developing countries*. FAO, Rome.

Leach, G., L. Jarass, G. Obermair, L. Hoffman 1986. *Energy and growth: a comparison of 13 industrial and developing countries*. Oxford: Butterworth.

Ledec, G. 1992a. New directions for livestock policy: an environmental perspective. In *Development or destruction: the conversion of forest to pasture in Latin America*, T. Downing, S.

REFERENCES

B. Hecht, S. Garcia-Downing (eds), Boulder, Co.: Westview Press.

Ledec, G. 1992b. *The role of bank credit for cattle raising in financing tropical deforestation.* PhD thesis, University of California at Berkeley.

Ledec, G., R. J. A. Goodland, J. W. Kirchner, J. M. Drake 1985. *Carrying capacity, population growth and sustainable development.* World Bank Staff Working papers No. 690, Population and Development Series, No. 15, World Bank, Washington DC.

Leonard, H. J. 1987. *Natural resources and economic development in Central America.* Washington DC: International Institute for Environment and Development.

Leslie, B. A. J. 1987. A second look at the economics of natural forest management systems in tropical mixed forests. *Unasylva* **39**, 46–58.

Lessard, D. R. & J. Williamson 1987. *Capital flight and Third World debt.* Washington DC: Institute for International Economics.

Lindsay, H. 1989. The Indonesian log export ban: an estimation of foregone export earnings. *Bulletin of Indonesian Economic Studies* **25**, 111–23.

Lisansky, J. 1990. *Migrants to Amazonia: spontaneous colonisation in the Brazilian frontier.* Boulder, Co.: Westview Press.

London, B. 1987. Structural determinants of Third World urban change: an ecological and political economic analysis. *American Sociological Review* **52**(1), 28–43.

Lopez-Gonzaga, V. 1987. *Capital expansion, frontier development and the rise of monocrop economy in Negros (1850–1898).* Occasional Paper, 1. LaSalle University, Bacolod, Philippines.

Low, P. 1992. *International trade and the environment.* Discussion Papers 159, World Bank, Washington DC.

Lugo, A. E. 1988. The future of the forest: ecosystem rehabilitation in the tropics. *Environment* **30**(7), 16.

Lugo, A. E. & S. Brown 1982. Conversion of tropical moist forests: a critique. *Interciencia* **7**(2), 89–97.

Lugo, A. E. & S. Brown 1986. Brazil's Amazon forest and the global carbon problem. *Interciencia* **11**(2), 57–8.

Lugo, A. E., J. A. Parrotta, S. Brown 1993. Loss in species caused by tropical deforestation and their recovery through management. *Ambio* **22**, 106–9.

Lugo, A. E., R. Schmidt, S. Brown 1981. Tropical forests in the Caribbean. *Ambio* **10**(6), 318–24.

Lutz, E. & H. Daly 1990. *Incentives, regulations, and sustainable land-use in Costa Rica.* Environment Working Paper, 34. Environment Department, World Bank, Washington DC.

MacArthur, R. H. & E. O. Wilson 1967. *The theory of island biogeography.* Princeton, New Jersey: Princeton University Press.

Mahar, D. J. 1989a. *Government policies and deforestation in Brazil's Amazon region.* Washington DC: World Bank.

Mahar, D. J. 1989b. Deforestation in Brazil's Amazon region: magnitude, rate and causes. In *Environmental management and economic development*, G. Schramm & J. Warford (eds). Baltimore: The Johns Hopkins University Press.

Malingreau, J. P. & C. J. Tucker 1988. Large-scale deforestation in the southern Amazon basin of Brazil. *Ambio* **17**(1), 49–55.

Mangurian, D. 1990. Enlisting nature's aid. *The IDB Newsletter* (September–October), 3.

Martine, G. R. 1988. Frontier expansion, agricultural modernization, and population trends in Brazil. In *Population, food and rural development*, R. Lee, B. Arthur, A. C. Kelley, G. Rodgers, T. N. Srinivasan (eds), 187–203. Oxford: Clarendon Press.

McClennan, M. S. 1980. *The Central Luzon Plain: land and society on the inland frontier.*

Quezon City, Philippines: Alemars.

McNeely, J. A. & R. J. Dobias 1991. Economic incentives for conserving biological diversity in Thailand. *Ambio* **20**(2), 86–90.

Melillo, J. M., C. A. Palm, R. A. Houghton, G. M. Woodwell, N. Myers 1985. A comparison of two recent estimates of disturbance in tropical forests. *Environmental Conservation* **12**(1), 37–40.

Mellor, J. W. 1988. The intertwining of environment problems and poverty. *Environment* **30**(9), 18–19.

Mendelsohn, R. 1990. *Market failure and tropical forest management.* Allied Social Science Associations Meeting, Washington DC.

Mendez, D. A. 1988. *Population growth, land scarcity, and environmental deterioration in rural Guatemala.* Unpublished paper, Universidad del Valle, Guatemala City.

Millikan, B. H. 1988. *The dialectics of devastation: tropical deforestation, land degradation and society in Rondônia, Brazil.* MA thesis, University of California at Berkeley.

Miron, P. A. 1984. Spatial autocorrelation in regression analysis: a beginners guide. In *Spatial statistics and models*, G. L. Gaile & C. J. Willott (eds). Reidel Publishing.

Molofsky, J., C. A. S. Hall, N. Myers 1986. *Comparison of tropical forest surveys.* Washington DC: US Department of Energy.

Moran, E. 1983. Government-directed settlement in the 1970's: an assessment of Transamazon Highway colonization. In *The dilemma of Amazonian development*, E. Moran (ed.). Boulder, Co.: Westview Press.

Moran, P. A. 1950. Notes on continuous stochastic processes. *Econometrica*

Moreddu, C., K. Parris, B. Huff 1993. Agricultural policies in developing countries and agricultural trade. In *Agricultural trade liberalisation: implications for developing countries,* I. Golden & O. Knudsen (eds), 115–58, Paris: Organization for Economic Co-operation and Development.

Mors, M. 1991. *The economics of policies to stabilize or reduce greenhouse gas emissions: the case of* CO_2. Economic Papers No. 87, CEE.

Myers, N. 1980. *Conversion of tropical moist forests.* National Research Council, Washington DC.

Myers, N. 1981. The hamburger connection: how Central America's forests become North America's hamburgers. *Ambio* **10**(1), 3–8.

Myers, N. 1984. *The primary source: tropical forests and our future.* New York: Norton.

Myers, N. 1986a. Economics and ecology in the international arena: the phenomenon of "linked linkages". *Ambio* **15**(5), 296–300.

Myers, N. 1986b. Tropical deforestation and a mega-extinction spasm. In *Conservation biology: the science of scarcity and diversity*, M. E. Soule (ed.), 394–409. Sunderland, Mass.: Sinauer Association.

Myers, N. 1986c. Tropical forests: patterns of depletion. In *Tropical rain forests and the world atmosphere*, G. T. Prance (ed.), Boulder, Co.: Westview Press.

Myers, N. 1988a. Tropical deforestation and remote sensing. *Forest Ecology and Management* **23**(2–3), 215–25.

Myers, N. 1988b. *Report on a visit to Jakarta for cabinet discussions.* Worldwide Fund for Nature, 12–15 September.

Myers, N. 1989. *Deforestation rates in tropical forests and their climatic implications.* London: Friends of the Earth.

Myers, N. 1990. *Population and environment: issues, prospects and policies.* Unpublished monograph, UN Population Fund, New York.

Myers, N. 1991. The world's forests and human populations: the environmental interconnections. In *Resources, environment and population*, K. Davis & M. S. Bernstam (eds), 237–51. New York: Oxford University Press.

Myers, N. 1992a. *Future operational monitoring of tropical forests: an alert strategy.* World Forest Watch Conference, Sao Jose dos Campos, Brazil, Joint Research Centre, Commission of European Communities.

Myers, N. 1992b. Tropical forests: present status and future outlook. In *Tropical forests and climate*, N. Myers (ed.), 3–32. Dordrecht: Kluwer Academic.

Myers, N. 1992c. *The primary source: tropical forest and our future.* New York: Norton.

Nectoux, F. & N. Dudley 1987. *A hard wood story.* London: Friends of the Earth.

Nectoux, F. & Y. Kuroda 1989. *Timber from the South Seas.* Gland: WWF International.

Nelson, M. 1973. *The development of tropical lands.* Baltimore: The Johns Hopkins University Press.

NESDB 1982. The fifth national economic and social development plan 1982–1986. Office of the Prime Minister, Bangkok, Thailand.

Netherlands Economic Institute (NEI) 1989. *An import surcharge on the import of tropical timber in the European Community: an evaluation.* Rotterdam: NEI.

Newcombe, K. & R. de Lucia 1993. *Mobilising private capital against global warming: a business concept and policy issues.* Unpublished paper, Global Environment Facility, Washington DC.

Ng, F. S. P. 1983. Ecological principles of tropical lowland rainforest conservation. In *Tropical rainforests: ecology and management*, S. L. Sutton, T. Whitmore, A. C. Chadwick (eds), 359–73. Oxford: Blackwell Scientific.

Nordhaus, W. 1991. Economic approaches to greenhouse warming. In *Global warming: the economic policy responses*, R. Dornbusch & J. M. Poterba (eds), 33–66. Cambridge, Mass.: MIT Press.

OECD 1993. *Agricultural policies, markets and trade: monitoring and outlook 1993.* Paris: Organization for Economic Co-operation and Development.

Olechowski, A. 1987. Barriers to trade in wood and wood products. In *The global forest sector: an analytical perspective*, M. Kallio, D. P. Dykstra, C. S. Binkley (eds), Chapter 28. Chichester, England: John Wiley.

Organization of American States (OAS) 1974. *Situacion y Perspectivas Alimentarias de America Latina.* Washington: OAS.

Oxford Forestry Institute (OFI) in association with the Timber Research and Development Association (TRADA) 1991. *Incentives in producer and consumer countries to promote sustainable development of tropical forests.* ITTO Pre-Project Report, PCM, PCF, PCI(IV)/1/1Rev.3, OFI, Oxford.

Ozbekhan, H. 1969. Toward a general theory of planning. In *Perspectives of planning*, E. Jantsch (ed.), 45–175. Paris: Organization for Economic Co-operation and Development.

Ozario de Almeida, A. L. 1992. *The colonisation of the Amazon.* Austin, Texas: University of Texas Press.

Page Jr, J. M., S. R. Pearson, H. E. Leyland 1976. Capturing economic rent from Ghanaian timber. *Food Research Institute Studies* **15**, 25–51.

Palo, M. 1984. Deforestation scenarios for the Tropics. In *Policy analysis for forestry development, Vol. II*, Proceedings of the International Conference held in Thessaloniki, 2–3 April, Thessaloniki, Greece: International Union of Forest Research Organizations.

Palo, M. 1987. Deforestation perspectives for the Tropics: a provisional theory with pilot applications. In *The global forest sector: an analytical perspective*, D. Dykstra, M. Kallio, C. Binkley (eds), Chapter 3: 57–89. Chichester, England: International Institute for Applied Systems Analyses and John Wiley.

Palo, M. 1988. The forest-based development theory revisited with a case study of Finland and prospects for developing countries. In Palo & Salmi (1988), vol. II, 13–156.

Palo, M. 1990. Deforestation and development in the Third World: roles of system causality and population. In *Deforestation or development in the Third World? Vol III*, M. Palo & G. Mery (eds), 155–72. Helsinki: Metsantutkimuslaitoksen Tiedonantoja (Research Bulletins of the Finnish Forest Research Institute).

Palo, M. & G. Mery 1986. *Deforestation perspectives in the Tropics with a global view: a pilot quantitative human population growth approach*. Congress Report, 18th International Union of Forest Research Organizations World Congress, Ljubljana: 552–85.

Palo, M. & J. Salmi (eds) 1988. *Deforestation or development in the Third World?* Helsinki: Finnish Forest Research Institute.

Palo, M., G. Mery, J. Salmi 1987. Deforestation in the Tropics: pilot scenarios based on quantitative analyses. In *Deforestation or development in the Third World? Vol. I*, M. Palo & J. Salmi (eds), 53–106. Helsinki: Metsantutkimuslaitoksen Tiedonantoja (Research Bulletins of The Finnish Forestry Institute).

Panayotou, T. 1991. *Population change and land-use in developing countries: the case of Thailand*. Workshop on Population Change and Land Use in Developing Countries. Committee on Population, National Academy of Sciences (5–6 December), Washington DC.

Panayotou, T. & P. S. Ashton, 1993. *Not by timber alone: economics and ecology for sustaining tropical forests*. Washington DC: Island Press.

Panayotou, T. & S. Sungsuwan 1989. *An econometric study of the causes of tropical deforestation: the case of Northeast Thailand*. Discussion Paper No. 284, Harvard Institute of International Development. Harvard University, Cambridge, Mass.

Paranjpye, V. 1988. *Narmada dam: a cost-benefit analysis*. New Delhi: INTACH.

Paris, R. & I. Ruzicka 1991. *Barking up the wrong tree: the role of rent appropriation in sustainable tropical forest management*. Occasional Paper No. 1, Environment Office, Asian Development Bank, Manila.

Paulino, L. A. 1986. *Food in the Third world: past trends and projections to 2000*. Research Report 52, The International Food Policy Institute, Washington DC.

Pearce, D. W., E. B. Barbier, A. Markandya 1990. *Sustainable development: economics and environment in the Third World*. London: Edward Elgar and Earthscan.

Pearce, D. W., D. Moran, E. Fripp 1992. *The economic value of biological and cultural diversity*. Report to International Union for the Conservation of Nature, Gland, Switzerland.

Pearce, D. W. & J. Warford 1993. *World without end: economics, environment and sustainable development*. New York and Oxford: Oxford University Press.

Peluso, N. L. 1983. *Markets and merchants: the forest products trade in East Kalimantan in historical perspective*. MA thesis, Department of Rural Sociology, Cornell University, Ithaca.

Pereira, A., A. Ulph, W. Tims 1987. *Socio-economic and policy implications of energy price increases*. Brookfield, Vermont: International Labour Office and Gower.

Perez-Garcia, J. M. & B. R. Lippke 1993. Measuring the impacts of tropical timber supply constraints, tropical timber trade constraints and global trade liberalization. In *The economic linkages between the international trade in tropical timber and the sustainable management of tropical forests*, E. B. Barbier, J. C. Burgess, J. T. Bishop, B. A. Aylward, C. Bann (eds), London: Environmental Economics Centre.

Peters, C. M., A. H. Gentry, R. O. Mendelsohn 1989. Valuation of an Amazonian rainforest. *Nature* 339 (29 June), 655–6.

Peters, W. F. & L. F. Neuenschwander 1988. *Slash and burn farming in Third World forests*. Moscow, Idaho: University of Idaho Press.

Peuker, A. 1992. *Public policies and deforestation: a case study of Costa Rica*. Regional Studies

Programme, 14. World Bank, Latin America and the Caribbean Technical Department. Washington DC: World Bank.

Pichon, F. J. & R. E. Bilsborrow 1992. Land tenure and land-use systems, deforestation, and associated demographic factors: farm-level evidence from Ecuador. In *Population and deforestation in the humid tropics*, R. Bilsborrow & D. Hogan (eds). Oxford: Oxford University Press.

Pinchot, G. 1947. *Breaking new ground*. New York: Harcourt Brace.

Pindick, R. S. 1979. *The structure of the world energy demand*. Cambridge, Mass.: MIT Press.

Pindyck, S. R. & D. L. Rubinfield 1981. *Econometric models and economic forecasts*. Tokyo: McGraw-Hill.

Pingali, P. L. & H. P. Binswanger 1987. Population density and agricultural intensification: a study of the evolution of technologies in tropical agriculture. In *Population growth and economic development: issues and evidence*, D. G. Johnson & R. D. Lee (eds), 27–56. Madison, Wisc.: University of Wisconsin Press.

Plotkin, M. & L. Famolare 1992. *Sustainable harvest and marketing of rain forest products*. Washington DC: Conservation International.

Poffemberg, M. (ed.) 1990. *Keepers of the forest: land management alternatives in Southeast Asia*. West Hartford, Conn.: Kumarian.

Poore, D., P. Burgess, J. Palmer, S. Rietbergen, T. Synnott 1989. *No timber without trees: sustainabilty in the tropical forest*. London: Earthscan.

Population Reference Bureau 1990. *World population data sheet*. Washington DC: Population Reference Bureau.

Population Reference Bureau 1992. *1992 World population data sheet*. Washington DC: Population Reference Bureau.

Poston, D. L., W. P. Frisbie, M. Micklin 1984. Sociological human ecology: theoretical and conceptual perspectives. In *Sociological human ecology: issues and application*, M. Micklin & H. Choldin (eds), 91–120. Boulder, Co.: Westview Press.

Principe, P. P. 1991. Valuing the biodiversity of medicinal plants. In *The conservation of medicinal plants*, O. Akerele, V. Heywood, H. Synge (eds) 79–124. Cambridge: Cambridge University Press.

Pritchett, L. 1991. *Measuring outward orientation in developing countries: can it be done?* PRE Working Paper Number 566, World Bank, Washington DC.

Pura, R. 1990. Battle over forest rights in Sarawak pits ethnic groups against wealthy loggers. *Asian Wall Street Journal*,26 February, p. A1.

Puri, G. S., V. M. Meher-Homji, R. K. Gupta, S. Puri 1983. *Forest ecology, phytogeography and forest conservation*. New Delhi: Oxford and IBH Publishing House.

Ragin, C. C. 1987. *The comparative method: moving beyond qualitative and quantitative strategies*. Berkeley, Los Angeles, London: University of California Press.

Rauscher, M. 1990. Can cartelization solve the problem of tropical deforestation. *Weltwirtschaftliches Arch* **126**, 380–87.

Redford, K. H. & C. Padoch (eds) 1992. *Conservation of neotropical forests, working from traditional resource use*. New York: Columbia University Press.

Reid, W.V. 1992. How many species will there be? In *Tropical deforestation and species extinction*, T. C. Whitmore & J. A. Sayer (eds), 55–74. London: Chapman & Hall.

Reid, W., S. Laird, C. Meyer, R. Gámez, A. Sittenfield, D. Janzen, M. Gollin, C. Juma (eds) 1993. *Biodiversity prospecting: using genetic resources for sustainable development*, Washington DC: World Resources Institute.

Reis, E. J. & R. M. Guzman 1992. *An econometric model of Amazon deforestation*. Paper presented at Statistics in Public Resources and Utilities, and in Care of the Environment

Conference (7–11 April) Lisboa, Portugal.

Reis, E. J. & S. Margulis 1990. *Economic perspectives on deforestation in the Brazilian Amazon.* Project Link Conference, Manila, Philippines.

Reis, E. J. & S. Margulis 1991. Options for slowing Amazon jungle-clearing. In *Global warming: the economic policy responses*, R. Dornbusch & J. Poterba (eds), 335–75. Cambridge, Mass.: MIT Press.

Repetto, R. 1987. Creating incentives for sustainable development. *Ambio* **16**(2–3), 94–9.

Repetto, R. 1988a. *The forest for the trees? Government policies and the misuse of forest resources.* Washington DC: World Resources Institute.

Repetto, R. 1988b. *Deforestation and government policy.* Occasional Papers No. 8, International Centre for Economic Growth, USA.

Repetto, R. 1989. *Government policies and deforestation in the Tropics.* Washington DC: World Resources Institute.

Repetto, R. 1990. Deforestation in the Tropics. *Scientific American* **262**(4), 36–45.

Repetto, R. & M. Gillis (eds) 1988. *Public policies and the misuse of forest resources.* Cambridge: Cambridge University Press.

Repetto, R. & T. Holmes 1983. The role of population in resource depletion in developing countries. *Population and Development Review* **9**(4), 607–32.

Revilla, A. V., J. A. Canonizado, M. C. Gregorio 1987. *Forest resource management report.* Manila: Forest Resources Management Task Force, Resource Policy Group.

Robison, R., K. Hewison, R. Higgot (eds) 1987. *Southeast Asia in the 1980s: politics of economic crisis.* Sydney and London: Allen & Unwin.

Rodhe, H. & R. Herrera (eds) 1988. *Acidification in tropical countries.* Chichester and New York: John Wiley.

Roemer, M. 1979. Resource-based industrialization: a survey. *Journal of Development Economics* **6**, 163–202.

Rudel, T. K. 1989. Population, development and tropical deforestation: a cross-national study. *Rural Sociology* **54**(3), 327–38.

Ruzicka, I. 1979. Rent appropriation in Indonesian logging: East Kalimantan 1972/73–1976/77. *Bulletin of Indonesian Economic Studies* **15**, 45–74.

Sadoff, C. W. 1992. *The effects of Thailand's logging ban: overview and preliminary results.* Unpublished paper, TDRI, Manila.

Salati, E., A. E. de Oliveira, H. O. R. Schubart, F. C. Novaes, M. J. Dourojeanni, J. C. Umana 1990. Changes in the Amazon over the last 300 years. In *The Earth as transformed by human action*, B. L. Turner (ed.), 479–93. Cambridge: Cambridge University Press.

Salter, R. E. 1990. *Letter of 29 March 1990.* Vientiane, Lao Pdr: Forest Resources Conservation Project, Lao/Swedish Forestry Cooperation Programme.

Salter, R. E. & B. Phanthavong 1989. *Needs and priorities for a protected areas system in Loa Pdr.* Stockholm: Swedish International Development Agency.

Saulei, S. M. 1989. *Letter.* Lae, Papua New Guinea: Director of Department of Forests.

Schmidt, R. 1990. *Sustainable management of tropical moist forests.* ASEAN Sub-Regional Seminar, Indonesia.

Schmink, M. & C. H. Wood 1987. The "Political Ecology" of Amazonia. In *Lands at risk in the Third World*, P. D. Little, M. M. Horowitz, A. E. Nyerges (eds), 38–57. Boulder, Co.: Westview Press.

Schneider, S. H. 1992a. Global climate change: ecological effects. *Interdisciplinary Science Review* **17**(2), 142–8.

Schneider, R. 1992b. *Brazil: an analysis of environmental problems in the Amazon.* Report, 9104-BR. World Bank.

REFERENCES

Schneider, R., J. Mckenna, C. Dejou, J. Butler, R. Barrows 1990. *Brazil: an economic analysis of environmental problems in the Amazon.* Washington DC: The World Bank.

Schuman, D. & W. L. Partridge 1989. *Human ecology of tropical land settlement in Latin America.* Boulder, Co.: Westview Press.

Scott, M. 1989. The disappearing forests. *Far Eastern Economic Review* **143** (12 January), 34–8.

Scotti, R. 1990. *Estimating and projecting forest area at global and local level: a step forward.* Rome: Forest Resources Assessment 1990 Project, FAO.

Sedjo, A., M. Bowes, C. Wiseman 1991. *Toward a worldwide system of tradeable forest protection and management obligations.* Washington DC: Resources for the Future.

Sedjo, R. A. 1987. *Incentives and distortions in Indonesian forest policy.* Unpublished paper, Resources for the Future, Washington DC.

Sedjo, R. A. & K. S. Lyon 1990. *The long-term adequacy of world timber supply.* Unpublished paper, Resources for the Future, Washington DC.

Serra-Vega, J. 1990. Andean settlers rush for Amazonia. *Earthwatch* **1**(39), 7–9.

Sesser, S. 1991. Logging the rainforest. *New Yorker* (27 May), 42.

Shafik, N. 1993. *Economic growth and environmental quality.* Forthcoming.

Shane, D. R. 1986. *Hoofprints in the forest: cattle ranching and the destruction of Latin America's tropical forests.* Philadelphia: Institute for the Study of Human Issues.

Sharma, R. A., J. F. Blyth, M. J. Macgregor 1990. The socio-economic environment of forestry development (since British period): a historical perspective. *The Indian Forester* **116**, 523–35.

Shaw, R. P. 1992. The impact of population growth on environment: the debate heats up. *Environmental Impact Assessment Review* **12**, 11–36.

Shilling, J. D. 1992. Reflections on debt and the environment. *Finance and Development* **30**, 28–30.

Shioya 1967. A short history of forestry and forestry research in East Asia. In *International Review of Forestry Research*, Romberger & Mikola (eds), 1–42. New York: Academic Press.

Shresta, N. R. 1987. Institutional policies and migration behaviour: a selective review. *World Development* **15**(3), 329–45.

Shrybman, S. 1990. International trade and the environment: an environmental assessment of present GATT negotiations. *Alternatives* **17**(2).

Simon, J. 1981. *The ultimate resource.* Oxford: Martin Robertson.

Simon, J. 1986. *Theory of population and economic growth.* Oxford: Basil Blackwell.

Singh, K. D. 1993. The 1990 tropical forest resources assessment. *Unasylva* **44**(174), 10–19.

Sinha, R. 1984. *Landlessness: a growing problem.* Economic and Social Development Series, 28. FAO.

Sittenfield, A. & R. Gámez 1993. Biodiversity prospecting in INBio. See Reid at al. (eds) (1993), 69–98.

Southgate, D. 1989. *Tropical deforestation and agricultural development in Latin America.* Unpublished paper, Quito, Instituto de Estrategias Agropecuarias.

Southgate, D. 1990. The causes of land degradation along "spontaneously" expanding agricultural frontiers in the Third World. *Land Economics* **66**(1), 93–101.

Southgate, D. 1991. *Tropical deforestation and agricultural development in Latin America.* LEEC Discussion Paper, 91–01, London Environmental Economics Centre, London.

Southgate, D. & D. W. Pearce 1988. *Agricultural colonization and environmental degradation in frontier developing economies.* Working Paper No. 9, World Bank Environment Department, Washington DC.

Southgate, D. & C. F. Runge 1990. *The institutional origins of deforestation in Latin America.* St

Paul Minnesota: University of Agricultural and Applied Economics, University of Minnesota.

Southgate, D., R. Sierra, L. Brown 1989. *The causes of tropical deforestation in Ecuador: a statistical analysis*. LEEC Discussion Paper 89–09, London Environmental Economics Centre, London.

Southgate, D., R. Sierra, L. Brown 1991. The causes of deforestation in Ecuador: a statistical analysis. *World Development* **19**(9), 1145–51.

Srivastrava, P. & W. Butzler 1989. *Protective development and conservation of the forest environment in Papua New Guinea*. Bonn: German Foreign Aid Programme.

Stonich, S. C. 1989. The dynamics of social processes and environmental destruction: a Central American case study. *Population and Development Review* **15**(2), 269–96.

Summers, R. & A. Heston 1991. The Penn World Table (Mark 5): an expanded set of international comparisons, 1950–88. *The Quarterly Journal of Economics* **106**(2), 327–68.

Sutter, H. 1989. *Forest resources and land-use in Indonesia*. Jakarta: Directorate General of Forest Utilization.

Swanson, T. 1993. *The international regulation of extinction*. London: Macmillan.

Tabah, L. 1992. *Population prospects with special reference to environment. Workshop 11: population growth of the Third World*. An International Conference on the Challenges to Systems Analysis in the Nineties and Beyond, International Institute for Applied Systems Analysis, Laxenburg, Austria.

Tata 1990. *Statistical outline of India 1989–90*. Tata Consultancy Services, Bombay.

Thampapillai, D. 1992. *Extensions of environmental accounting: internalizing the environment in a Keynesian framework*. Working Paper 9201, Macquarie University, Sydney, Australia.

Thiesenhusen, W. C. (ed.) 1989. *Searching for agrarian reform in Latin America*. Boston: Unwin Hyman.

Thirgood, J. V. 1981. *Man and the Mediterranean forest: a history of resource depletion*. London: Academic Press.

Toffler, A. 1980. *The third wave*. London: Collins.

Tongpan, S., T. Panayotou, S. Jetanavanich, K. Faichampa, C. Mehl 1990. *Deforestation and poverty: can commercial and social forestry break the vicious circle?* Bangkok: Thailand Development Research Institute.

Uhl, C. 1987. Factors controlling succession following slash-and-burn agriculture in Amazonia. *Journal of Ecology* **75**(2), 377–407.

Uhl, C., R. Buschbacher, E. A. S. Serrao 1988. Abandoned pastures in eastern Amazonia, I: patterns of plant succession. *Journal of Ecology* **76**(3), 663–81.

Uhlig, H. (ed.) 1984. *Spontaneous and planned settlement in Southeast Asia*. Hamburg: Fischer.

United Nations 1988. *World demographic estimates and projections, 1950–2025*. A report prepared jointly by the United Nations, the International Labour Organization and the Food and Agricultural Organization of the United Nations. Department of International Economic and Social Affairs, UN, New York.

United Nations 1990. *World population trends: 1989 monitoring report*. UN: New York.

United Nations Development Programme (UNDP) 1991. *Human development report 1990*. Oxford: Oxford University Press.

USDA Forest Service 1990. *An analysis of the timber situation in the United States, 1989–2040*. General Technical Report RM–199, USDA Forest Service.

Valdes, A. 1986. Impact of trade and macroeconomic policies on agricultural growth: the South American experience. In *Economic and social progress in Latin America* Washington DC: Inter-American Development Bank.

Vincent, J. R. 1989. Optimal tariffs on intermediate and final goods: the case of tropical forest products. *Forest Science* **35**(3), 720–31.

Vincent, J. R. 1990a. Rent capture and the feasibility of tropical forest management. *Land Economics* **66**, 212–23.

Vincent, J. R. 1990b. Don't boycott tropical timber. *Journal of Forestry* **88**(4), 56.

Vincent, J. R. 1992a. A simple nonspatial modelling approach for analysing a country's forest-products trade policies. In *Forestry sector analysis for developing countries*, R. Haynes, P. Harou, J. Mikowski (eds), 43–54. Seattle: Centre for International Trade in Forest Products, University of Washington.

Vincent, J. R. 1992b. The tropical timber trade and sustainable development. *Science* **256**, 1651–5.

Vincent, J. R. & C. S. Binkley 1991. *Forest based industrialization: a dynamic perspective*. Development Discussion Paper No. 389, Harvard Institute for International Development (HIID), Cambridge, Mass.

Vincent, J. R., D. J. Brooks, A. K. Ganadpur 1991. Substitution between tropical and temperate sawlogs. *Forest Science* **37**(5), 1484–91.

Vincent, J. R., A. K. Gandapur, D. J. Brooks 1990. Species substitution and tropical log imports by Japan. *Forest Science* **36**(3), 657–64.

Vincent, J. R. & M. R. Newmark 1992. *Deforestation and economic growth when food is nontradable*. Development Discussion Paper No. 420, Harvard Institute for International Development, Cambridge, Mass.

Von Moltke, K. 1990. *International economic issues in tropical deforestation*. Unpublished Paper, Workshop on Climate Change and Tropical Forests, Sao Paulo, Brazil.

Walker, R. T. 1990. *A behavioral model of tropical deforestation under the system of concession logging*. Paper presented at ISEE Conference on The Ecological Economics of Sustainability (21–23rd May), Washington DC, USA.

Westoby, J. C. 1978a. *Forest industries for socio-economic development*. FID/GS. Eighth World Forestry Congress, Jakarta.

Westoby, J. C. 1978b. Forest industries for socio-economic development. *Commonwealth Forestry Review* **58**, 107–116.

Westoby, J. C. 1989. *Introduction to world forestry: people and their trees*. Oxford: Basil Blackwell.

Whitaker, M. 1990. The human capital and science base. In *Agriculture and economic survival: the role of agriculture in Ecuador's economic development*, M. Whitaker & D. Colyer (eds), Boulder, Co.: Westview Press.

Whitaker, M. & J. Alzamora 1990. In *Agriculture and economic survival: the role of agriculture in Ecuador's economic development*, M. Whitaker & D. Colyer (eds), Boulder, Co.: Westview Press.

Whitmore, T. C. 1984. *Tropical rain forests of the Far East*. Oxford: Clarendon Press.

Wilson, E. O. (ed.) 1988. *Biodiversity*. Washington DC: National Academy Press.

Winpenny, J. T. 1990. *Development research: the environmental challenge*. London: Overseas Development Institute.

Winterbottom, R. 1990. *Taking stock: the tropical forestry action plan after five years*. Washington DC: World Resources Institute.

Wood, D. 1993. Forest to fields: Restoring tropical lands to agriculture. *Land Use Policy* **10**(2), 91–107.

Woods, R. 1987. Malthus, Marx and the population crisis. In *A world in crisis? Geographical perspective*, R. J. Johnston & P. J. Taylor (eds), 127–49. Oxford: Basil Blackwell.

World Bank 1985. *World Development Report 1984: population change and economic development*. Oxford and London: Oxford University Press.

REFERENCES

World Bank 1989a. *World tables*. Baltimore, Maryland: The Johns Hopkins University Press.

World Bank 1989b. *Indonesia – forest and water: issues in sustainable development*. World Bank, Washington DC.

World Bank 1990a. *Population pressure: the environment and agricultural intensification in the Philippines*. Washington DC: Environment Department, World Bank.

World Bank 1990b. *World development report: focus on poverty*. Washington DC: World Bank.

World Bank 1990c. *World development report 1990*. Oxford and London: Oxford University Press.

World Bank 1991a. *The forest sector: a World Bank policy paper*. Washington DC: World Bank.

World Bank 1991b. *World development report 1991*. Oxford and London: Oxford University Press.

World Bank 1992a. *World development report 1992: development and the environment*. Washington DC: World Bank.

World Bank 1992b. *World debt tables*. Washington DC: World Bank.

World Bank 1992c. *World tables*. Washington DC: World Bank.

World Bank various years-a. *World tables*. Baltimore, Maryland: The Johns Hopkins University Press.

World Bank various years-b. *World debt tables*. Baltimore, Maryland: The Johns Hopkins University Press.

World Commission on Environment and Development (WCED) 1987. *Our common future*. London: Oxford University Press.

World Resources Institute 1985. *Tropical forests: a call to action*. Washington DC: WRI and the World Bank.

World Resources Institute 1990. *World resources, 1990–1991*. Washington DC: World Resources Institute.

World Resources Institute 1992. *World resources, 1992–1993*. Oxford: Oxford University Press.

Youn, Y. C. & S. C. Yum 1992. *A study on the demand and supply of timber in South Korea*. Paper presented at the Forest Sector, Trade and Environmental Impact Models: Theory and Applications Symposium (30 April–1 May), Centre for International Trade in Forest Products, University of Washington, Seattle, USA.

Zellner, A. 1962. An efficient method of estimating seemingly unrelated regression and tests for aggregation bias. *Journal of the American Statistical Association* **57**, 348–68.

Index

Bold entries, tables; italic entries, illustrations

329